Severe Plastic Deformation Technology

Edited by
Andrzej Rosochowski

Reader in Light Metals Advanced
Technology, University of Strathclyde, UK

WHITTLES PUBLISHING

Published by
Whittles Publishing,
Dunbeath,
Caithness KW6 6EG,
Scotland, UK

www.whittlespublishing.com

ISBN 978-184995-091-6

Printed by
Gomer Press Limited, Llandysul, Wales

Contents

Preface

In the past, improvements in the properties of metallic materials were achieved mainly by adding more alloying elements and/or applying complex thermomechanical treatments. The former is a costly solution and may involve rare or strategic elements while the latter is usually limited in terms of the achievable improvements. Recently, a more unconventional approach has been tried, which is based on refining grain structure of metals below the level normally achievable by traditional techniques, that is below an average grain size of 1 μm. Such metals are referred to as ultrafine grained (UFG) metals.

The most feasible method of producing UFG metals is based on so-called severe plastic deformation (SPD), a new branch of metal forming technology, in which very large plastic deformation of the material is not accompanied by any substantial change of its shape and dimensions.

UFG metals are characterised by improved mechanical, physical and technological properties. The most prominent one is much higher yield strength, which enables the design of substantially lighter structures. Ultimate tensile strength and high cycle fatigue are also improved. Typically, ductility is reduced but for some naturally brittle metals such as magnesium alloys ductility at room temperature can be increased. This opens up an opportunity to form these materials without heating. UFG metals exhibit increased fracture toughness at low temperatures, which makes them ideal candidates for cryogenic and outer space applications. There are claims of more uniform corrosion and improved biocompatibility through faster integration of biological tissue with UFG metal surfaces. The diffusion rate is increased due to the larger surface of grain boundaries, which is beneficial in technological processes such as superplastic forming and diffusion bonding since they can be performed at a lower temperature and/or higher strain rate. For micro-components, the UFG structure helps avoiding so-called scale effects. The above benefits of UFG metals should make them very popular in a variety of applications spread over many industrial sectors such as transport (aerospace, automotive, rail), energy (traditional, nuclear, renewable), medical (implants and devices) and micro-manufacturing. However, progress in practical applications of UFG metals is very slow due to three main reasons: limited availability of UFG metals, few examples of industrial applications which could be followed and lack of awareness among designers and engineers.

The fundamental knowledge of UFG metals is well developed because of the work of numerous groups of researchers, thousands of publications and many dedicated conferences. The problem is that only a small percentage of this research effort refers to methods of producing UFG metals and when it does, it only briefly describes laboratory equipment used to produce small size UFG samples for structural investigations. The challenge still remains to convert the laboratory SPD processes to scaled-up processes, which are industrially viable. One element of this challenge is the lack of detailed knowledge regarding engineering aspects of different SPD processes. This book aims to provide this sort of insight into the most popular SPD processes of ECAE (equal channel angular extrusion)/ECAP (equal channel angular pressing), I-ECAP (incremental equal channel angular pressing), HPT (high pressure torsion), CEC (cyclic extrusion–compression), TE (twist extrusion) and ARB (accumulative roll-bonding). The historical background of these processes is followed by explaining their principles, engineering implementations in terms of machines, tooling and process parameters, some structural results and material properties as well as new process developments. We hope the book, written by a team of international experts, will be useful to researchers trying to understand the background of the main SPD processes and practical challenges in their implementation as well as to engineers interested in transferring these processes from laboratory to industry.

Andrzej Rosochowski

The Authors

Yan Beygelzimer is Dr.Sc., professor and principal research scientist of Donetsk Institute for Physics and Engineering named after O. O. Galkin of the Ukrainian National Academy of Sciences. In 1999 he suggested the idea of twist extrusion and subsequently was a research leader for the development of this process. His research interests lie in the area of severe plastic deformation, metal forming, materials sciences and mathematical modelling. He proposed a continuum model of grain refinement and damage of polycrystalline materials during severe plastic deformation and developed a continuum theory of plastic deformation in structurally-inhomogeneous porous bodies. Based on this theory, he proposed a mathematical model of metal forming of porous and powder materials and developed a new model for predicting the ductility of materials under deformation.

Anton Hohenwarter studied materials science at the Montanuniversität of Leoben and undertook his PhD at the Erich Schmid Institute of Materials Science of the Austrian Academy of Sciences. Currently, he is a group leader at the Department of Materials Physics of the Montanuniversität Leoben. His research interests concern mainly the fracture and fatigue behaviour of ultrafine-grained and nanocrystalline metals and alloys. He has been working with Prof. Reinhard Pippan for many years and has been strongly involved in the further development of high pressure torsion as a severe plastic deformation method.

Roman Kulagin has a Ph.D. degree in materials science and engineering. For over 10 years, he has been working on industrially relevant problems in metal forming and the theory of plasticity. Between 2005 and 2014, he worked at the Donetsk Institute for Physics and Engineering, Ukraine. He is presently a postdoctoral research fellow at the Karlsruhe Institute of Technology, Germany. His work has led to many improvements in metal forming processes already in use as well as to the development of novel forming techniques (e.g., hot section rolling on a continuous rolling mill, continuous extrusion of copper (CONFORM), drawing of high precision copper profiles, direct extrusion of aluminium profiles, and forging of titanium alloys for gas-turbine engine applications).

Lech Olejnik is professor at the Department of Metal Forming and Casting in the Faculty of Production Engineering at Warsaw University of Technology, the biggest technical university in Poland. Since 2005, he has been chair of the UFGbySPD Group which carries out both research work and production of engineering components made of UFG metals and alloys. The UFGbySPD laboratory occupies now more than 200 square meters and is equipped with eight presses (from 100kN to 10000kN) for ECAP and post-processing, and special SPD machines for I-ECAP of bars and plates.

Dmytro Orlov, PhD, is professor and Head of the Division of Materials Engineering at the Faculty of Engineering (LTH) in Lund University, Lund, Sweden. Among other professional activities, he has a joint appointment as Senior Scientist at the University of Nova Gorica in Slovenia and serves as a Vice-Chair of a Magnesium Committee at The Minerals, Metals and Materials Society (TMS), USA. His PhD project at Donetsk National Technical University under the supervision of Professor Beygelzimer was the first work dedicated to the development of twist extrusion. His present research interests are focused on the design of Mg alloys for biomedical and light-weight mobility applications, multi-scale architectured structures with topological control of their heterogeneity, and the development of relevant in-situ characterisation techniques at large-scale facilities.

Professor Reinhard Pippan is the vice director and group leader of the Erich Schmid Institute of Materials Science at the Austrian Academy of Sciences. He studied physics at the Technical University in Graz and received his PhD degree in 1982 at the Montanuniversität Leoben. Pippan's research activities are focused on mechanical properties of metals, alloys and composites. The improvement of the basic understanding of the relations between the mechanical behaviour, the deformation processes, the fracture processes, and the micro- and nano-structure of the material is the main goal of his research. The pioneering work of his group on severe plastic deformation (SPD) has made them one of the leaders in the international SPD community.

Jan Richert is now emeritus professor in the Department of Plastic Working and Physical Metallurgy of Non-ferrous Metals at the AGH University of Science and Technology Kraków, Poland. He is a member of the Theory of Metalforming Processes Section at the Metallurgy Committee of Polish Academy of Science. He is the author/co-author of 133 publications, the inventor/co-inventor of 19 patents and the author of the book *Innovative Methods of Metal Plastic Forming* (in Polish). His main scientific fields are experimental investigations, mathematical analysis and computational modelling of various unconventional plastic working processes of metals. Many of these investigations are devoted to manufacture of nanostructural metals by using the original CEC method, which enables the achievement of a very large plastic deformation in the presence of a high hydrostatic pressure.

Andrzej Rosochowski graduated and received his PhD degree from Warsaw University of Technology in Poland. In 1991, he moved to University of

Strathclyde in Glasgow, UK. He is currently reader at the Department of Design, Manufacture and Engineering Management of this university, leader of the Light Metals Advanced Technology Research Group and Co-Director of Centre for Precision Manufacturing in this department. He was a member of Board of Directors of ESAFORM for twelve years and a Fellow of the Institute of Nanotechnology for ten years. He has been Fellow of IMMM and member of its Light Metals Division Board since 2007. His research interests cover all aspects of metal forming technology. Since 2000, his main research field has been severe plastic deformation technology and its applications. He has invented an incremental version of equal channel angular pressing.

Professor Vladimir Segal D.Sc. is now retired. He worked for Physical – Technical Institute of Belorussian Academy of Sciences (USSR, 1962–1986), Texas A&M University, College Station, TX, (USA, 1992–1995), Honeywell International, Inc. (USA, 1996–2002). In 2005 he founded Engineering Performance Materials Co. His research interests include metal forming, materials processing for properties, and applied plasticity. He has pioneered research on special techniques of severe plastic deformation since 1970. In 1972 he invented and developed equal channel angular extrusion, and the first commercialized the process at Honeywell. He is author and co-author of more than 150 papers, eight technical books (in Russian and English), and a number of patents.

Nobuhiro Tsuji has been a professor at the Department of Materials Science and Engineering, the Graduate School of Engineering, Kyoto University, Japan, since March 2009, after working in Osaka University from 1994 to 2009 as an assistant professor and then associate professor. He developed the ARB process with his colleagues including Professor Yoshihiro Saito in Osaka University around 1998. He has been working on process-microstructure-property correlations of structural metallic materials, and his current research interest includes unique mechanical properties of bulk nanostructured metals (ultrafine grained metals), microstructure and properties of high entropy alloys, and dynamic structure evolution during thermomechanical processing of steels and nonferrous alloys.

Viktor Varyukhin, D.Sc., is Correspondent Member of the Ukrainian National Academy of Sciences, and director of Donetsk Institute for Physics and Engineering named after O. O. Galkin of the Ukrainian National Academy of Sciences. He organized institute activities involving severe plastic deformation, including twist extrusion. He is an expert in metal physics and physics of high pressures. His areas of interest include SPD, sub-microcrystalline and nano-structured materials. He has made significant contributions to research on the effect of high pressures on the mechanisms of plastic deformation and metal fracture.

Equal channel angular extrusion (ECAE)

CHAPTER 1

Vladimir M. Segal

1.1 Introduction

The origins of severe plastic deformation (SPD) may be traced to ancient times. Since the introduction of special deformation techniques, it has become an important field in processing materials to enhance their properties. During the last 20 years, extensive academic activity has developed around SPD—an avalanche of publications and numerous conferences. As a part of modern materials science, SPD raises specific problems, which are still far from adequately resolved at different structural scales. However, our understanding of many details of SPD is continually improving. In contrast, technological developments in SPD and related industrial applications remain at the initial stages. There is a clear disproportion between academic and technical activities in the field.

Despite many papers on structure and materials characterisation, there are only a few on the engineering and practical aspects of SPD. To some extent, this is quite surprising. On the one hand, new deformation techniques result in discoveries of unusual opportunities to control materials structure and properties, and various concepts of SPD are tied to specific methods of deformation. On the other hand, the effects of SPD depend strongly on the mechanics of the processing and it is known that modifications in tool design and boundary conditions of different SPD techniques may lead to diverse results. This decisive characteristic of SPD is often overlooked. Practical applications are the final goal of any development in materials science and engineering. Therefore, to succeed as a new processing technology, significant efforts should be focused on developing industrial processes using SPD. The lack of such works and related information is the main reason why SPD remains a laboratory tool and industry remains sceptical about SPD despite numerous promises. Hopefully, the present book will help to overcome these problems.

This chapter starts by briefly considering a correlation between structure modification and the mechanics of SPD. Some general requirements for deformation techniques are outlined. The main part of the chapter deals with equal channel angular extrusion (ECAE), the most developed and practical SPD technique. Stress/strain states are analysed for a plastic zone, inlet and outlet channels. The emphasis is on ECAE as a materials processing technology for properties. The data that is presented would be useful for R&D on ECAE and to design production tools.

1.2 Materials processing for properties

1.2.1 Processing mechanics

Plastic deformation significantly influences the structure and properties of materials. At the meso scale, grains change their shape and crystallographic orientation, second phases and hard particles are crushed and refined, volume defects are generated or healed. At the micro scale, deformation produces numerous dislocations which form sub-structures and patterns, induces new boundaries and a high concentration of vacancies as well as stalking faults and twins in some metals. These defects increase the internal energy and diffusivity, and reduce structural stability, thus promoting recrystallisation and phase transformations during plastic deformation and heat treatment. The evolution of the final structure depends on the processing conditions and the mechanics of deformation. Processing conditions are defined by temperature and strain rate. Their overall effect is described by the Zener–Hollomon parameter. The choice of processing temperatures for particular materials depends on the mechanisms of plastic flow and structural modification during cold, warm and hot deformation whereas the strain rate provides conditions for their dynamic or static realization. The most important parameters of the processing mechanics are the stress and strain states. It is always desirable to have uniform stress/strain distributions under sufficiently large compressive hydrostatic pressures to prevent material fracture at the prescribed effective strain, strain rate and temperature.

A general description of different plastic states can be given in terms of the directions of the principal shear stresses, known as α and β slip lines (Hill, 1950). For the plane (two-dimensional) and uniform strain states, slip lines form a Cartesian coordinate system (α, β). In this system, stress tensor components can be expressed through the material shear stress k and the hydrostatic pressure p. As hydrostatic pressure does not affect the transition to plastic flow, the ratios of deviator components of stress tensors to k are identical for any state. For the fixed effective strain rate $\eta = (\eta_\alpha + \eta_\beta)/2$, the deformation mode depends on the distribution of the strain rate components η_α, η_β between the α and β slip lines. Then, all possible strain/strain rate states are described by parameter

$$C = \frac{2}{\left(1 + \dfrac{\eta_\alpha}{\eta_\beta}\right)} \tag{1.1}$$

This parameter was introduced by Segal (2002) as a single characteristic of the deformation mode. A coefficient $C = 0$ when straining is performed along one family of slip lines ($\eta_\beta = 0$). For $C = 1$ the strain rates are equally distributed between both families of slip lines ($\eta_\alpha = \eta_\beta$). In mechanics of solids, the first case is defined as simple shear and the second case is defined as pure shear.

These two limit situations are of special interest because all possible modes of deformation are confined between them. Therefore, coefficient C is the main characteristic of the mechanics of processing during plastic deformation.

1.2.2 Structure evolution and deformation mode

Depending on circumstances, distribution of strains and strain rates during plastic deformation may indicate the normal level of non-uniformity in microstructures or strong localisation in shear bands (SBs). The first case of continuous evolution was described by a hypothesis of low energy dislocation structures (Kuhlmann-Wilsdorf, 1989). In accordance with that hypothesis, under monotonic loading, dislocations form dense dislocation walls (DDWs) which subdivide original grains and rotate them to a direction of the principal elongation with increased angles of misorientation between subdivided areas. Simultaneously, dislocations that are trapped inside the subdivided areas organise cells and cell blocks separated by incidental dislocation boundaries (IDB). Finally, they form structures of cells and sub-grains. For large strains during rolling (Hughes and Hansen, 1997), cell blocks are readily transformed into flat lamellae with high-angle boundaries parallel to a rolling plane. Identical 'onion-like' structures are observed after deep drawing of wire. Such very fine two-dimensional structures cannot be considered to be new grains.

In the second case of discontinuous evolution or strain localisation, totally different structures of extremely fine equaxial grains down to 100 nm were found within SBs during rolling with large reductions (Hirsch *et al.*, 1988). Similar sub-micron grained structures were also observed in surface layers of metals exposed to dry friction (Rigney *et al.*, 1986), in adiabatic SBs under dynamic loading (Meyer *et al.*, 2006), in thin discs twisted and compressed between Bridgman's anvils (Smirnova *et al.*, 1986), and other situations.

A clear difference between the two kinds of structures can be explained in terms of processing mechanics. For typical forming operations such as rolling and drawing at fixed processing conditions (temperature, strain, strain rate) it relates to the deformation mode. Hill (1967) introduced a concept of the 'ideal' forming operations which provide strain uniformity and minimum pressure p_{id} and load P_{id} during reduction of the material cross-section area from F to f

$$p_{id} = \sqrt{3}k\varepsilon,\ P_{id} = p_{id}F \tag{1.2}$$

Where $\varepsilon = \ln(F/f)$ is the von Mises effective strain and k is the material yield shear stress. In 'ideal' forming operations, such as extrusion or drawing through 'sigmoidal' dies without friction, flow lines form angles ±45° with slip lines, strain rates $\eta_\alpha = \eta_\beta$, and the deformation mode conforms to pure shear. For rolling and drawing with low friction and slowly convergent plastic flow, the deformation mode also approximates to pure shear. This mode provides a stable continuous evolution of dislocation structures until very large strains have accumulated, when localisation starts to operate into SBs along two families of slip lines. After their first appearance at the micro-scale, SBs penetrate all material and accommodate the larger part of the straining. Within SBs, the strain rates $\eta_\alpha >> \eta_\beta$ (or $\eta_\beta >> \eta_\alpha$), $C \to 0$ and the deformation mode changes to simple shear. However, in these cases SBs cover a small part of the whole material and do not contribute noticeably to the evolution of the structure.

In processing of well-annealed materials by simple shear, continuous evolution takes place at the initial stages of deformation. During subsequent loading, simple shear provides a few significant advantages in comparison to pure shear (Segal, 2002; Segal *et al.*, 2010). Plastic deformation is concentrated along one family of slip lines instead of two families of slip lines. At the same effective strain/strain rate, it is twice as large in one direction for simple shear than for pure shear. Under simple shear, grains and their crystallographic orientations are rotated to the principal shear direction instead of the principal stress direction as for pure shear. That increases the Schmid factor, reduces textural hardening and promotes an earlier transition to localisation with the formation of SBs in one direction throughout the material. When localisation proceeds, there are no restrictions on strain accommodation along SBs in the case of simple shear in contrast to pure shear. During extended straining, simple shear promotes rotational localisation of the material elements with development of three-dimensional high-angle boundaries. Finally, as simple shear retains a cross-section of the processed material, a change in shear directions during multi-step processing allows the spatial control of structures in bulk materials. These peculiarities provide exceptional potentials for structure fragmentation and refinement of grains to sub-micron and, sometimes, to nano sizes. It has also been found that simple shear is an effective deformation mode for many other modifications of structures such as: development of various textures, geometrical control of structural elements, refinement of second phases and precipitates, phase transformations, breakdown of cast structures, consolidation and bonding of powder materials, processing of specialty and functional materials. Therefore, no matter what mechanisms are operating in each case, simple shear can be considered as a near 'ideal' deformation mode in material processing for properties. This concept was first formulated by Segal *et al.* (1981) and has received numerous experimental confirmations.

1.3 SPD processes

There are various options to exploit plastic deformation in materials processing. Such opportunities are restricted by the available deformation techniques with specific deformation modes and attainable strains. Most of them were developed as forming operations to change the shape and dimensions of different products. From Equation (1.2) for 'ideal' forming, one can see that large effective strains ε are limited by the capabilities of tool and equipment to withstand high stresses p_{id} and loads P_{id}, and they can only be developed in thin final products. In real forming operations, contact friction significantly increases the 'ideal' values p_{id}, P_{id} from Equation (1.2), induces strain non-uniformity and applies much stronger restrictions on attainable strain but retains a near pure shear deformation mode in most of the cross-section. For that reason multi-pass cold rolling and drawing to large strains usually leads to ordinary heavily deformed lamellae structures of cells and sub-grains (Gil Sevillano *et al.*, 1980).

Problems arise when realizing uniform simple shear within a finite volume of material because of the need to satisfy specific static and kinematic boundary conditions, however, this can easily be achieved in thin layers. The first such technique—twist–compression—was introduced by Bridgman (1952), in the 1930s for thin discs. In this technique, large friction shear stresses and strains developed by rotating one of the anvils under compressive load are transferred throughout the material thickness δ, if $\delta < 0.3$ mm. Later, intensive simple shear was observed in thin surface layers during dry plastic friction (Rigney *et al.*, 1986) where severe deformation was generated by multiple micro-asperities of rigid tools. Similarly, simple shear in thicker layers can be induced by surface burnishing with a special local tool or during surface indentation by numerous small high-energy balls. Segal *et al.* (2010) have reviewed a few modifications of surface shear techniques.

It has also been suggested that contact friction should be used to induce additional shears within materials. In the accumulative roll bonding process (Saito *et al.*, 1998), surface layers are embedded into the material by cumulative stacking and rolling of sheets. In differential speed rolling (Ji *et al.*, 2007), extensive surface shear is transferred through thin sheets. However, in forward extrusion of the bulk material through a rotating die, friction shear was localised near a contact surface and changes to the strain distribution were insignificant within the plastic zone (Ma *et al.*, 2004).

The greatest practical interest concerns internal simple shear in areas with high gradients of strain and strain rate. These areas, penetrating through the material and mathematically identified as lines of velocity discontinuities along slip lines (Hill, 1950), are the general characteristic of plastic flows. They are observed in most technological problems with sharp corner tools and contact friction provoking non-uniform flow. When crossing a line AA of velocity discontinuity (Figure 1.1), material particles experience simple shear

$$\gamma = \frac{[v]}{v_n} \tag{1.3}$$

Where $[v]$ is the discontinuity of the tangent velocity component, and v_n is the normal velocity component at AA. Along curved slip lines, the velocity discontinuity $[v]$ = constant whereas v_n and γ may vary. In steady plastic flows with stationary slip lines having velocity discontinuities, large von Mises strains can accumulate during multi-pass processing. Such an approach can be applied to different forming operations with higher pressure and load but smaller attainable reductions than for 'ideal' forming operations.[1] (See chapter end for notes.) One example is an application of hydro-extrusion for SPD (Kurzydlowski, 2004). Another example is cyclic extrusion–compression (Richert *et al.*, 1979) where

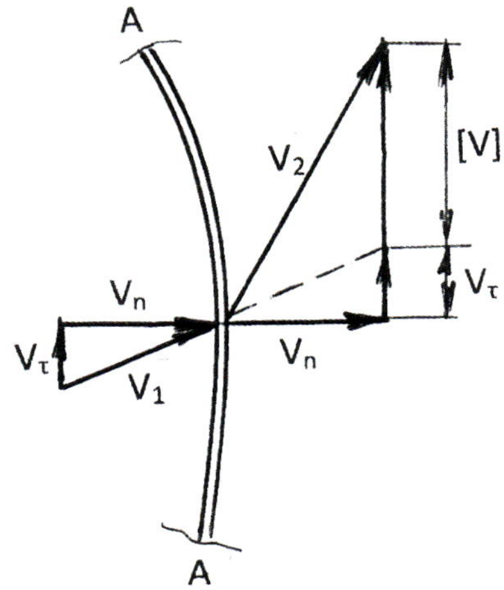

Figure 1.1 Velocity discontinuity along slip line

extrusion is followed by a disturbance in the identical channel to restore the cross-section of the material after each step. This technique can be modified (Pardis *et al.*, 2011). For forming operations with non-steady plastic flow, like multi-directional forging between flat dies, a material shape can be periodically distorted and restored during the numerous processing steps which are necessary to compensate for strain non-uniformity and accumulate large shears in bulk products (Imayev *et al.*, 1992).

Other processing techniques for properties have also been suggested, and the search for new ones continues. They have been defined as SPD processes. According to Valiev *et al.* (2006), SPD is

> ... any metal forming method under extensive hydrostatic pressure that may be used to impose a very high strain on a bulk solid without the introduction of any significant change in the overall dimensions of the sample and having the ability to produce exceptional grain refinement.

This definition of SPD is controversial. Refinement is not the only goal of SPD and many other effects of plastic deformation have been discovered with new processing techniques. Ultra-fine grained structures produced by SPD can also be obtained at relatively moderate strains, without extensive hydrostatic pressure and with a significant change of overall dimensions. Experiments and analyses show that SPD processes depend on both the strain intensity and the special deformation mode. The general characteristic of these techniques is the modification of pure shear deformation, typical for ordinary forming operations, to the simple shear deformation mode that used not to be used in material processing. Such an approach has been consistently advocated by the present author (Segal *et al.*, 1981; Segal *et al.*, 2010). Therefore, SPD techniques can be classified as processes of predominantly simple shear deformation mode irrespective of strain intensity including moderate, large or severe strains, hydrostatic pressure and overall shaping. The performance of various simple shear techniques may be evaluated by different criteria: stress–strain homogeneity, approximation to simple shear deformation mode, attainable strains per pass in thin and bulk products, required pressure and load, hydrostatic pressure, and, as the ultimate goal of any technological development, simplicity of industrial realization, low cost and operational productivity. An important technological parameter of SPD processes is a dimensionless ratio

$$K = \frac{P}{(\sqrt{3}k\gamma f)} \tag{1.4}$$

Where P is the processing load, k is the material yield shear stress, γ is the shear strain per pass, and f is the cross-sectional area of the product (Segal *et al.*, 1981). Then the lower is K the more energetically effective is the SPD processing.

1.4 Concept and history of ECAE

1.4.1 Concept of ECAE

Figure 1.2(a) demonstrates a concept of ECAE. A well-lubricated billet 1 is inserted into tool 2 having two crossing channels of identical cross-sections (a) with the billet. A punch 3 then extrudes the billet from one channel into the second channel 4 till point O. For zero friction, the material moves as a rigid whole within each channel and a plastic zone is localised around a single

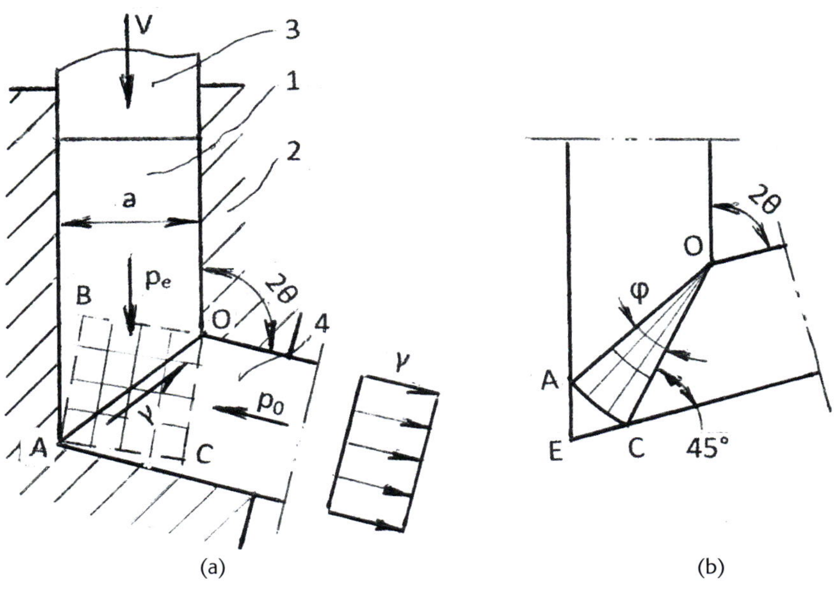

Figure 1.2 Frictionless ECAE: (a) with tool angle $2\theta \geq 90°$; (b) with tool angle $2\theta < 90°$

α-slip line AO. Along AO, velocity discontinuity $[v]$, normal velocity component v_n and shear strain $\boldsymbol{\gamma}$ acquired by material particles during the crossing AO are

$$[v] = 2V \cos \theta, \, v_n = V \sin \theta, \, |\boldsymbol{\gamma}| = 2 \cot \theta \tag{1.5}$$

Where 2θ is an angle between channels and V is the extrusion speed. Shear strain $\boldsymbol{\gamma}$ is uniformly distributed through a thickness a.

For very low friction in channels, the extrusion pressure p_e, extrusion load P_e and compressive hydrostatic pressure p along AO are

$$\frac{p_e}{2k} = \frac{p_0}{2k} + \cot \theta, \, P_e = p_e f, \, p = p_0 + k \cot \theta \tag{1.6}$$

Where p_0 is a back pressure applied to the material into the second channel and f is the cross-sectional area of the billet. Usually, $p_0 = 0$. In this case, a comparison of Equations (1.5), (1.6) and (1.2) for the same cross-section areas $f = F$ of the final product and $\gamma = \sqrt{3}\varepsilon$, $\sigma_s = \sqrt{3}k$ shows that one pass of ECAE requires the minimum extrusion pressure $p_e/2k = 1$ and plastic work as in 'ideal' forming operations. However, coefficient K in Equation (1.4) for ECAE has the absolute minimum $K = 1$ against $K = \exp(\varepsilon)$ for 'ideal' forming operations. Therefore, the required load for ECAE is $\exp(\varepsilon)$ times lower than for 'ideal' forming operations. Additionally, ECAE can be repeated many times in the same tool to accumulate extremely high strains without any change in the dimensions of the billet. Different orientations of shear planes may be also induced by rotating the billet between successive passes. Together with uniform strain distribution and simple shear deformation mode ($C = 0$), these properties provide optimal characteristics of ECAE for material processing. Other technological advantages of ECAE will be considered in Section 1.8.

1.4.2 History of ECAE

In the scientific literature ECAE is known under various names (ECA extrusion, ECA pressing, ECA forging, ECA rolling), many publications propose different dates for the introduction of ECAE ranging from the 1970s to the 1990s. Its introduction has been attributed to different authors, and there are examples of duplication and reinvention. Related information is often concealed and confusing. So, it is time to tell the true story.

The first part of the story is the motivation for introducing ECAE. The author has to return to early 1970 when he worked at the Physical Technical Institute (FTI), Minsk, USSR. His scientific interest was in the mechanics of metal forming, in particular slip line theory. This theory, unfortunately almost forgotten in the present age of finite element (FE) analysis, gives a clear understanding of plastic deformation but requires intuition and refinement of mind. To extend the slip line approach, we developed a sophisticated technique of physical modelling of plastic flow using the Moiré effect and computer analysis to calculate experimental slip line fields (Segal *et al.*, 1974). That turned mathematical abstractions such as rigid zones, velocity discontinuities and singular points into reality and provided us with a means to correlate the local processing mechanics with the evolution of the structure. Being involved in different technical projects, the author realized that existing forming operations were not optimal for modifying materials and recognised the general problem of finding the 'ideal' deformation mode(s) in processing for properties.

The subsequent research revealed that the deformation mode during plastic flow can be varied from pure shear to simple shear. For pure shear, extensive information was available for a wide range of strains and processing conditions. For simple shear at moderate temperatures the information was restricted to some results obtained during twist tests. However, Bridgeman's technique of unrestricted shear by twist–compression of thin discs, which was applied in the USSR by Enikolopov to organic and non-organic materials discovered unusual effects and reactions in quantum-mechanics chemistry.[2] These works encouraged the development of simple shear processing for bulk metals. After numerous experiments with different approaches, partly described by Segal *et al.* (1994), a stable and uniform simple shear was finally attained in so-called 'cross-extrusion' using an idea of a steady flow through straight lines of velocity discontinuity (Figure 1.3). Multi-pass processing in tool 1 was performed by alternative extrusion of a billet 2 by punches 3 and 4 from vertical channels into horizontal channels and vice versa.

Thus, the first sub-micron grained bulk materials by SPD were produced in 1971 (Segal *et al.*, 1981). They demonstrated superior properties such as a combination of high strength and good ductility. However, because of a flow singularity in the vicinity of the centre of the cross, streamlines near an axis of symmetry *a–a* experience a split and rebounding with opposite streamlines resulting in infinitely large strains along an axis *b–b*. After multi-pass processing, the central area was very hard and had a low ductility that

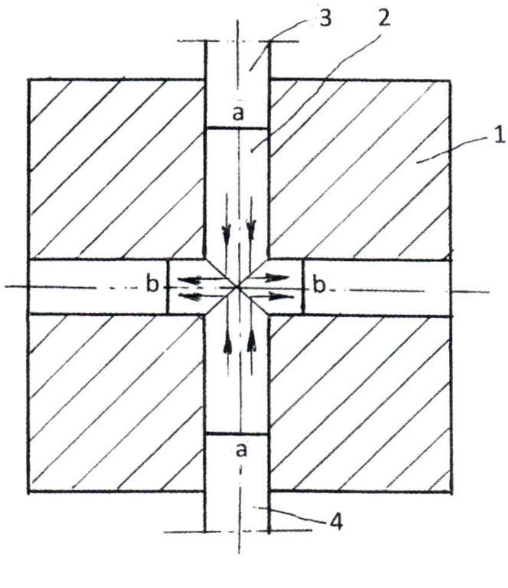

Figure 1.3 Simple shear in cross-extrusion

often resulted in cleavage of the material during mechanical testing. As the next stage of development, the problem was resolved by separating the plastic zone shown in Figure 1.3 into four similar zones and introduction of an additional parameter, angle 2θ between channels, as shown in Figure 1.2. This technique was first introduced in 1972 as a new materials processing technology by simple shear in an unpublished report from FTI.[3] Later, in 1975, the ECAE project was awarded a USSR Science and Technical Committee Grant, and an R&D group was formed to develop ECAE and related techniques. The group included young and enthusiastic scientists (V. I. Reznikov, V. I. Kopylov, D. A. Pavlik and V. F. Malyshev) and existed until 1985. It was an exciting time of everyday discoveries.

Our first goal was to develop reliable tools for ECAE of different materials including those that were of high strength and difficult to deform. A few sets of dies were created for cold, warm and hot ECAE with movable channel walls. They significantly reduced pressure, prevented material sticking and tool fracture. Processing of alloys with an ultimate strength up to 1600 MPa became possible. The first Conform mill for continuous ECAE was built at FTI in 1976. Multi-pass processing, changing the orientation of the billet between passes (now known as routes) was also invented. Many structural effects of simple shear deformation were demonstrated: refinement of grains to the submicron scale, refinement of second phases and precipitates; control of textures and morphology of micro-structure; consolidation and bonding of particulate materials; healing of volume defects including large cavities and pores. ECAE was applied to numerous metals and alloys: Al, Cu, Ni, Armco-iron, steels, dispersion hardening and maraging steels, magnets, elinvar and invar alloys, and superconductor alloys. In most cases, there were clear improvements in mechanical, functional or special properties even after one or two passes.

The results that were obtained confirmed the huge potential for simple shear in materials processing. A few projects were initiated in the period 1976–1986 on practical applications of ECAE. Multi-pass continuous ECAE was developed for structure refinement of Cu stocks for deep drawing of superfine wires. Thermo-mechanical processes comprising two passes of ECAE and heat treatment were developed for high-carbon and high-speed tool steels used for piercing punches. A special 800 t ECAE press was designed at Ural's Machine Building Plant in Yekaterinburg, Russia (URALMASH) to break down the structure of continuous cast steel billets. However, these projects were not completed because of the economic and political situation in the late Soviet Union.

This extensive research was confidential and most of the results have never been published. Information on ECAE was available to a narrow circle of specialists in the USSR and remained unknown abroad. In 1986, the author left FTI (Minsk) and in 1989 emigrated to the USA. Later, in 1989–1990, under a national (USSR) program 'Superplasticity', FTI transferred information on ECAE to the Institute of Metals Superplasticity Problems (IMSP) at Ufa, and delivered a large number of samples of different materials to IMSP. Soon after the collapse of the Soviet Union, in 1991, Dr R. Z. Valiev, then the Deputy Director of IMSP, Ufa, travelled to many countries giving numerous presentations, providing foreign laboratories with these samples and promoting international interest in SPD and ECAE (Hashimoto, 2005). Unfortunately, many of his papers published in the 1990s did not refer to the earlier work performed at Minsk. It is to be hoped that the information provided here will help to complete the picture.

The research on SPD and ECAE is extensive and well recognised in contrast to its still insignificant industrial impact. The author's activity on ECAE in USA is mostly focused on practical applications and commercialisation. The ECAE technology of plate billets was licensed to Johnson Matthey Electronic Materials (now Honeywell Electronic Materials) in 1996 and commercialised for sputtering targets in 2002. This product has remained the only bulk sub-micron grained material which is produced by SPD and is currently available on the market. Presently, there are three small companies (Metallicum, Shear-Form and Engineered Performance Materials) in the USA that have been founded with a mission of industrial development and commercialisation of ECAE.

1.5 Plastic zone during ECAE

During ECAE the material passes through three zones (Figure 1.2(a)): (i) inlet channel 1, (ii) plastic zone (around a crossing plane AO) and (iii) outlet channel 4. Processing effects depend on the stress–strain state within the plastic zone which has been extensively analysed by slip lines, upper-bound and FE methods as well as experimental techniques using coordinate grids and Moiré fringes (Segal *et al.*, 2010). For optimal processing, strain within the plastic zone should be intensive, uniform and approximating simple shear whereas the stress state should provide sufficient compressive hydrostatic pressure and low extrusion load. Among the analytical methods that have been mentioned, the slip line theory (Hill, 1950) presents a simple and natural way to evaluate the deformation mode and accumulated shear strains. This approach is especially suitable for ECAE because of plane plastic flow and an opportunity to neglect material elasticity and hardening during severe deformation. A slip line analysis describes the correct stress–strain states as function of the tool geometry and boundary conditions such as friction in inlet and outlet channels. As we will see later in the following subsections of Section 1.5, this requires an additional consideration of the contact interaction between material and channels. Different boundary conditions lead to different solutions and results. For an optimal design of ECAE processing, it is informative to consider a slip line analysis for some characteristic cases.

1.5.1 Frictionless conditions

A slip line field for zero friction at tool walls is shown in Figure 1.2(a). The plastic zone is a single slip line AO. Stress–strain states are described by Equations (1.5) and (1.6). A shear strain vector $\boldsymbol{\gamma}$ is uniformly distributed and oriented along AO. Triangles AOB and AOC constitute an extension of the plastic state into a rigid material. For a tool angle $2\theta > 90°$, they do not touch the channel walls. Plastic contact with channels is established at $2\theta = 90°$ when the slip line forms an angle 45° with the walls of the tool. For $2\theta < 90°$ (Figure 1.2(b)), the slip line field comprises a central fan AOC with an angle $\varphi = (90° - 2\theta)$ and a dead metal zone ACE. Maximum friction, $\tau = k$, acts along the boundary AC of this zone and in some cases the zone ACE may be separated from the rest of the material. Therefore, the tool angle 2θ controls all the processing parameters. It should be noted that simple shear γ from Equation (1.5) is performed at an angle θ to the axis of the billet. In practice, near zero friction is attained by the application of very good lubricants and solid coatings.

1.5.2 Uniform friction in channels

To satisfy the prescribed friction τ in channels, β slip lines should form an angle with channel walls equal

$$\eta = \frac{\left[\frac{\pi}{2} + \arcsin\left(\frac{\tau}{k}\right)\right]}{2} \tag{1.7}$$

The corresponding slip line field (Figure 1.4(a)) comprises a central fan AOC with an angle $\varphi = 2(\eta - \theta)$, $\phi \geq 0$, a dead metal zone ACE and continuations of the plastic state into the rigid material ABO and COD. Plastic friction τ acts along AB and CD. If $\eta \leq \theta$, the angle $\phi = 0$ and the slip field becomes identical to Figure 1.2(a). In another limit case of maximum friction $\tau = k$ slip lines AO and CO are perpendicular to the channel walls (shown in Figure 1.18(b)). The extrusion pressure p_e applied to a boundary AO is

$$\frac{p_e}{2k} = \cot\eta + 2(\eta - \theta) + \frac{\tau}{2k}[\sin\eta(\sin\eta + \cos\eta)] + \frac{p_0}{2k} \tag{1.8}$$

A strain analysis (Segal, 2003) shows that after crossing the plastic zone material particles experience three successive simple shears

$$\begin{aligned} |\gamma_1| &= \cot\eta \text{ along a slip line AO with velocity discontinuity} \\ |\gamma_2| &= \phi \text{ into a flow direction within the central fan AOC} \\ |\gamma_3| &= \cot\eta \text{ along a slip line CO with velocity discontinuity} \end{aligned} \tag{1.9}$$

Accumulated von Mises shear $\gamma = |\gamma_1| + |\gamma_2| + |\gamma_3|$ is uniformly distributed throughout the material and shows the dissipated plastic work but not element distortions. Within the central fan AOC flow lines coincide with circular α slip lines. Strain rates along these lines are

$$\eta_\alpha = V\left(\frac{\sin\eta}{r}\right),\ \eta_\beta = 0 \tag{1.10}$$

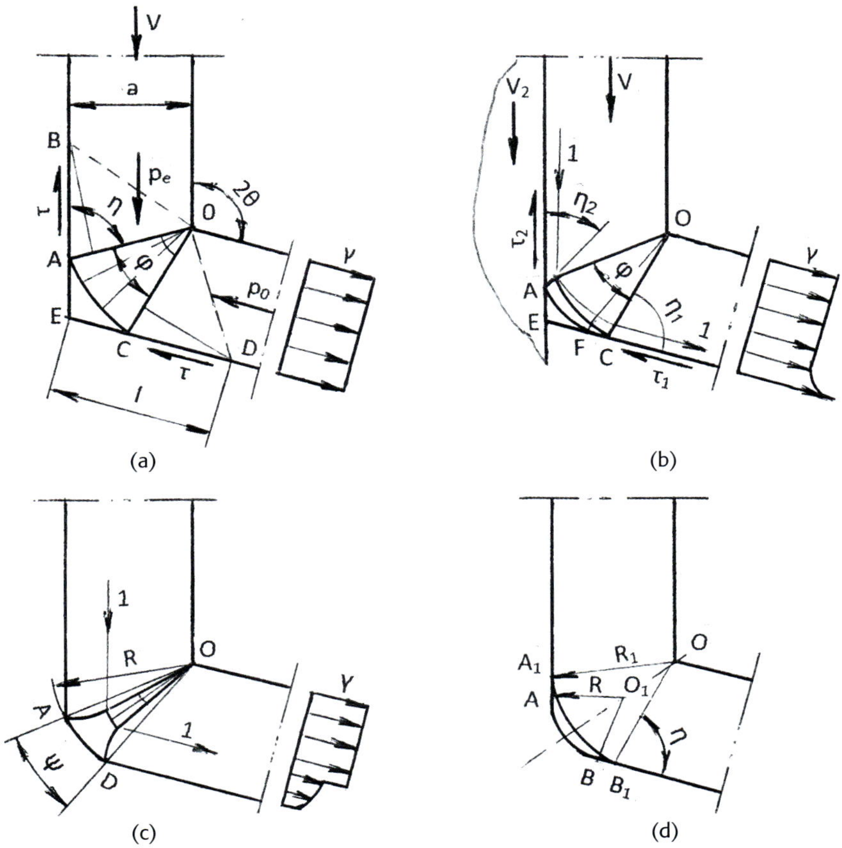

Figure 1.4 Slip line fields for particular cases: (a) equal friction in channels; (b) different friction in channels; (c) round corner channels without friction; (d) dead metal zone at round corner with friction

Where V is the extrusion speed, r is a radius of the considered material particle from the centre O. For maximum friction $\tau = k$, shears $\gamma_1 = \gamma_3 = 0$, velocity discontinuity along slip lines disappears and shear $|\gamma_2| = (\pi - 2\theta)$ only occurs within the fan AOB in the extrusion direction. It should be also noted that a velocity discontinuity along a boundary AC of the dead metal zone develops additional intensive shear at the bottom surface of the processed material.

1.5.3 Non-uniform friction in channels

In reality, it is difficult to provide the same low friction τ_1 in the outlet channel as the friction τ_2 in the inlet channel. Figure 1.4(b) shows a slip line field for $\tau_1 > \tau_2$. Angles η_1 and η_2 are found from Equation (1.7) for friction τ_1 in the outlet channel and friction τ_2 in the inlet channel. An additional friction area FC is also introduced to satisfy these conditions. The solution includes a dead metal zone AFE and continuations of the slip line field in rigid zones (not shown in Figure 1.4(b)). Simple shears γ_1, γ_3 occur along the slip lines AO and OC with velocity discontinuities. In other plastic areas, the deformation mode is not exactly simple shear. Accumulated strains are uniformly distributed above a flow line 1–1 but increase sharply to the bottom tool wall below line 1–1.

The slip line field of Figure 1.4(b) relates to passive friction τ_2 when a boundary EA is stationary. However, this boundary can be moved with a speed V_2. If $V_2 = V$, where V is the extrusion speed, friction $\tau_2 = 0$. If $V_2 > V$, friction along BE is active. It acts in an opposite direction to the direction shown in Figure 1.4(b), and should be taken with a minus sign in Equation (1.7). The largest length of a contact area AE occurs if $\tau_2 = k$. When friction $\tau_2 \rightarrow 0$, then point A shifts to point E and receives some intermediate position for $\tau_2 = 0$. For active friction $\tau_2 < 0$, it moves closer to point E and coincides with E for $\tau_2 = -k$. The related cases of movable tool walls and continuous ECAE will be considered in Subsection 1.7.3 and Section 1.9.

1.5.4 Round corner channels

Round corner channels were suggested by Iwahashi *et al.* (1996) to improve lubrication of the outlet channel and reduce an extrusion load. Figure 1.4(c) shows a slip line field for frictionless channels connected by radius R with an angle ψ from a centre O. It comprises a fan BOC and circular slip lines AB and DC. An area ABCD rotates about the centre O. To eliminate plastic overloading in points A and D, angles between slip lines AO, DO and a boundary AD should be greater than 45°. As for the uniform friction in both channels (Figure 1.4(a)), in the area above a flow line 1–1, accumulated shear γ is composed by shears γ_1 along a boundary AO, γ_2 along round slip lines and γ_3 along a boundary DO. In an area below the flow line 1–1, shearing occurs along slip lines AO and DO. The distribution of γ shows strong non-uniformity extended through a large part of the cross-section of the material.

In practical applications, the round corner centre O_1 conjugates channels and friction τ is not zero (Figure 1.4(d)). As the area adjacent to AB cannot rotate as a rigid whole about O, there is a high probability that a thin dead metal zone will develop along a boundary A_1B_1 formed by radii A_1O and B_1O from the centre O under angles η to the tool walls in accordance with Equation (1.7).

These examples of correct analytical solutions demonstrate that deviations in contact friction or the tool geometry from the optimal frictionless case with sharp corners, extend plastic zone, reduce strain, develop dead metal zone and strain non-uniformity, redistribute shear between a few directions and modify the simple shear deformation mode. These changes are undesirable for structure formation and should be avoided or minimised.

1.5.5 Additional effects

Under correct processing conditions, the single slip line model of Figure 1.2(a) provides a realistic description for plastic flow during ECAE of metals. Figure 1.5

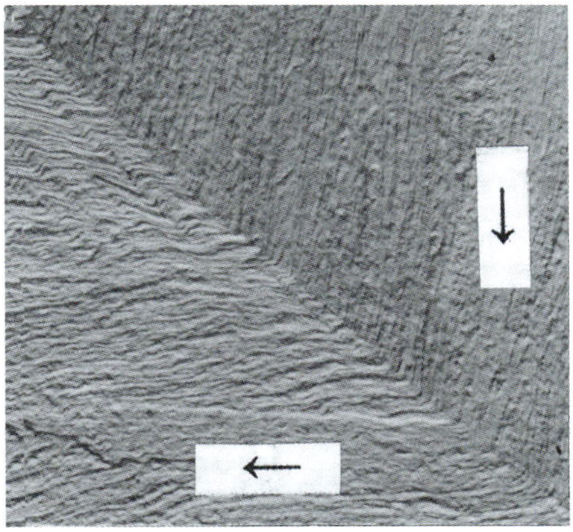

Figure 1.5 Experimental flow lines during crossing of shear plane (Al0.5%Cu alloy, frictionless conditions, $2\theta = 90°$, room temperature) (× 50)

shows macro-etched flow lines in Al0.5Cu alloy after two passes of ECAE with a tool angle $2\theta = 90°$. However, effects of material hardening and elasticity are also observed for a low number of passes. They develop a small-scale plastic flow above the plastic zone AO, which typically appears as some broadening and curvature of the shear plane AO and a characteristic gap AB between material and tool (Figure 1.6(a)). Although the 'gap' was a concern in many publications, it changes localised shear along AO insignificantly and can be eliminated by the application of sufficient back pressure p_0. More important is a tendency for the rigid material to rotate after crossing the shear plane AO. Additional reasons for such rotation are: the singularity of the stress/strain states in the

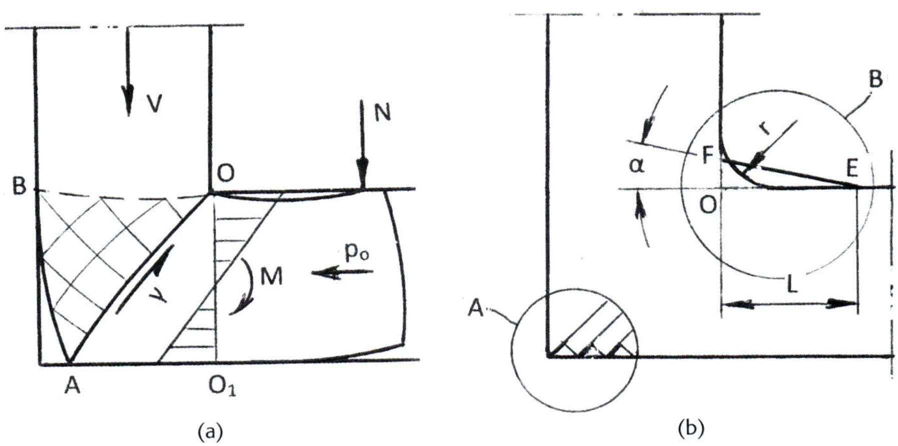

Figure 1.6 (a) Corner gap and material rotation in outlet channel; (b) micro-scale localisation (area A) and balance of rotation (area B)

vicinity of a sharp corner O and the strain rate sensitivity of real metals. As it is impossible to fully eliminate contact friction in channels, there is always a small central fan ϕ as shown in Figure 1.4(a). From Equation (1.10) one can see that shear strain rate η_α becomes infinitely large in the vicinity of O when $r \to 0$. Therefore, the material yield stress significantly increases near O developing plastic heterogeneity, non-uniform flow and a tendency to rotate.

1.6 Outlet channel

1.6.1 Contact friction in outlet channel

The main function of the outlet channel is to balance an extrusion load at the bottom contact area of the billet. From a slip line field of Figure 1.4(a) and Equations (1.6) and (1.8), one can see that for a tool angle $2\theta = 90°$ and $p_0 = 0$, the contact length l and extrusion pressure $p_e/2k$ change from $l = a$, $p_e/2k = 1$ for zero friction to $l = 2a$, $p_e/2k = 2.07$ for maximum friction. Therefore, the channel length ED should be $l > 2a$, where a is the thickness of a billet, and the contact pressure along ED is compared to the material flow stress $\sqrt{3}k$. If $p_0 > 0$, the contact pressure is even larger. Intensive material slip under this high pressure during ECAE squeezes out the lubricant applied to the bottom channel walls whereas lubricant applied to the billet in the inlet channel cannot penetrate onto the bottom wall of the outlet channel where the contact pressure is larger than that in the inlet channel, especially if ECAE is accompanied by the formation of a dead metal zone. An additional complication arises due to a specific characteristic of simple shear processing. Usually, simple shear is not uniformly distributed and small-scale localisation is observed at micro and meso scales as shows in Figure 1.6(b) for an area A (with large magnification). Such localisation unveils the atomically clean material which contacts tool and forms sites of galling. As a result, despite lubricants, sticking is often observed during ECAE that damages material and tool. The bottom channel wall is one of the critical parts of an ECAE die, and there is a significant problem in providing low friction in the outlet channel to reduce stresses, load, improve product quality and increase tool life.

1.6.2 Material flow

Another function of the outlet channel is to preserve the thickness and straightness necessary for multi-pass processing of a billet without machining or reshaping. If the billet moves in the channel as a rigid whole, its dimensions are defined by the cross-section of a channel and the volume of the billet. As can be seen, there is a tendency for rigid billet rotation rather than for progressive movement. Therefore, an additional reaction N arises at the top channel wall to compensate rotation and to straighten the billet in the channel (Figure 1.6(a)). At the start of extrusion, the front end of the billet rotates up and forms a local contact area with the top channel wall. Following extrusion, a channel reaction N along this area develops a bending moment

M in a cross-section OO_1 which unbends the billet and induces a clearance at the top surface of the billet. A small area of material in the vicinity of the centre O is subjected to simple shear along AO and tensile stresses induced by bending. Assuming, for simplicity, frictionless ECAE, a tool angle $2\theta = 90°$ and that the bending stress near O attains the material yield stress $\sigma_s = \sqrt{3}k$ and acts as the additional tensile back pressure $p_0 = -\sigma_s$, then Equation (1.6) gives the tensile hydrostatic pressure around O, $p = -0.42\sigma_s$ against compressive hydrostatic pressure $p = 0.58\sigma_s$ in the identical case of the ideal plastic material with progressive flow into the outlet channel. This notorious difference conforms to the results of FE modelling (Semiatin *et al.*, 2000) and explains the development of local porosity and cracks at the top billet surface during multi-pass ECAE.

One possible solution to the problem is to apply a sufficient radius r at the inner channel corner (Luri *et al.*, 2005) to reduce the area of stress/strain singularity around O and balance the rotation of the material by relatively small elastic strains in the outlet channel (Figure 1.6(b)). In another solution (Wang *et al.*, 2009), progressive flow is provided by a tilted area FE having a small angle $\alpha = 2°–5°$ and sufficiently large length $L \sim 1.5a$ at the top channel wall next to the centre of singularity O (Figure 1.6(b)). This area also balances rotation and develops an additional compressive back pressure around O to compensate for tensile stresses.

1.6.3 Movable channel walls

A radical solution for friction and galling problems is provided by a slider acting as a movable bottom channel wall. In the most important case of ECAE with a tool angle $2\theta = 90°$,[4] two designs found practical applications. In the first (Figure 1.7(a)), a front billet end is open and slider 1 is supported by a guide

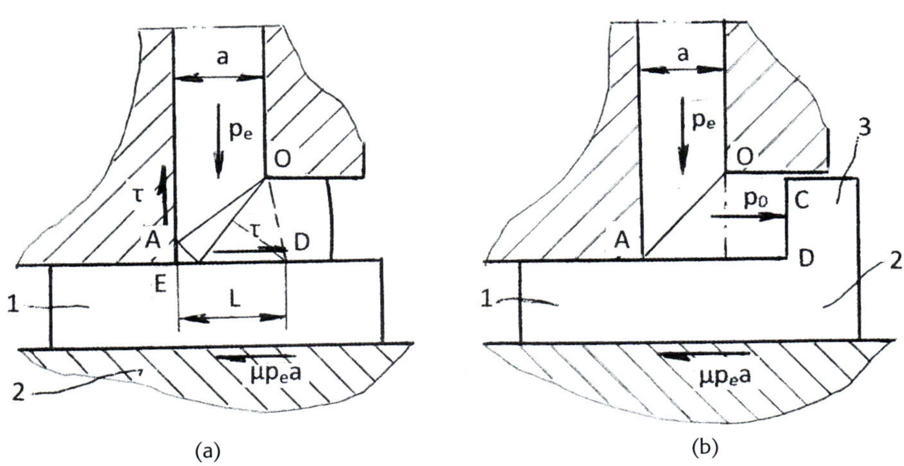

Figure 1.7 Outlet channels with slider: (a) open; (b) locked

plate 2. The extrusion pressure p_e applied to a boundary AO of the plastic zone is transferred to slider and guide plate developing a unit friction force $\mu p_e a$ where a is a billet thickness and μ is the elastic (Coulomb) friction coefficient between the slider and the guide plate. This force is balanced by friction τ along a contact area L between the plastic zone and the slider

$$\tau L = \mu p_e a \tag{1.11}$$

Where the friction τ is not the sliding friction that usually occurs during steady material flow along tool but is self-adjustable friction under insignificant slip. Figure 1.8 is a general diagram of friction τ as a function of a contact displacement (or slip) $[s]$. In area of very small elastic displacements $s < [s]_e$ contact friction changes near linearly with $[s]$. The maximum friction τ_{st} at $[s] = [s]_e$ is known as static friction. For $[s] >> [s]_e$, so-called dynamic or sliding friction τ_s remains stable. Therefore, the contact displacement $[s]$ controls friction τ between material and slider and defines slip line fields in the plastic zone. Assuming that the same friction τ acts in both channels and using slip line field of Figure 1.4(a) together with Equations (1.7) and (1.8) for $p_0 = 0$ and $2\theta = 90°$, Equation (1.11) finally gives

$$\mu = \frac{A}{B} \tag{1.12}$$

Where A and B only depend on the dimensionless ratio (τ/k)

$$A = \left(\frac{\tau}{k}\right)\frac{\sin\eta}{(\sin\eta + \cos\eta)},\ B = \cot\eta + 2\left(\eta - \frac{\pi}{4}\right) + \frac{2\left(\frac{\tau}{k}\right)}{[\sin\eta(\sin\eta + \cos\eta)]} \tag{1.13}$$

Equations (1.12) and (1.13) establish that the friction τ at the channel walls depends on a friction coefficient μ between slider and guide plate. From

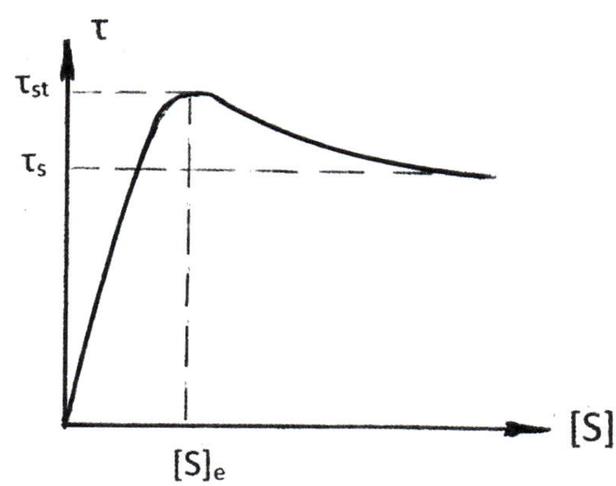

Figure 1.8 Friction τ versus contact slip $[s]$

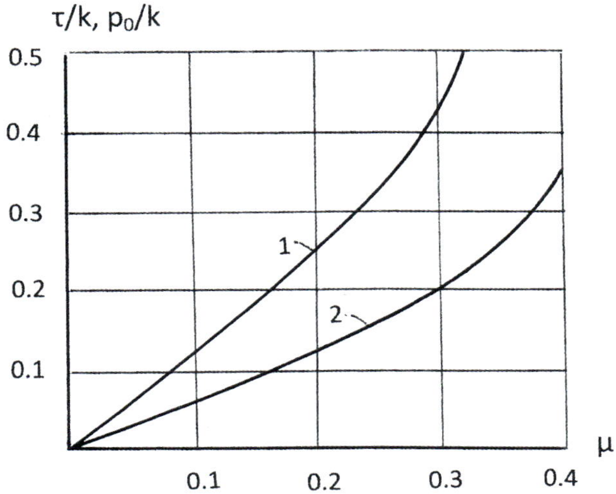

Figure 1.9 Effect of friction coefficient μ on plastic friction τ/k (plot line 1) and back pressure $p_0/2k$ (plot line 2) in outlet channel with slider

numerical results (Figure 1.9, plot line 1) it is obvious that $\tau = 0$ when $\mu = 0$ and τ increases with μ. However, it cannot exceed the static friction τ_{st} (Figure 1.8) for specific processing conditions such as lubricant, material, temperature, etc. If a slider/guide plate friction coefficient μ is sufficiently large and $\tau = \tau_{st}$, after a small amount of slippage, ordinary plastic friction τ_s is established in channels, the slider does not move and ECAE is performed as in stationary channels. This effect conforms to experimental observations. Therefore, good lubrication of channels and the guide plate is important for effective ECAE with low friction.

The situation is different when slider 1 closes the outlet channel (Figure 1.7(b)). In the original position, the material is touching the slider surface CD and maintains tight contact at any moment during extrusion. Because a relative displacement between slider and material is not possible, $\tau = 0$ along AD and ECAE is performed under frictionless conditions. The material develops a pressure p_0 that moves the slider and balances the Coulomb friction $\mu p_e a$ between slider 1 and guide plate 2

$$p_0 = \mu p_e$$

Using the slip line field of Figure 1.2(a) and Equation (1.6) for $2\theta = 90°$, the equation immediately above gives

$$\frac{p_0}{2k} = \frac{\mu}{(1 - \mu)} \tag{1.14}$$

The pressure p_0 acts on the material as the back pressure. A diagram of $p_0/2k$ versus a friction coefficient μ (plot line 2, Figure 1.9) is similar to a considered diagram (τ/k) versus μ. In comparison with the previous case, this design provides four important advantages

- optimal frictionless ECAE irrespective of friction between slider and guide plate
- conversion of adverse effects of tool friction into the compressive back pressure during ECAE which is very desirable in most cases. A sufficiently large friction coefficient μ may be of special interest. For example, with a friction coefficient $\mu = 0.2$ for lubricated hardened steels under large compressive stresses, the back pressure is $p_0/2k = 0.25$
- preservation of the correct rectangular billet shape after each pass of ECAE without reshaping
- stable processing with repeatable material properties

It has also been suggested that the bottom channel wall could act as a roll (Stecher *et al.*, 2003; Park *et al.*, 2006). Such a complicated design, especially for bulk billets, does not find practical applications.

1.7 Inlet channel

Although the quality of the product depends to a large extent on the outlet channel, the inlet channel defines the processing pressure and load. Minimisation of these characteristics is critical for ECAE of high-strength materials and bulk billets, as well as to reduce the wasting of material and to increase the life of the tools. The main function of the inlet channel is to feed material and lubricant into the plastic zone (Figure 1.10). The billet length L should be

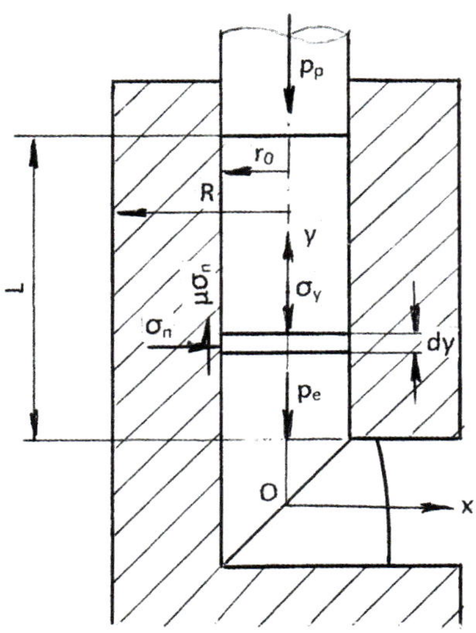

Figure 1.10 Elastic stresses in inlet channel

sufficiently larger than a channel thickness a, that is $L/a >> 1$. As the channel is straight and has the same cross-section as the billet, the movement of the billet within the channel does not induce plastic deformation. Some deformation is observed at the start of extrusion until a pressure at the front billet end attains the value p_e which is necessary for extrusion. If $p_e < 2k$, deformation of long billets in the channel takes place by buckling, which forms irregular local contacts with the channel walls. Such small contact areas do not develop noticeable friction, and extrusion is performed at a punch pressure p_p comparable to p_e. According to Equation (1.6), this is a low friction case with $2\theta > 90°$ and $p_o = 0$. When friction is noticeable and $p_e = 2k$, the billet, is first upset in some areas and then comes into tight contact with the channel walls. Due to friction, the pressure above these areas consistently increases and promotes the upsetting of the entire billet. Finally, the full billet–channel contact area is formed before the extrusion commences (Figure 1.10). This is the undesirable case of ECAE. Therefore, an analysis of punch pressure and contact stresses on the channel walls is of considerable practical interest when selecting the equipment and designing the tools.

1.7.1 Friction in the inlet channel

If $p_e < 2k$ and there is no full contact between material and channel, friction conditions in the inlet channel are much more favourable than those in the outlet channel because: the contact pressure at the channel walls is significantly lower, there is no dead metal zone, lubricant is deposited directly on billet and channel, and the movement of the billet (as the rigid body) does not destroy the lubricant. However, large displacements may deplete a lubrication layer with low viscosity. Heavy greasing with powder graphite and MoS_2 provide good lubrication at cold and warm temperatures. Application of commercial anti-seize lubricants was also successful for many materials. Solid films and coatings are preferable for high-strength and hard-to-deform materials. With a correct choice of lubricants, load–displacement diagrams show a steep drop in proportion to decreasing billet contact area from the beginning to the end of ECAE, confirming that the friction in the inlet channel is low.

1.7.2 Stress distribution

When $p_e >> 2k$, there is full contact between the material and channel walls while the billet–tool interaction remains elastic. We now consider ECAE with a cylindrical inlet channel of length L and radius r_0 used as a tool with an external radius R (Figure 1.10). Neglecting friction between material and channel, for an inner channel radius r, elasticity theory gives (Segal *et al.*, 1994)

$$r = r_0\left[\frac{1+\sigma_n(m^2 - \nu_1 m^2 + \nu_1 + 1)}{(E_1 - E_1 m^2)}\right] \tag{1.15}$$

Table 1.1 Normal to axial tool pressure ratio c for different materials processed by ECAE (tool is made of hot working steel H13 and its geometry characterised by inner to outer radius ratio of 0.33)

Material	Al	Cu	Ti	Mg	Fe
Coefficient c	0.275	0.19	0.26	0.356	0.13

Where σ_n is a normal pressure at the channel wall, $m = r_0/R$, v_1 and E_1 are the Poisson coefficient and the Young modules of the tool, respectively. Similarly, the radius of the material element dy is

$$r = r_0 \frac{1 + (v_2\sigma_y - \sigma_n(1 - v_2))}{E_2} \tag{1.16}$$

Where σ_y is the axial stress, v_2 and E_2 are, respectively, the Poisson coefficient and the Young modulus of the material. Comparing Equations (1.15) and (1.16), one finds

$$\sigma_n = c\sigma_y \tag{1.17}$$

Where

$$c = \frac{[E_1 v_2(1 - m^2)]}{[E_2(m^2 - v_1 m^2 + v_1 + 1) + E_1(1 - m^2)(1 - v_2)]} \tag{1.18}$$

Coefficients c calculated for typical metals and tool fabricated from steel H13 with $m = 0.33$ are presented in Table 1.1.

Equations (1.17) and (1.18) correlate the normal pressure σ_n at the channel wall with axial stresses σ_y along the channel, tool ratio r/R and the elastic constants of the material and tool. Then, a balance of forces acting on the element dy with friction $\tau = \mu\sigma_n = \mu c\sigma_y$, where μ is the friction coefficient, is

$$\frac{d\sigma_y}{\sigma_y} = \frac{2\mu c dy}{r}$$

Integration of this equation at the boundary condition $\sigma_y = p_e$ at $y = 0$ gives a distribution of stresses along the channel length

$$\sigma_y = p_e \exp\left(\frac{2\mu c y}{r}\right), \ \sigma_n = c\sigma_y \tag{1.19}$$

The punch pressure at $y = L$ is

$$p_p = p_e \exp\left[2\mu c\left(\frac{L}{r}\right)\right] \tag{1.20}$$

For a square channel of a thickness a, the radius r in Equation (1.19) should be substituted by $r \sim a/\sqrt{\pi}$. For billets with a length-to-diameter ratio $L/r \leq 10$,

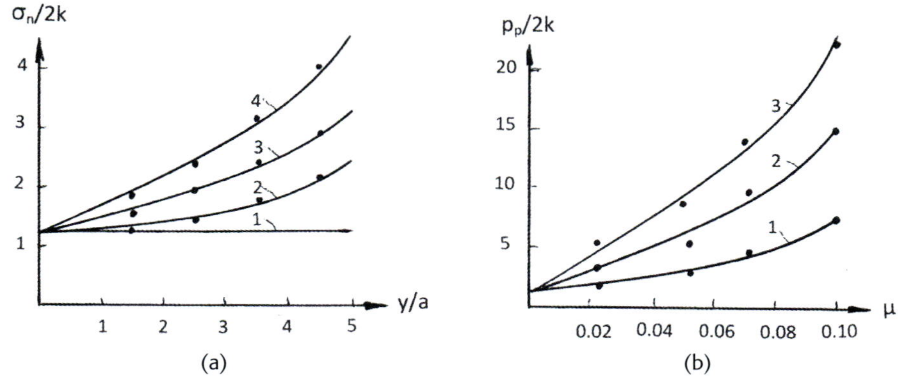

Figure 1.11 (a) Distribution of contact stress $\sigma_n/2k$ along inlet channel for friction: plot 1 – $\mu = 0$, plot 2 – $\mu = 0.05$, plot 3 – $\mu = 0.075$, plot 4 – $\mu = 0.1$; (b) effect of friction μ on punch pressure $p_p/2k$ for channels with $L/r = 10$ (plot 1); $L/r = 16$ (plot 2); $L/r = 20$ (plot 3). Material is Pb

$c < 1$, and low friction coefficient $\mu << 1$, approximate formulae can be used instead of Equations (1.19) and (1.20)

$$\sigma_y \simeq p_e\left(1+\frac{2\mu cy}{r}\right), p_p \simeq p_e\left(1+\frac{2\mu cL}{r}\right) \tag{1.21}$$

These solutions have been experimentally verified (Segal *et al.*, 1994). Full contact with different lubricants was provided by a preliminary pressing of the billet in the channel before ECAE. Solid curves in Figures 1.11(a, b) show, respectively, analytical distributions of stresses $\sigma_n/2k$ along a channel length y/a and punch pressure $p_p/2k$ versus friction coefficient μ calculated from Equations (1.19) and (1.20) which fit well with experimental data. These results demonstrate that even low friction in the inlet channel with full plastic contact leads to a significant increase of stresses on the channel walls and increased extrusion load. Development of a reliable ECAE tool with stable and cost-effective processing is an important engineering problem.

1.7.3 Movable channel walls

Movable walls can be used for inlet channels having rectangular cross-sections. They eliminate friction along certain channel walls and provide a multi-fold increase of punch cross-section and stability. A few possible designs are shown in Figures 1.12–1.14. The largest effect provides a die with three movable walls incorporated into a slider 1 moved by a punch 2 (Figure 1.12). Such a design was first described by Segal *et al.* (1994) and later by Mathieu (2006) and Azushima (2007).

Punch 2 is a massive block that overlaps the channel and slider. The length of the billet can therefore be, at least, tripled in comparison with an identical

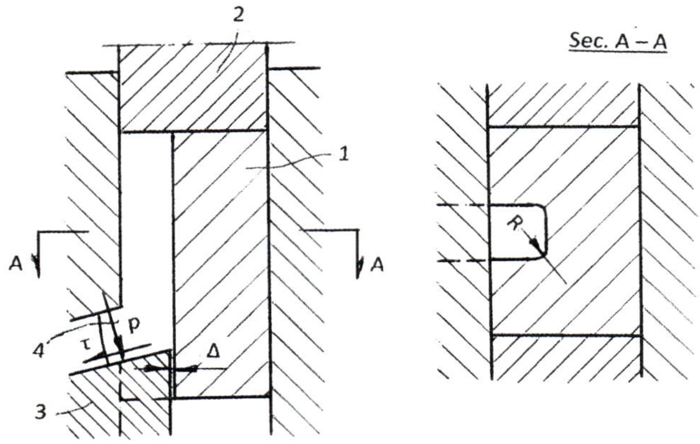

Figure 1.12 ECAE tool with three movable walls of inlet channel

solid channel shown in Figure 1.10. Some problems occur due to material sticking and flashes along block 3, which forms a bottom wall of the outlet channel 4. Side flashes may be prevented by sufficient radii R at the channel corners. However, it is quite difficult to eliminate a clearance Δ between slider and block 3 through which lubricant flows out and does not come onto block 3. Large friction τ at this block increases the clearance Δ during extrusion. The clearance may be controlled by the balance of normal p and tangential τ stresses acting on block 3

$$\frac{p}{\tau} \sim -\tan 2\theta$$

As $p > 0$, $\tau > 0$, it can only be achieved for tool angles $2\theta > 90°$. For the tool angle $2\theta = 90°$ and effective high-temperature lubricants, this design also showed good performance during hot ECAE of steels (Segal *et al.*, 1994).

Another design which eliminates friction along channel walls has been proposed (Segal *et al.*, 1997) (Figure 1.13). In this case, two identical billets 1, 2 are extruded simultaneously into opposite directions without friction along their mutual boundary ab. Side plates 3, 4 are moved together with billets by punch 5 which overlaps the channel and side plates and provides frictionless conditions along two other channel walls. The lubricant seized between billets is fed directly to the top surface of block 6. To reduce a normal pressure at this surface, the channel angle is $2\theta > 90°$. Flashes are eliminated by a chamfer on the edges of the billet. In comparison with the previous case, this design allows high-strength materials with large billet length-to-thickness ratios to be processed.

For ECAE of batch billets, a tool with two side sliders 1, 2 may be combined with a bottom slider 3 in the outlet channel (Figure 1.14). Punch 4 overlaps the channel cross-section and sliders. This design removes 50% friction in the inlet channel, increases punch stability and eliminates material galling in the outlet

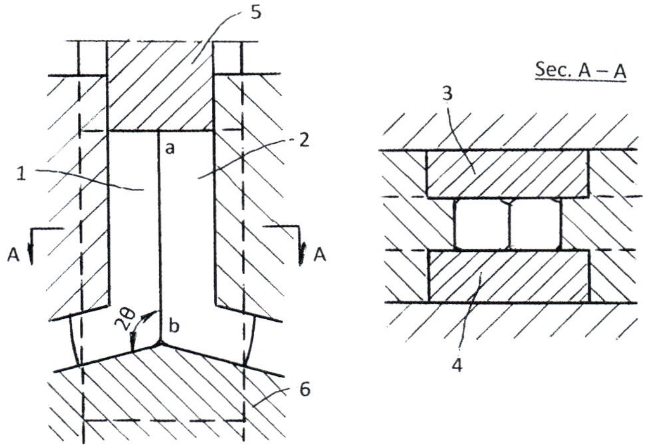

Figure 1.13 ECAE tool with two billets and two side sliders

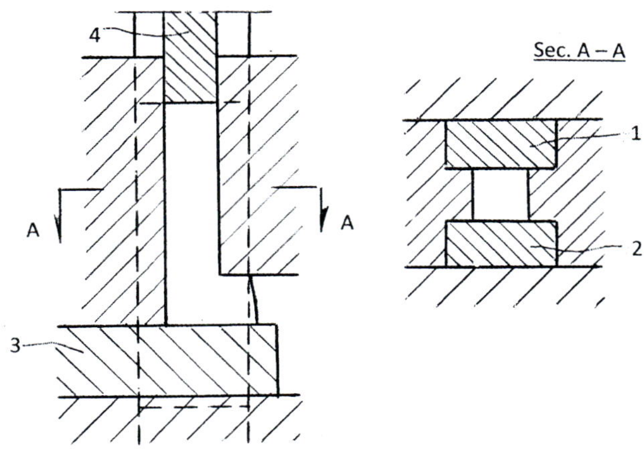

Figure 1.14 ECAE tool with two side sliders and a bottom slider

channel providing a good compromise between optimal conditions in both channels. Its disadvantages include complexity and the possibility of forming flashes in clearances at the bottom slider 3.

1.8 ECAE of batch billets

1.8.1 Shapes of billets

ECAE is usually performed for separated batch billets such as bars, rods and plates. To use multi-pass processing in the same dies, bars have square cross-sections ($a \times a$) and plates have a square shape ($L \times L$) (Figure 1.15). In order to achieve punch stability, the billet length-to-thickness ratio is usually

$L/a < 6$. For long square billets ($a \times a \times L$) (Figure 1.15(a)), the most advanced is a tool with two movable walls into the inlet channel and a movable bottom wall into the outlet channel (Figure 1.14). In this case, depending on material and lubricant, the billet length-to-thickness ratio can be $L/a \approx 10$. With good lubricants, the maximum punch pressure at the beginning of extrusion is about $p_p/2k \approx 1.5$, that is much less than $p_p/2k \approx 7–9$ in dies with solid channels. With the movable bottom wall in the outlet channel, the billet can easily be removed from the die to clean and lubricate the channels after each pass. The thickness of the inlet channel is slightly larger whereas the thickness of the outlet channel is slightly smaller than the channel width in order to provide the clearance necessary to reinsert the billet for subsequent passes.

As movable die walls cannot be applied to cylindrical billets, solid dies are used in this case (Figure 1.10). Such a simple tool presents a few problems

- The billet ejection from the outlet channel should be performed by the following billet or a dummy billet that distorts both billet ends.
- Billet ends should be cut off after each pass.
- The outlet channel remains unavailable for lubrication at each pass.
- There is a high probability of the material sticking and galling to the channel walls.

Therefore, ECAE of cylindrical billets leads to considerable waste of material waste and a short tool life. References to high extrusion pressures and punch fracture can be found in the literature. Nevertheless, owing to simplicity, cylindrical dies are widely used as a laboratory tool.[5]

For plate billets ($L \times L \times a$) (Figure 1.15(b)) movable walls can be used in both channels. However, for $L/a >> 1$, friction along the side walls ($L \times a$) of the inlet channel is insignificant in comparison with friction along the top and bottom billet surfaces ($L \times L$),[6] and movable side walls in the inlet channel are ineffective. In this case, a simple die design with a single slider in the outlet channel (Figure 1.7) provides only a slightly higher punch pressure than the

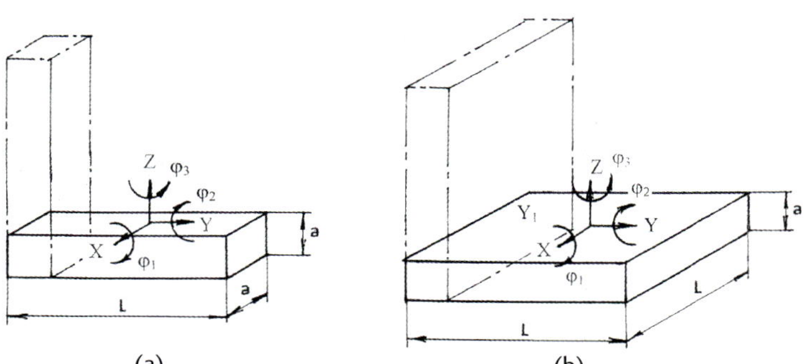

Figure 1.15 Orientations of billets during ECAE: (a) long square billets; (b) plate billets

die with three sliders (Figure 1.14). Together with easy billet ejection, lubrication of both channels after each pass and maintenance of the rectangular billet shape, multi-pass ECAE of plate billets can be performed pass-by-pass without reshaping the billet and preheating between passes. Additionally, the weight of the plate billets is L/a times larger than the weight of identical bar billets and the required load and stroke for plate billets fit the existing presses better.

1.8.2 Multi-pass processing

In most cases, ECAE should be repeated a few times to accumulate strains and modify the structure of the material. The number of passes should be as low as possible to increase productivity and reduce cost. The obtained structure is usually used as a criterion of strain-induced material modifications. The related changes are large after the first few passes and are gradually damped during subsequent passes. Therefore, control of the attainable properties is also useful to establish the optimal number of passes. The processing with low friction and better approximation to simple shear, providing stronger structural effects, is preferable to ECAE which has noticeable friction and a larger number of passes.

The control of the orientation of the billet between passes, which has been proposed for long and plate billets (Segal, 1996, 1998), presents additional options. Numerous combinations of possible billet rotations ϕ_1, ϕ_2, ϕ_3 about coordinate axes X, Y, Z (Figure 1.15) are reduced to four basic routes A–D as shown in Table 1.2.

These routes induce different shear planes and distortions of material elements shown in Figure 1.16 for four passes. Correspondingly, numerous micro-SBs arise in the material along shear planes at each pass. After multi-pass ECAE, different routes produce specific patterns of SBs which affect microstructures, textures and properties. In particular, because the boundaries of the cell blocks are oriented along micro-SBs and are transformed into high-angle boundaries, the spacing and intersections of SBs define the size and morphology of strain-induced micro structures. Other structural characteristics depend on the number of ECAE passes and routes in various ways. Their selection should be considered and optimised in particular cases. More details related to the structural effects of ECAE can be found elsewhere (Meyers *et al.*, 2006; Valiev *et al.*, 2006; Beyerlein *et al.*, 2009).

Table 1.2 Definition of billet rotation for routes A–D for long billets and plate billets

Route	Long billet			Plate billet		
	$\phi 1$	$\phi 2$	$\phi 3$	$\phi 1$	$\phi 2$	$\phi 3$
A	0	0	0	0	0	0
B	0	±90°	0	0	0	±90°
C	0	0	180°	0	0	180°
D	0	90°	0	0	0	90°

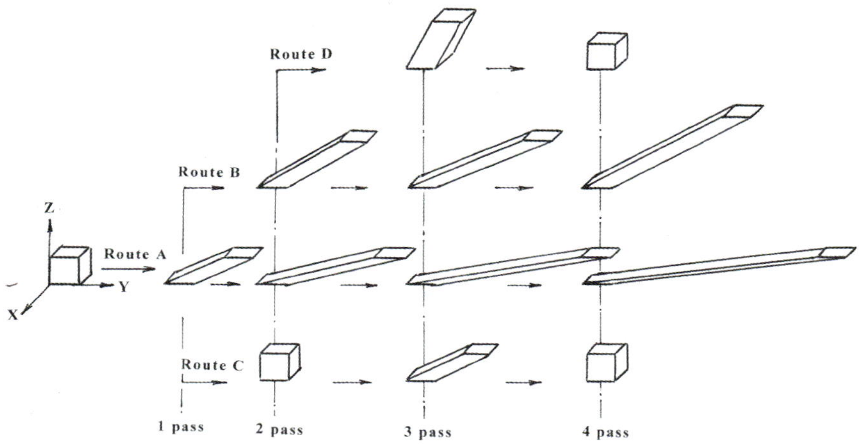

Figure 1.16 Material distortions for basic ECAE routes

With correct processing, ECAE provides uniform strains in most of the billet volume except the front and back triangle areas resulting in the so-called 'end effect'. During multi-pass ECAE via different routes, these areas change orientation, loading history and properties. The 'end effect' becomes insignificant with the increase of passes and the billet length-to-thickness ratio (L/a). Special combinations of the basic routes have been suggested to optimise ECAE in particular cases (Barber *et al.*, 2004; Bian *et al.*, 2011, Segal *et al.*, 2010).

1.8.3 Processing parameters

As a processing technology for properties, ECAE can be applied to numerous materials and problems. In any case, near zero friction, tool angle $2\theta = 90°$ and sharp corner channels provide intensive and uniform simple shear with effective structure modification. Similarly to the number of passes and routes, the processing temperature and strain rate depend on the material and character of modifications with different combinations of hardening, recovery and recrystallisation. Temperatures can be selected within ranges of cold, warm and hot deformation. The general requirement is that the material is able to withstand the necessary strains without fracturing. Materials with high ductility can be processed at room temperature. Materials with moderate or low ductility are preheated below their recrystallisation temperature for warm ECAE or above their recrystallisation temperature for hot ECAE. There are some restrictions on strain rate. Because there are large strains per pass within a narrow shear zone, the strain rates during ECAE are of an order of magnitude larger than for ordinary forming operations. Thus, the processing speeds for ECAE should be lower. Some modifications to the structure, such as dynamic recrystallisation, dynamic aging and phase transformation controlled by diffusion are time sensitive and require a precise selection of strain rate and temperature during ECAE.

The structural effects of ECAE are often established in small-scale experiments. These results can be transferred to a large-scale processing of the same material for geometrically similar dies and billets, identical lubricant, processing temperature and time

$$t = \frac{L}{V} = \frac{l}{v}$$

Where L and l are lengths and V and v are processing speeds at large and small scales, respectively. With an increase of the billet size from l to L, the extrusion speed increases in proportion to (L/l), the extrusion pressure remains the same and the processing load increases in proportion to $(L/l)^2$. When the processing time t reaches 10 seconds and more, ECAE, especially of small billets, should be performed at isothermal conditions in dies equipped with heating and temperature control devices. On the other hand, adiabatic heating during sufficiently fast multi-pass ECAE, especially for large billets, may increase the temperature by a few hundred degrees Celsius and should be taken into account when considering structural changes. Extrusion temperature and speed are the main factors used to control structures and produce defectless materials using ECAE.

1.8.4 Defects associated with ECAE

The unusual mechanics of ECAE predispose it to specific processing defects. These defects are not inevitable and can usually be eliminated by fixing the process. The most common defect is fracture along a shear plane that is at an angle θ to the direction of extrusion. Materials with low ductility develop deep cracks or undergo separation of the front billet end and following parts as simple shear periodically restarts. Materials with deformation softening may exhibit coarse macro-localisation within small areas along the shear plane with a highly inhomogeneous strain distribution and a tooth-like top surface. Shear cracks sometimes grow from sharp tooth edges into the billet. Another type of surface cracking is observed at the top billet surface in the outlet channel (Figure 1.6(a)) because of additional tensile stresses in this area induced by rotation of the billet. These cracks are not oriented along the shear plane but are generated by the ordinary mechanisms of pore nucleation, growth and coalescence (Semiatin *et al.*, 2000). Both types of cracks can be eliminated by raising the extrusion temperature and reducing the speed. Additional back pressure can be applied in some cases (Lapovok, 2005).

Surface defects also arise at the bottom billet surface (Figure 1.6(a)) due to irregularities in the shape of the billet and contact friction. At the start of ECAE, some area around A contacts the bottom die wall whereas an area above A forms a gap. Their superposition develops a shallow surface lap at A. As contact friction is a naturally unstable phenomenon, small dead metal zones and gaps may be periodically originating and disappearing at the die corner during steady ECAE. After transition to the bottom billet surface, they also form shallow laps. These surface defects can be removed by scalping or machining after ECAE.

1.9 Continuous ECAE

Continuous ECAE (CECAE) is an attractive alternative to ECAE of batch billets owing to its high productivity, low wastage of material and the fabrication of long semi-finished products. Instead of the punch pressure in ordinary ECAE, an extrusion load in CECAE may be developed by a pulling force at the outlet billet end, pushing force at the inlet billet end, and pushing and pulling forces at both ends of the billet. The first concept was realized by Chukkingal *et al.* (1998) in equal channel angular drawing (ECAD). Lakota *et al.* (2002) showed that the stress–strain state during ECAD is similar to tensile bending around a channel corner that leads to material thinning, near pure shear deformation mode and local fracture. The pushing forces developed by contact friction at the front billet end before entering the inlet channel or by the moving walls of the inlet channel, as in the Conform process, are more effective. Different designs were used to develop pushing forces (Chung *et al.*, 2003; Chaudhury *et al.*, 2005; Saray *et al.*, 2010; Utsunomiya *et al.*, 2001; Lee *et al.*, 2002; Han *et al.*, 2002; Stecher *et al.*, 2003; Rusz *et al.*, 2010). Conform CECAE is the most practical. It was first realized in 1976 at FTI, Minsk (Segal *et al.*, 1994) and later by others (Han *et al.*, 2002; Huang, 2005; Zhu *et al.*, 2006). However, in none of these papers were investigated the specifics of CECAE that often led to material slippage and overheating, surface defects, increased power and tool wear. As in other SPD techniques, an analysis of the mechanics is necessary to optimise processing, structure and properties. A simple analytical model of Conform CECAE has been suggested in (Segal, 2010).

In Conform ECAE, the inlet channel works as a friction press that grabs the material and develops an extrusion pressure. In contrast to ordinary Conform extrusion, the required pressure is significantly lower and processing can be performed with lubricants, at low speeds and without intensive heating. The advanced design of Conform ECAE is shown in Figure 1.17. The inlet channel is formed between a stationary block 3 and a driven roll 1 with a groove 2. An auxiliary roll 8 coins the billet 5 into the groove 3 to a thickness which is slightly smaller than a gap between roll 1 and block 3. That provides a tight contact between the billet and groove walls. When the billet is advanced to a support 4, slip and friction between billet and roll increase and develop sufficient pressure to upset the front billet end. The material fills the inlet channel and forms a contact area with block 3, slippage and friction continue to grow until extrusion starts into an outlet channel 7 formed between block 3 and support 4. This design provides billet grasping, feeding and automatic control of pressure by self-adjustment of slip, friction and contact length L_b during unsteady extrusion with variable parameters. Related analysis will be considered further in the following subsections of Section 1.9.

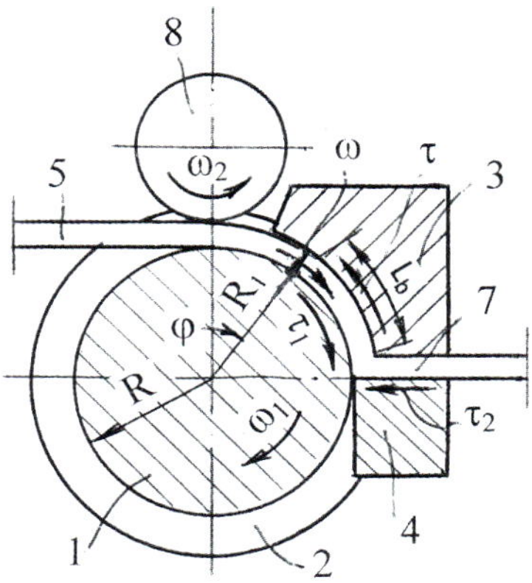

Figure 1.17 Conform ECAE with a coining roll

1.9.1 Friction conditions

There are three different areas of friction between material and tool (Figure 1.17): τ_2 along support 4, τ along block 3 and τ_1 along roll 1. Passive friction τ, τ_2 can be described by Prandtl's equations $\tau = mk$, $\tau_2 = m_2k$ where m and m_2 are related plastic friction factors. Active friction τ_1 depends on the slip $[s]$ between material and roll in accordance with the general diagram shown in Figure 1.8. It may be described by the approximate equations

$$\tau_1 = \frac{\tau_s[s]}{[s]_e} \text{ in an area of small displacements } [s] \leq [s]_e \quad (1.22)$$

$$\tau_1 = \tau_s = mk \text{ in area of large displacements } [s] > [s]_e \quad (1.23)$$

Consider Conform ECAE of a rectangular billet ($a \times b$) assuming that the coining roll 8 bends the billet along a roll radius R and provides full plastic con-

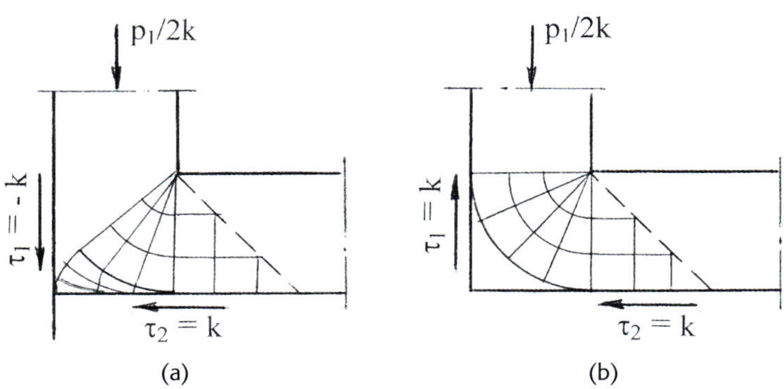

Figure 1.18 Plastic zones for ECAE with: (a) maximum active friction; (b) passive friction in both channels

tact between billet and roll. The material within the groove rotates about the roll centre O with an angular speed ω that is slightly slower than an angular speed ω_1 of the roll. For $a << R$ and $\omega \approx \omega_1$, the local slip between material and roll is

$$[s] \approx \left(1 - \frac{\omega}{\omega_1}\right) l \tag{1.24}$$

Where $l = r\phi$ is a distance from the coining area. Equations (1.22)–(1.24) describe a friction distribution between material and roll

$$\tau_1 = \frac{\tau_s\left(1 - \frac{\omega}{\omega_1}\right) l}{[s]_e} \quad \text{for} \quad [s] \leq [s]_e,\ \tau_1 \leq \tau_s \tag{1.25}$$

$$\tau_1 = \tau_s \quad \text{for} \quad [s] > [s]_e \tag{1.26}$$

Depending on slip $[s]$, there are three possible situations:

- Friction increases linearly with l but maximum friction τ_s is not attained at any point of the whole contact length L (Figure 1.19)

$$0 \leq \tau_1 \leq \tau_s \quad \text{and} \quad \frac{\omega}{\omega_1} > 1 - \frac{[s]_e}{L} \tag{1.27}$$

- Friction increases from zero to maximum at some point l^* and sliding friction τ_s acts along the remaining part of the contact length L

$$0 \leq \tau_1 \leq \tau_s \quad \text{for} \quad 0 \leq l \leq l^*,\ \tau_1 \sim \tau_s \quad \text{for} \quad l^* \leq l \leq L, \frac{\omega}{\omega_1} = 1 - \frac{[s]_e}{l^*} \tag{1.28}$$

- Sliding friction τ_s acts along the entire contact length

$$\tau_1 = \tau_s = m_1 k \quad \text{for} \quad 0 \leq l \leq L \tag{1.29}$$

For the friction distributions given in Equations (1.27) and (1.28), a ratio ω/ω_1 is definitive with low slip $[s]$. These conditions describe stable processing. For friction (Equation (1.29)), the ratio ω/ω_1 is indefinite and slip $[s]$ may be large.

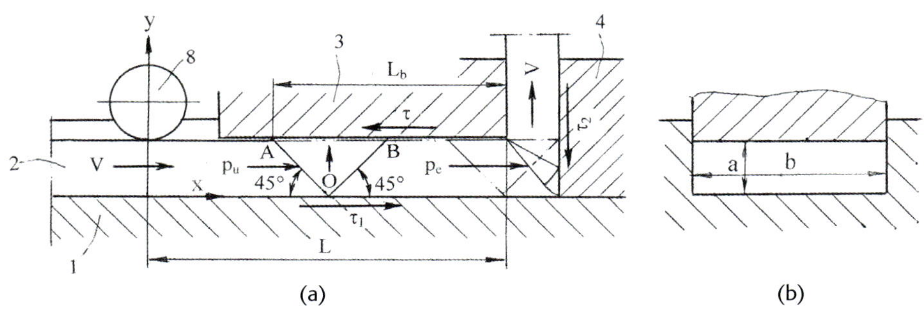

Figure 1.19 Balance of forces into an inlet channel

Once it has started, it leads to intensive surface heating, a decrease in contact friction and skidding between material and roll.

1.9.2 Plastic zone

Equations (1.27)–(1.29) define friction τ_1 in a plastic zone. The slip line field in this zone can then be found from Figure 1.4(b) for particular values τ_1 and τ_2. For example, Figure 1.18 shows slip line fields for $2\theta = 90°$ in the limit cases of maximum active (Figure 1. 18(a)) and passive (Figure 1.18(b)) friction $|\tau_1| = \tau_2 = k$ in both channels. It should be noted that active friction in the inlet channel eliminates a dead metal zone at the die corner, in contrast to the extensive dead zone for passive friction. Calculations of extrusion pressure and accumulated shear give $p_e/2k = 1.96$, $\gamma = 1.92$ in the first case against $p_e/2k = 2.07$, $\gamma = 1.57$ in the second case. Although extrusion pressures p_e are almost identical, punch pressures p_p and required work are significantly larger in the second case because of contact friction at the channel walls, especially for long billets. At the same time, the shear strain γ is larger for active friction and occurs near a bisect channel plane instead of shear into an extrusion direction for passive friction. This comparison demonstrates that the processing mechanics of CECAE are closer to ideal frictionless ECAE than those of ordinary ECAE with the same friction.

1.9.3 Analysis of CECAE

As $a << R$, a balance of forces within the inlet channel (Figure 1.17) can be considered for a rectilinear diagram shown in Figure 1.19 where a coordinate $x = r\phi$. Initially (dashed lines), the material 2 moves together with roll 1 without slip ($[s] = 0$) and approaches a wall 4 of the outlet channel. Then slip $[s]$ and friction τ_1 increase until the axial pressure p_u becomes sufficient to upset the front billet end. A simple upper-bound solution assuming lines of velocity discontinuity AO and BO and friction $\tau_2 = m_2 k$ at groove walls gives the upsetting pressure

$$\frac{p_u}{2k} = 1 + \frac{m_2 a}{2b}$$

For $a \leq b$ this pressure is lower than the extrusion pressure $p_e/2k$. Therefore, slip $[s]$ increases further to start extrusion that is accompanied by the increase in the contact area with block 3. Finally, steady extrusion into the outlet channel is attained at some ratio ω/ω_1 and a contact length L_b. Assuming, for simplicity, that axial pressure $p_x = 0$ for $x = 0$, friction factors m, m_1, m_2, critical slip $[s]_e$ and the extrusion pressure p_e are known, then different boundary conditions (1.27)–(1.29) can be considered separately. The minimum contact length L providing steady extrusion is of practical interest. Obviously, the minimum length L_{min} conforms to the distribution given in Equation (1.29) when $m_1 = m_2 = m$. An equilibrium of forces along an axis x is then

$$m_1 k L_{min}(2a + b) - m_1 k L_b b - p_e ab = 0$$

An additional equation is condition $p_x = p_u$ for $x = (L - L_b)$

$$m_1 k(L_{min} - L_b)(2a + b) - p_u ab = 0$$

The solution of both of the equations immediately above gives

$$\frac{L_{min}}{a} = \frac{np_e}{(2km_1)} - \frac{n(2n + m_1)}{(2(2 + n)m_1)} \tag{1.30}$$

$$\frac{L_b}{a} = \frac{np_e}{(2km_1)} - \frac{(2n + m_1)}{(2m_1)}$$

Where $n = b/a$ is the width-to-thickness ratio for a billet. However, as the contact friction during processing may vary, this case is unstable and uncontrollable slip may occur. To compensate for possible variations in friction, material properties and temperature, the actual channel length L should exceed L_{min} by a stability factor ν that is $L = \nu L_{min}$, $\nu > 1$. In these cases, distribution τ_1 along L includes areas of the linear and constant friction (Equation (1.28)).

Diagrams of (L_{min}/a) calculated from Equation (1.30) versus the friction factor m_1 are shown in Figure 1.20 for billet width-to-thickness ratios n = 1, 2, 4, 8 and an extrusion pressure $p_e/2k$ = 1.97 corresponding to the maximum friction in the outlet channel. The characteristic length (L_{min}/a) plays an important role in the optimisation of continuous ECAE and depends almost linearly on the ratio n and inversely on friction τ_1. For a square cross-section with n = 1,

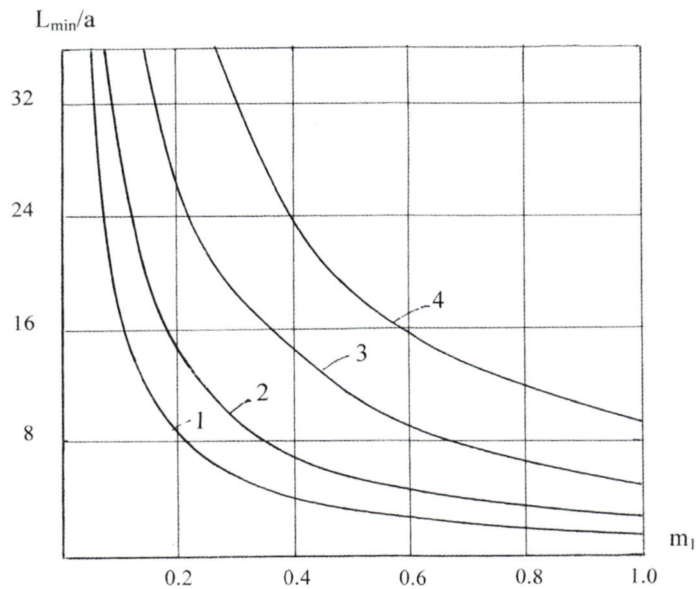

Figure 1.20 Effect of plastic friction factor m_1 on minimum length of inlet channel (L_{min}/a) for $n = 1$ (plot line 1), $n = 2$ (plot line 2), $n = 4$ (plot line 3) and $n = 8$ (plot line 4)

a relatively short channel provides the necessary extrusion pressure even for low contact friction, and CECAE can be performed with effective lubricants. However, CECAE of strips with ratio $n > 4$ requires sufficiently high friction $m_1 > 0.5$. Continuous processing of thin sheets with large ratios n is especially difficult because $(L_{min}/a) \rightarrow \infty$ when $n \rightarrow \infty$. In such cases, differential friction at top and bottom billet surfaces or additional auxiliary rolls are necessary.

1.10 Modifications of ECAE

ECAE was originally introduced as a deformation technique for simple shear processing for properties. If contact friction can be reduced to near zero, ECAE also provides exceptional technological characteristics such as minimum stresses and process work together with stress–strain uniformity. Practical realization of these opportunities presents an engineering challenge requiring sophisticated operations and tools. Additional shortcomings are low productivity, wastage of material and product cost. Modifications of ECAE were therefore proposed to resolve these problems. Multi-pass ECAE without ejecting the billet from the tool was suggested in a rotary die (Nisida *et al.*, 2001) or a special press (Azushima *et al.*, 2002). In multi-turn ECAE, a few billets are simultaneously extruded through a few successively located ECAE channels in a semi-continuous manner billet-by-billet (Schukin and Segal, 1976; Liu *et al.*, 1998; Nakashima *et al.*, 2000; Rosochowski *et al.*, 2002; Varyukhin *et al.*, 2002). Both approaches reduce the processing time and eliminate operations of lubrication, billet reshaping and reinserting after each pass. However, significant disadvantages are: high extrusion pressure, material sticking, and the impossibility of lubricating and inspecting a die which leads to poor quality, excessive load and a short tool life.

Some improvements have been oriented towards reducing the friction in the channels. A die with low friction in the inlet channel similar to that shown in Figure 1.13 was used by Canta *et al.* (2008). An inversion of this scheme into a two billet–two punch design eliminates friction at the bottom wall of the outlet channel (Rosochowski *et al.*, 2011). Rosochowski (2006) also introduced an incremental ECAP in which an approximation to simple shear was attained by a cyclic movement of the bottom wall of the outlet channel into the shear direction with periodic material feeding. Such processing is performed semi-continuously, in small steps, and potentially eliminates friction in both channels. It will be considered in detail in Chapter 2 of this book.

A few attempts were also made to increase strain intensity during ECAE and to reduce the number of passes. For example, it was suggested that additional deformation could be induced by changing the cross-section of the channels (Kocisko *et al.*, 2009) or by twisting the material in the inlet channel (Kosich *et al.*, 2010) as well as non-ECAE (Toht *et al.*, 2009; Nagasekhar *et al.*, 2008). In these cases, in addition to significant technological complications, the processing mechanics deviates from the simple shear deformation mode, and is probably less effective for structure formation.

Most modifications of ECAE are at the early stage of R&D. At present, it is not clear what practical and technological benefits they may provide because particular improvements are achieved at the expense of other characteristics such as strain uniformity, low stresses and loads, simplicity, etc.

1.11 Applications

Similarly to other SDP techniques, ECAE results in various changes in structure and properties which are usually difficult or impossible to obtain with ordinary forming operations. One remarkable effect is structure refinement to sub-micron and, sometimes, to nano scale. For certain conditions, mechanisms of continuous and discontinuous recrystallisation, phase separation in duplex alloys, phase transformation and geometric recrystallisation may operate during ECAE. The specific mechanism is strain-induced new high-angle boundaries organised into three-dimensional configurations such as sub/grains and fragments. Although such structures differ from ordinary structures of recrystallisation, in many aspects they behave like grains and are identified as ultra-fine grained (UFG) structures. Fabrication of UFG materials is only one of the many possible applications of ECAE. Less acknowledged applications include: modification of material texture, consolidation and bonding of particulate materials, breakdown and homogenisation of cast alloys, topological control of structural elements, *in situ* composites, phase transformation, processing of specialty and functional materials. ECAE may be applied to different materials including: metals, ceramics, semiconductors and polymers; polycrystals and single crystals; solids, semisolids and powders. An additional area is ECAE processing to improve manufacturing characteristics such as: high strain rate/low temperature superplasticity, advanced formability and machinability to fabricate complicated and precision components at large and miniaturised scales in a productive and cost-effective manner.

As a consequence, there are potentially numerous opportunities for applications of ECAE as a material processing technology (Segal *et al.*, 2010). However, 40 years since the invention of ECAE in the USSR and about 20 years after its introduction to the West, that has still not happened. Experience at Honeywell Electronic Materials, Spokane, WA with commercialisation of ECAE for bulk Al alloy billets up to 32 kg of weight has proved that, at least in special applications, the process is scalable, practical and cost-effective whereas UFG structures provide superior performance for sputtering targets (Ferrasse *et al.*, 2008). This is still the only success story, demonstrating that a key factor is the development of reliable and operative industrial tools as well as the optimisation of preliminary and post-extrusion steps. It is also known that some attempts to transfer simple experimental dies into production have failed. A lack of engineering developments of ECAE (i.e. the relevant processing technology for different products), is the main reason why ECAE has remained as a laboratory tool. Such a conclusion is now recognised in the research community (Filho *et al.*, 2011).

With the shift of interest to ECAE as an industrial technology, new applications may first be expected in areas where the performance of a material is most important, the cost of the material comprising only a small part of the total cost of the product. Examples include high-tech products such as: medical implants, materials for micro-machining and functional alloys or critical components for special techniques. More broad applications of ECAE in the manufacturing of structural components may require a high-volume production at a large metallurgical scale, with a significant decrease in the material and operational costs.

References

Azushima, A., Aoki, K., Tsukagoshi, H. and Yamagida, A. (2007) 'Development of ECAE process for reducing friction', in *Proceedings of Japanese Spring Conference on Technology of Plasticity*, 2007, 5–6 (in Japanese).

Azushima, A. and Aoki, K. (2002) 'Properties of ultrafine-grained steel by repeated shear deformation extrusion process', *Materials Science and Engineering A*, 377: 45–49.

Barber, R. E., Dudo, T., Yasskin, P. B. and Hartwig, K. T. (2004) 'Product yield for ECAE processing', *Scripta Materialia*, 51: 373–377.

Beyerlein, I. J. and Toth, L. S. (2009) 'Texture evolution in equal-channel angular extrusion', *Progress in Materials Science*, 54: 427–510.

Bian, L., Liang, W., Xie, G., Zhang, W. and Xue, J. (2011) 'Enhanced ductility in an Al–Mg2Si *in situ* composite processed by ECAP using a modified B_C route', *Materials Science and Engineering A*, 528: 3463–3467.

Bridgman, P. W. (1952) *Studies in Large Plastic Flow and Fracture*, McGraw-Hill, New York.

Canta,T., Frunza, D., Szilagyi, E. and Lungu, M. (2008) 'A new low friction die design for equal channel angular extrusion', *Key Engineering Materials*, 367: 215–220.

Chakkingal, P. K., Suriadi, A. B. and Thomson, P. F. (2009) 'Microstruture development during equal channel angular drawing of Al at room temperature', *Scipta Materialia*, 39: 677–684.

Chaudhury, P. K., Srinivasan, R. and Viswanathan, S. (2005) 'Continuous severe plastic deformation process for metallic materials', US Patent 6895795.

Chung, Y.-H., Park, J.-W., Moon, I.-G. and Shin, M.-C. (2003) 'Continuous severe plastic deformation process for metallic materials', US Patent 6571593.

Ferrasse, S., Segal, V. M., Alford, F., Kardocus, J. and Strothers, S. (2008) 'Scale up and application of equal-channel angular extrusion for the electronics and aerospace industries', *Materials Science and Engineering A*, 493: 130–140.

Filho, A. de A. M., Prados, E. F., Valio, G. T., Rubert, J. B., Sordi, V. L. and Ferrante, M. (2011) 'Severe plastic deformation by equal channel angular pressing: Product quality and operational details', *Materials Research*, 14: 335–339.

Gil Sevillano, J., van Houtte, P. and Aernoudt, E. (1980) 'Large strain work hardening and textures', *Progress in Materials Science*, 25: 69–134.

Hashimoto, S. (2005) 'Latest frontiers of nanomaterials research – perspective on ECAP method', *Koyo Engineering Journal English Edition, No. 167E*, 2–8.

Han, J.-H., Seok, H.-K., Chung, Y.-H., Shin, M.-C. and Lee, J.-C. (2002) 'Texture evolution of the strip cast 1050 Al alloy processed by continuous confined strip shearing and its formability evaluation', *Materials Science and Engineering A*, 323: 342–347.

Hill, R. (1950) *The Mathematical Theory of Plasticity*, Oxford University Press, Oxford, UK.
Hill, R. (1967) 'Ideal forming operations for perfectly plastic solids', *Journal of the Mechanics and Physics of Solids*, 15: 223–227.
Hirsch, J., Lucke, K. and Hatherlay, M. (1998) 'Overview No. 76: Mechanism of deformation and development of rolling textures in polycrystalline fcc metals –III. The influence of slip inhomogeneities twinning', *Acta Metallurgica*, 36: 2905–2927.
Huang, Y. (2005) 'Method and apparatus for extrusion', International Patent Publication, WO 2005/065856.
Hughes, D. A. and Hansen, N. (1997) 'High angle boundaries formed by grain subdivision mechanism', *Acta Materialia*, 45: 3871–3886.
Imayev, R. M., Imayev, V. M. and Salischev, G. A. (1992) 'The development of the submicrocrystalline structure in intermetallic TiAl during hot deformation', *Journal of Materials Science*, 27: 4465–4470.
Iwahashi, Y., Wang, J., Horita, Z., Nemoto, M. and Langdon, T. G. (1996) 'Principle of equal-channel angular pressing for the processing of ultra-fine grained materials', *Scripta Materialia*, 36: 143–148.
Ji, Y. H., Park, J. J. and Kim, W. J. (2007) 'Finite element analysis of severe deformation in Mg–3Al–1Zn sheets through differential-speed rolling with a high-speed ratio', *Materials Science and Engineering A*, 454/5: 570–574.
Kocisko, R., Kvackaj, T., Bidulska, J. M. and Molnarova, L. (2009) 'New geometry of ECAP channel', *Acta Metallurgica Slovaca*, 15: 228–233.
Kocich, R., Greger, M., Kursa, M., Szurman, I. and Machackowa, A. (2010) 'Twist channel angular pressing (TCAP) as a method for increasing the efficiency of SPD', *Materials Science and Engineering A*, 257: 6386–6392.
Kopylov, V. I. (1983) 'Development and research on a cold metal forming process of simple shear', PhD thesis, Physical-Technical Institute of National Academy of Sciences of Belarus, Minsk, Belarus (in Russian).
Kuhlmann-Wilsdorf, D. (1989) 'Theory of plastic deformation: properties of low energy dislocation structures', *Materials Science and Engineering A*, 113: 1–41.
Kurzydlowski, K. J. (2004) 'Microstructural refinement and properties of metals processed by severe plastic deformation', *Bulletin of the Polish Academy of Sciences, Technical Sciences*, 52: 301–311.
Lapovok, R. (2005) 'The role of back-pressure in equal channel angular extrusion', *Journal of Materials Science*, 40: 341–346.
Lee, J.-C., Seok, H.-K., Park, J.-W., Chung, Y.-H. and Lee, H.-I. (2002) 'Continuous shear deformation device', US Patent 6370930.
Liu, Z. Y., Liang, G. X., Wang, E. D. and Wang, Z. R. (1998) 'The effect of cumulative large plastic strain on the structure and properties of a Cu–Zn alloy', *Materials Science and Engineering A*, 242: 137–140.
Luri, R., Luis, C. J., Leon, J. and Sebastian, M. A. (2006) 'A new configuration for equal channel angular extrusion dies', *Journal of Manufacturing Science and Engineering*, 128: 860–865.
Ma, X., Barnett, M. R. and Kim, Y. H. (2004) 'Forward extrusion through steadily rotation conical dies: Part II: theoretical analysis', *International Journal of Mechanical Sciences*, 46: 449–464.
Mathieu, J.-P., Suwas, S., Eberhardt, A., Toth, L. S. and Moll, P. (2006) 'A new design for equalchannel angular extrusion', *Journal of Materials Processing Technology*, 173: 29–33.

Meyers, M. A., Mishra, M. and Benson, D. J. (2006) 'Mechanical properties of nanocrystalline materials', *Progress in Materials Science*, 51: 427–556.

Nagasekhar, A. V. and Kim, H. S. (2008) 'Analysis of T-shaped equal channel angular pressing using the finite element method', *Metals and Materials International*, 14: 565–568.

Nakashima, K., Horita, Z., Nemoto, M. and Langdon, T. G. (2000) 'Development of a multi-pass facility for equal-channel angular pressing to high total strains', *Materials Science and Engineering, A*, 281: 82–87.

Nishida, Y., Arima, H., Kim, J. C. and Ando, T. (2001) 'Rotary-die equal channel angular processing of an Al-7 mass% Si-0.35 mass% Mg alloy', *Scripta Materialia*, 45: 261–266.

Park, J. W. and Chung, Y. H. (2006) 'Apparatus and method for uniform shear deformation', US Patent 7140222.

Pardis, N., Talebanpour, B., Ebrahimi, R. and Zomorodian, S. (2011) 'Cyclic expansion–extrusion (CEE): A modified counterpart of the cyclic extrusion–compression (CEC)', *Materials Science and Engineering A*, 528: 7537–7540.

Pobogyi, U. (1982) 'Pressure plus shear', *Science and Life*, 8: 17–23 (in Russian).

Richert, J., Richert, M., Zasadzinski, J. and Korbel, A. (1984) 'Method and apparatus for plastic deformation of metals and alloys' (in Polish), Polish Patent 123026.

Rigney, D. A., Naylor, M. G. S., Divakar, R. and Ives, L. K. (1986) 'Low-energy dislocation-structures caused by sliding and by particle impact', *Materials Science and Engineering*, A81: 409–425.

Rosochowski, A. (2006) 'Severe plastic deformation of metals', International Patent Application WO/2006/100448.

Rosochowski, A. and Olejnik, L. (2002) 'Numerical and physical modelling of plastic deformation in 2-turn equal channel angular extrusion', *Journal of Materials Processing Technology*, 125/6: 309–316.

Rosochowski, A. and Olejnik, L. (2011) 'Equal channel angular pressing with converging billets - FE simulation', Paper presented at *10th International Conference on Technology of Plasticity*, held 25–30 September 2011, Aachen, Germany, pp. 235–240.

Rusz, S., Malanik, K., Dutkiewicz, J., Cizek, L., Donic, T., Kedron, J. and Tylsar, S. (2010) 'New design of the forming equipment DRECE for obtaining UFG structure in strip of sheet', *Archives of Materials Science and Engineering*, 42: 111–118.

Saray, O., Purcek, G. and Karaman, I. (2010) 'Principles of equal-channel angular sheet extrusion (Ecase): application to IF-steel sheets', *Reviews on Advanced Materials Science*, 25: 42–51.

Schukin, V. I. and Segal, V. M. (1976) 'Apparatus for strengthening materials by plastic deformation' (English translation), Invention Certificate of USSR, 515968.

Segal, V. M. (1996) 'Plastic deformation of crystalline materials', US Patent 5513512.

Segal, V. M. (1998) 'Method and apparatus for intensive plastic deformation of flat billets', US Patent, 5850755.

Segal, V. M. (2002) 'Severe plastic deformation: simple shear versus pure shear', *Materials Science and Engineering A*, 338: 331–344.

Segal, V. M. (2003) 'Slip line solutions, deformation mode and loading history during equal channel angular extrusion', *Materials Science and Engineering A*, 345: 36–46.

Segal, V. M. (2010) 'Mechanics of continuous equal-channel angular extrusion', *Journal of Materials Proccessing Technology*, 210: 542–549.

Segal, V. M. and Segal, L. V. (1997) 'Method of and apparatus for processing tungsten heavy alloys for kinetic energy penetrators', US Patent 5600989.

Segal, V. M., Beyerlein, I. J., Tome, C. N., Chuvil'deev, V. N. and Kopylov, V. I. (2010) *Fundamentals and Engineering of Severe Plastic Deformation*, Nova Science Publishers, Hauppauge, NY.

Segal, V. M., Makushok, E. M. and Reznikov, V. I. (1974) '*Investigation of Plastic Flow by Moiré Method*', Metallurgy, Moscow (in Russian).

Segal, V. M., Reznikov, V. I., Drobyshevsky, A. E. and Kopylov, V. I. (1981) 'Plastic treatment of metals by simple shear', *Russian Metallurgy*, 1: 115–123.

Segal, V. M., Reznikov, V. I., Kopylov, V. I., Pavlik, D. A. and Malyshev, V. F. (1994) *Processes of Plastic Structure Formation*, Science and Engineering, Minsk (in Russian).

Semiatin, S. L., Delo, D. P. and Shell, E. B. (2000) 'The effect of material properties and the tooling design on deformation and fracture during equal channel angular extrusion', *Acta Materialia*, 48: 1841–1851.

Smirnova, N. A., Levit, V. I., Pilyugin, V. P., Kuznetsov, R. I., Davydova, L. S. and Sazonova, V. A. (1986) 'Evolution of structure of FCC single crystals during strong plastic deformation', *Physics of Metals and Metallography*, 61: 127–134.

Stecher, G. and Thomson, P. (2003) 'Improved angular channel processing', International Patent Application WO2003027337.

Toth, L. S., Lapovok, R., Hasani, A. and Gu, C. (2009) 'Non-equal channel angular processing of aluminum alloy', *Scripta Materialia*, 61: 1121–1124.

Utsunomiya, H., Saito, Y., Suzuki, H. and Sakai, T. (2001) 'Development of continuous shear deformation process', *Proceedings of the Institution of Mechanical Engineers, Part B*, 215: 947–957.

Valiev, R. Z., Estrin, Y., Horita, Z., Langdon, T. G., Zehetbauer, M. J. and Zhu, Y. T. (2006) 'Producing bulk ultrafine-grained materials by severe plastic deformation', *JOM Journal of the Minerals, Metals and Materials Society*, 58(4): 33–39.

Valiev, R. Z. and Langdon, T. G. (2006) 'Principles of equal-channel angular processing as a processing tool for grain refinement', *Progress in Materials Science*, 51: 881–981.

Wang, S., Liang, W., Wang, Y., Bian, L. and Chen, K. (2009) 'A modified die for equal channel angular pressing', *Journal of Materials Processing Technology*, 209: 3182–3186.

Zhu, Y. T., Lowe, T. C., Valiev, R. Z. and Raab, G. J. (2006) 'Continuous equal-channel angular pressing', US Patent 7152448.

Notes

1. In 'ideal' forming operations velocities are continuous.
2. An interesting and popular description of this work can be found in Pobogyi (1982).
3. The abbreviation ECAE (or ECAP) was introduced later (Kopylov, 1983).
4. For tool angles $2\theta > 90°$ or $2\theta < 90°$, projections of the extrusion load onto an extrusion direction are not zero. These projections act on the slider and develop tensile back stresses in the first case and a compression back pressure in the second case.
5. It is interesting to note that the high back pressure in such cases increases material workability during ECAE. Some results on processing of hard-to-deform materials at low temperatures cannot be reproduced in dies with movable walls.
6. For example, for $L/a = 5$, friction along side walls is five times lower than friction along top and bottom walls.

Incremental ECAP

CHAPTER 2

Andrzej Rosochowski

In this chapter, the severe plastic deformation (SPD) process of equal channel angular extrusion (ECAE) is referred to as equal channel angular pressing (ECAP). The reason for this is that ECAE is a registered trademark of Honeywell International Inc. so ECAP has increasingly become the term of choice for researchers and engineers.

2.1 Concept of incremental ECAP (I-ECAP)

The concept of incremental ECAP (I-ECAP) addresses the issue of continuous feeding of the workpiece material into a classical ECAP die, which proves to be problematic if attempted using only friction forces (Segal, 2010). I-ECAP is a variant of ECAP, which enables billets of infinite length to be processed into the form of bars, plates and sheets. This is possible due to a dramatic reduction of the billet feeding force, which is achieved by separating the feeding and deformation stages of the process. The process has been patented in Europe (Rosochowski, 2012) and the USA (Rosochowski, 2014).

2.1.1 First idea

The concept of I-ECAP was formulated in April 2004. It originated from the thought that there is substantial shear deformation preceding fracture in the metal-forming process of cutting. If cutting could be stopped before fracture and the consecutive portions of the material 'pre-cut' in the same way, the whole length of a workpiece would be subjected to shear deformation. To check this idea a finite element (FE) simulation was carried out using Abaqus Explicit. Figure 2.1 shows the distribution of the equivalent plastic strain resulting from a single stroke of the cutting tool, which stopped before fracture. The cutting edges were rounded in order to delay fracture. A solid two-sided arrow in Figure 2.1 indicates the reciprocating movement of the cutting tool while the dashed one-sided arrow represents the incremental nature of material feeding.

Problems with consecutive cycles of this process led to the idea of changing the tool geometry, as shown in Figure 2.2. Only then, looking at this new geometry, did it become clear that the new process resembled classical ECAP with two channel turns (Rosochowski and Olejnik, 2002). Many more FE simulations were required to find the acceptable tool geometry and process parameters. Generally, the FE technique is not without its own problems but it has proved

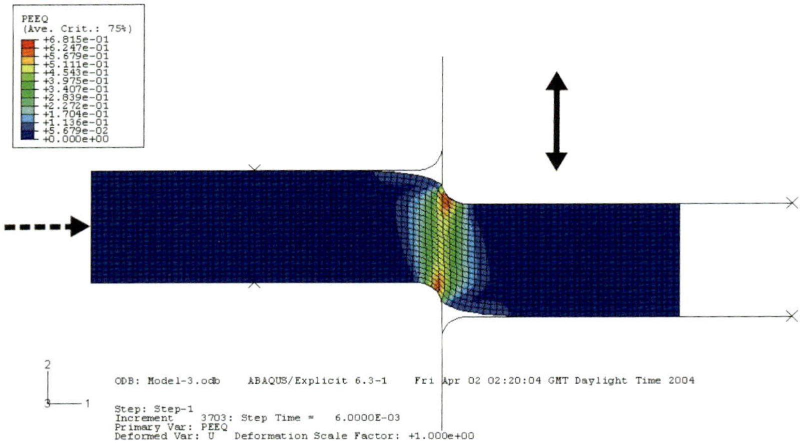

Figure 2.1 Distribution of equivalent plastic strain during 'pre-cutting' with rounded tools

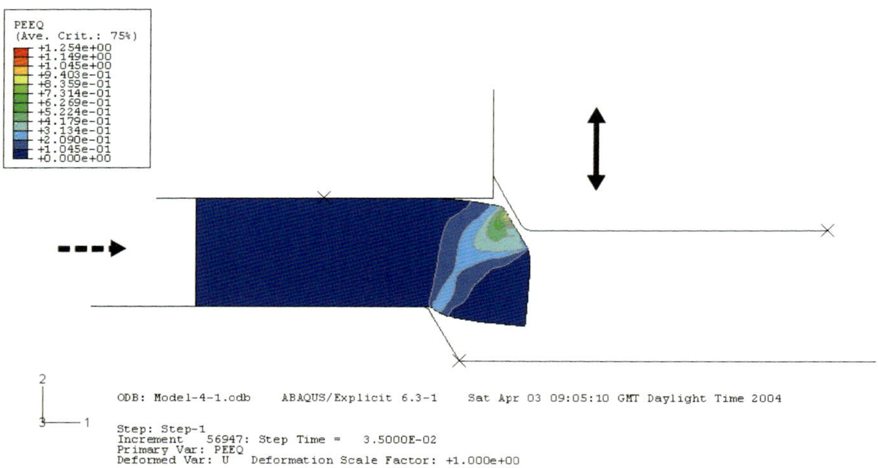

Figure 2.2 Distribution of strain at early stage of next version of process with different tool geometry

to be very useful in providing insights into material flow in the first pass of this and other versions of I-ECAP discussed throughout this chapter.

2.1.2 Process definition

As shown in the right-hand side of Figure 2.3, dies A and B define the input channel while die A and punch C define the output channel of the tooling. Dies A and B can be fixed or be movable to enable clamping of the billet. Punch C

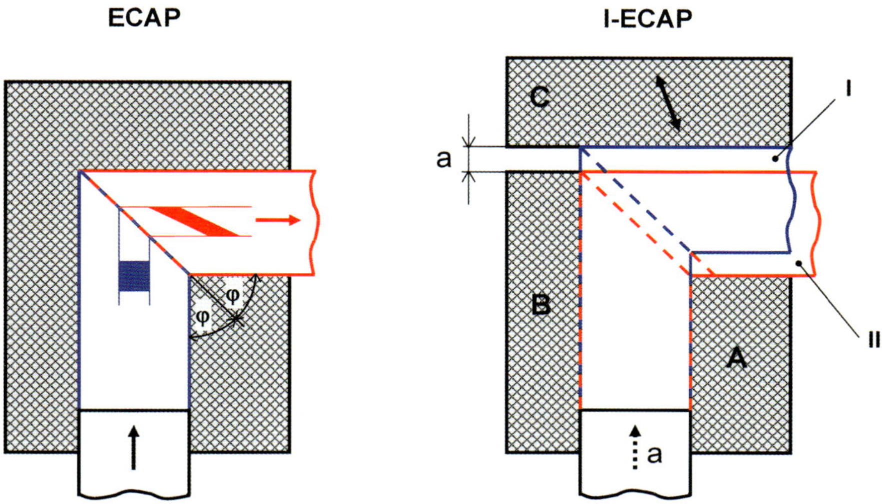

Figure 2.3 Schematics of classical ECAP and I-ECAP

moves in a reciprocating manner at an appropriate angle to the billet. The movement of the punch is synchronised with the feeding movement of the billet. The sequence of operations is as follows

1. Punch C is moved away from the billet by a distance of at least *a*.
2. This enables the billet 'I' to be fed forward by a distance *a* so that it sticks out by the same distance beyond die B. Without contact with punch C, feeding of the billet requires only a small force.
3. The billet is fixed in position to prepare it for the next stroke of the punch, for example, by supporting it by a feeding punch located between dies A and B or clamping it with dies A and B.
4. Punch C moves in a predetermined way towards the billet causing plastic deformation of the billet in the narrow zone marked by the dashed lines. The billet assumes the form 'II.'
5. The whole sequence described above is repeated as many times as necessary to process the entire length of the billet. Due to its incremental nature, I-ECAP can process billets of infinite length.

2.1.3 Relation to ECAP

Initially the new process was known as incremental shear (Rosochowski and Olejnik, 2007) but since its similarity to ECAP (left-hand side of Figure 2.3) was so obvious and most people were familiar with the term ECAP, it was decided to call it incremental ECAP or I-ECAP. Nevertheless, some sort of proof that ECAP and I-ECAP were similar was required and this was provided by simple geometrical considerations.

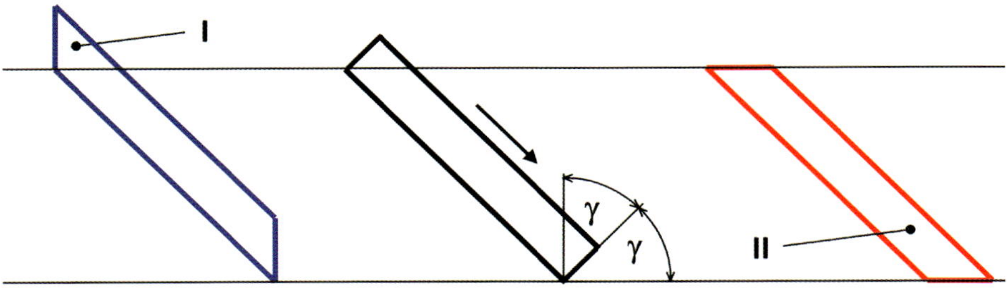

Figure 2.4 Simple shear mode of deformation in I-ECAP

The geometrical analysis of the material flow in I-ECAP, in the zone marked by the two parallel dashed lines in Figure 2.3, leads to the conclusion that the mode of deformation is that of simple shear. For the purpose of strain analysis, the deformation stage during I-ECAP is split into two steps as shown in Figure 2.4. Shearing the initial parallelepiped 'I' by an angle γ produces a rectangle, for which the shear strain is $tg\gamma$ and the equivalent von Mises strain is $\varepsilon = tg\gamma/3^{0.5}$. Continuing this shearing by another angle γ converts the rectangle to the final parallelepiped 'II' and doubles the equivalent strain to $\varepsilon = 2tg\gamma/3^{0.5}$. For $\gamma = 45°$, the total equivalent strain is $\varepsilon = 1.155$, which is the value known from classical ECAP with a channel angle of 90°. This value of equivalent strain has been confirmed by FE analysis (shown in Section 2.2.1) so, in terms of the type and value of the strain produced, I-ECAP is equivalent to classical ECAP.

2.1.4 Place of I-ECAP in the metal-forming discipline

All SPD processes are related to the known metal-forming processes, the main difference being the lack of dimensional change of a billet. For example, ECAP is an equal channel version of side extrusion. The incremental nature of I-ECAP places it in a large family of incremental bulk metal-forming processes. The definition used in (Groche *et al.*, 2007) states that

> In an incremental bulk forming process, regions of the workpiece experience more than one loading and unloading cycle due to the action of one set of tools within one production stage.

This assumes that the tool's geometry is simple and the final shape of the workpiece depends on the multiple hits/passes of the tool performed on the same small portion of the material. Examples of such processes include: open die forging, ring rolling, flow forming, swaging (also known as radial forging) and rotary forging. Nominally, the consecutive portions of the material in I-ECAP are deformed only once in one pass. However, the real deformation zone in I-ECAP is substantially wider than indicated by the pair of dashed lines in Figure 2.3 and analysed in Figure 2.4. It can reach 50% of the thickness of the workpiece. This, together with usually a small feeding increment of the billet material, causes each material point passing the deformation zone to be subjected to multiple strain increments, which added together give the final value of strain at the zone exit (Figure 2.9(c)). In this respect, I-ECAP is similar to radial forging. On the other hand it is very different from radial forging because it realises a different mode of deformation, which is simple shear. The related geometrical aspect is the lack of change in the workpiece cross-section, which means creating the plain strain state in the direction of the channel width and ensuring that the workpiece thickness remains the same by controlling the punch's lowest position.

2.2 I-ECAP of square bars

I-ECAP of square10 mm × 10 mm cross-section billets was the first process tested experimentally (Rosochowski *et al.*, 2008) and it remains the basis for process improvements and trying new ideas. Because of the limited space for tooling, the length of billets was not more than 200 mm. Sometimes, to save on material and shorten the processing time, that length was shorter but it was never less than 100 mm, which seems to be the upper limit for classical ECAP (typically it would be 60 mm for classical ECAP).

For the sake of load symmetry and higher productivity, the version of I-ECAP realised experimentally was one with two parallel billets fed into the input channel as proposed by Segal (1997). It is shown schematically in Figure 2.5. Compared to the basic process in Figure 2.3, there is only die A, punch C extends over two billets and the direction in which it reciprocates is along the input channel. The punch has a central spike whose role is to help direct material in the output channels. According to this configuration, there are two zones of simple shear. This will increase the punch force compared to the one-billet configuration but by less than twice because the friction surface in the input channel is less than doubled.

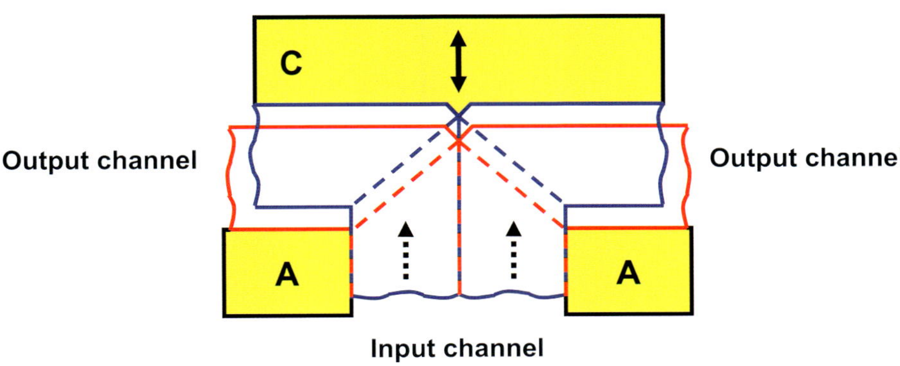

Figure 2.5 I-ECAP of two billets fed in parallel in input channel

2.2.1 FE simulation

Nearly all FE simulations presented in Chapter 2 have been carried out using a commercial program (Abaqus/Explicit). Taking into account that initial experiments usually involve a soft material deformed at room temperature, the only material used in these simulations was a commercial purity aluminium, Al1070. It was modelled as an isotropic elastic–plastic von Mises strain hardening material, which was not temperature or strain rate sensitive. Its strain hardening curve was described as $\sigma = 159(0.02 + \varepsilon)^{0.27}$. The curve was established experimentally by straining the material to approximately $\varepsilon = 1.5$ in a compression test (Rosochowski *et al.*, 2006). This allowed only the first pass of I-ECAP to be simulated, however, this was usually sufficient to assess the effect of the tool geometry and process parameters on the material flow.

The cross-section of the input channel was a 10 mm × 20 mm rectangle which forked into two 10 mm × 10 mm output channels. The angle of each channel turn was 90°. The channel corners were rounded with a radius of 2 mm. Billets that were 100 mm in length were chamfered to match a 2 mm high spike. The feeding stroke was 1 mm and it was synchronised with a reciprocating movement of the punch. The peak-to-peak amplitude of this movement was 1.2 mm and the frequency was 100 Hz. This high frequency as well as a high feeding speed was used to shorten the calculation time. Figure 2.6 shows the kinematics of tools in the first three cycles of the simulation.

Each 100 mm long billet was divided into about 1900 plane strain, bilinear, quadrilateral elements with reduced integration. No remeshing was attempted so the material flow was easier to observe and interpret. Friction was assumed to follow Coulomb's law with friction coefficient $\mu = 0.1$.

Figure 2.7 shows a few snapshots of I-ECAP, which illustrate the incremental character of the process. Some of these snapshots refer to a stage of the cycle in which the punch reaches its lowest position while others show the stage in

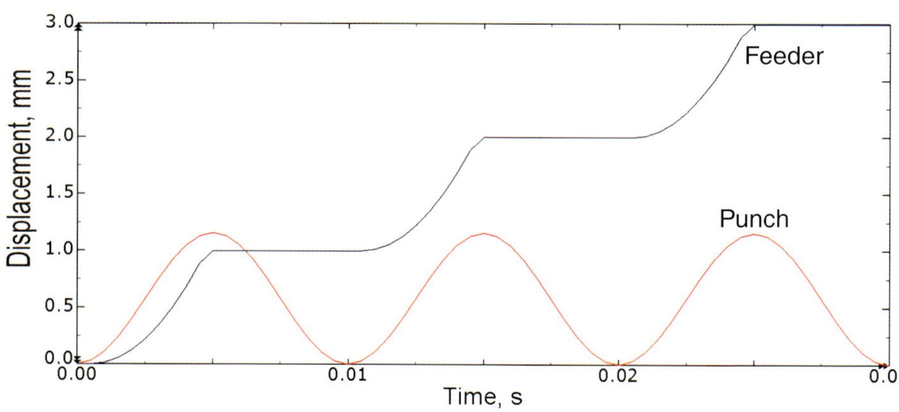

Figure 2.6 Kinematics of feeder and punch used in FE simulation of I-ECAP

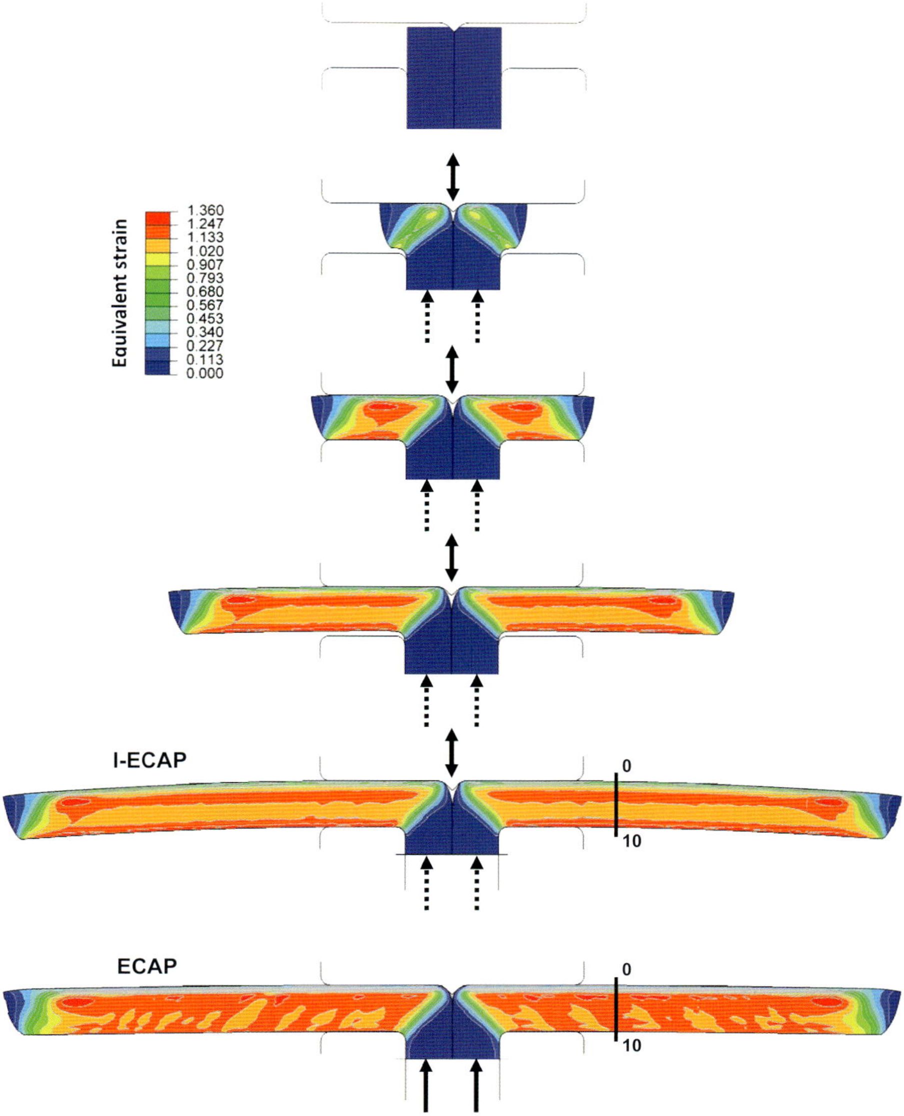

Figure 2.7 Progression of deformation during I-ECAP and comparison with ECAP

which the punch is in its highest position. The latter stage is accompanied by a gap between the die and the billets in the output channel, which is equal to the material feeding increment. The equivalent strain contours give evidence of a non-uniform strain distribution with no strain at both ends of the billet (so-called end effects) and some strain gradient in the transverse direction. A similar strain distribution can be observed in the classical ECAP process,

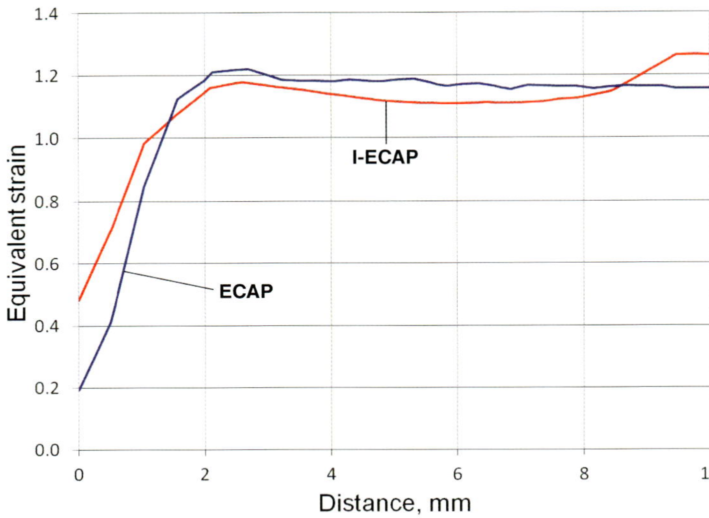

Figure 2.8 Strain distribution along path 0–10 in Figure 2.7 for I-ECAP and ECAP

which supports the opinion formulated earlier that I-ECAP and ECAP are nearly identical. Figure 2.8 enables one to compare the strain distribution in the transverse direction, along a 0–10 path shown in Figure 2.7, for both processes. There appear to be only small differences but it should be noted that I-ECAP results in a higher strain on the punch side (distance 0). There is some bending of the billets in I-ECAP, which can be removed by reducing the width of the punch.

Another insight into the material flow can be achieved by plotting the history of the equivalent strain for a few representative material points. Figure 2.9 displays

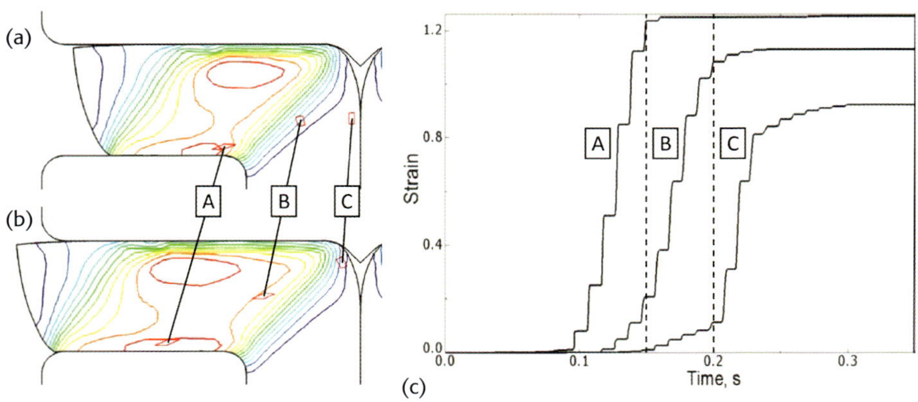

Figure 2.9 (a, b) Snapshots of equivalent strain distribution; (c) history of equivalent strain for three material points A, B and C in I-ECAP

two snapshots of strain distribution for 0.15 s (Fig. 2.9(a) and 0.2 s (Figure 2.9(b)) into the process, with the material points A, B and C shown as finite elements. Figure 2.9(c) shows the whole strain history for these points with the snapshot times indicated by the dashed lines. In addition to confirming the lower strain towards the top surface of the billet, Figure 2.9(c) gives an indication of the strain rate. Since each of the material points is subjected to a stepwise pattern of strain increments, the strain rate in each increment is higher compared to classical ECAP with a smooth strain rise over the deformation zone. For some brittle materials, the strain rate should be reduced to avoid fracture of the material. In I-ECAP, this can be achieved by reducing the cycle frequency and the feeding stroke.

2.2.2 Tooling

The term tooling, in the version of I-ECAP presented here, refers to the punch and the die, their holders and plates as well the prestressing ring and the pusher (Figure 2.10). The punch, shown in Figure 2.11, has a cylindrical base fitting into a punch holder. The lower end of the punch has a rectangular cross-section, which fits the output channels of the die. The width of this section has to be slightly smaller than the width of the output channels to enable the punch to move in the channel but not too small to avoid flow of the material into gaps between the punch and the die (so-called flash). The geometry of the spike, visible on the punch face, should facilitate material flow but also take into account that it influences the strain on the punch side of the billet (Figure 2.8).

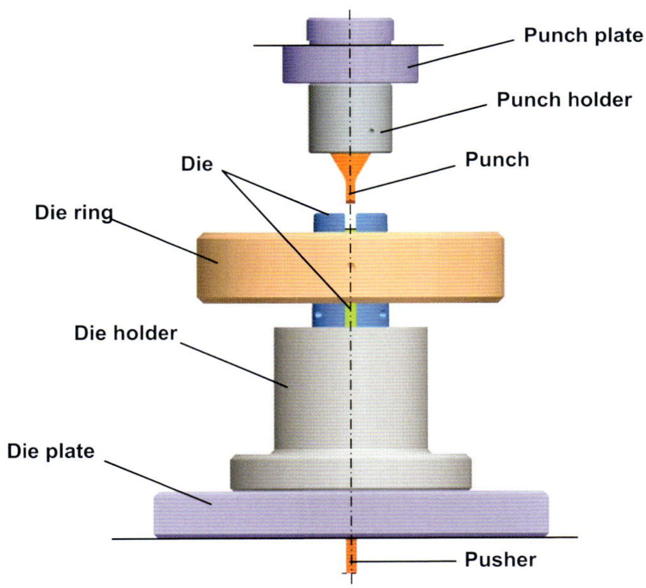

Figure 2.10 Tool elements in I-ECAP of two billets

The die (Figure 2.11) is made of four segments to avoid stress concentration along the corner edges of the input and output channels. The segments are kept together by dowel pins. The lower part of the segmented die is placed in a die holder while the upper part is prestressed with a massive ring. The ring is situated below the output channel to enable the processed material to exit the die. The middle segments of the die are 15 mm shorter than the outer segments, which provides 10 mm for the billet and 5 mm for guiding the punch.

The punch holder is attached to a punch plate, which in turn, is attached through a load cell to a press actuator. The die holder is attached to a die plate clamped to the press table. To keep the die inside the die holder, the prestressing ring has four vertical holes near its edge to accommodate four threaded pins connecting it to the die plate (not shown in Figure 2.10). The pusher is a long rod with rectangular cross-section matching the input channel of the die. It comes through a hole in the press table. All stressed tools, which come in contact with the processed material (punch, die and pusher) are made of high-quality high-speed steel, heat treated and polished.

Figure 2.11 Punch, die and ring for I-ECAP of two billets

2.2.3 Machine

The machine used to realise I-ECAP of two billets, as well as any other I-ECAP process reported in this chapter, is a laboratory servo hydraulic press (Figure 2.12). The press is rigid and lightweight because of the hollow structure of its frame. It is equipped with a hydraulic actuator capable of generating 1000 kN force with a stroke up to 300 mm. This large stroke is obviously not required for I-ECAP but the press is designed as a general purpose metal-forming machine. The actuator is attached to an adjustable cross head. This cross head, together with the press table, connect two side pillars of the frame. Additionally, the pillars are connected with two top bars and two bottom bars. The bottom bars are employed to support an additional plate placed under the table to provide a

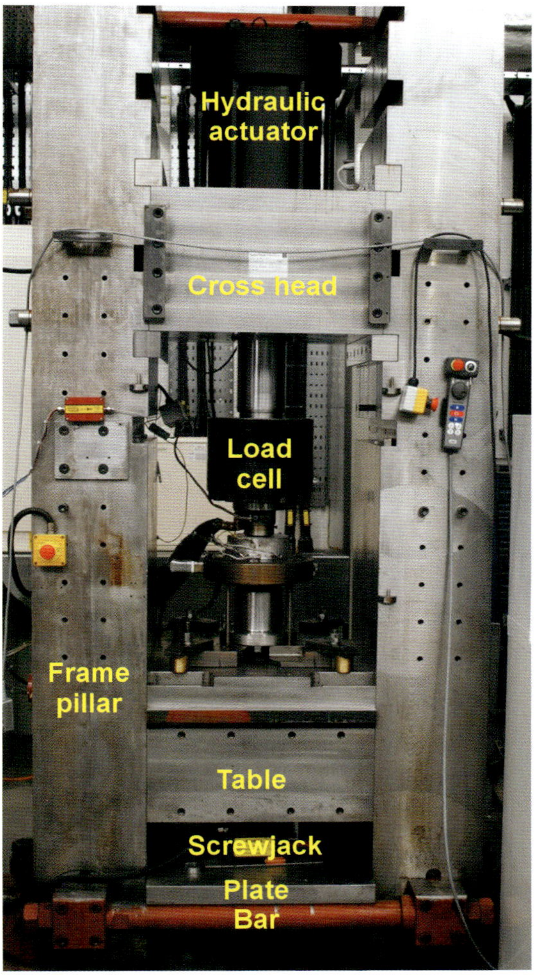

Figure 2.12 Laboratory press used for I-ECAP experiments

Figure 2.13 Heating elements

base for a 200 kN screw jack. The purpose of the screw jack, and other elements attached to it, is to drive the pusher up and to support it during deformation. The hydraulic actuator is connected to a 11 kW hydraulic power pack delivering 23 litre/min of oil at 25 MPa. The temperature of the oil is maintained by a closed-circuit water-cooling system comprising an industrial chiller and a heat-exchanger. The same chiller is used to provide cooling for the punch plate, which incorporates cooling channels. Cooling in this area is required when processing materials at elevated temperature. In this case, the die and the punch are heated up to the same temperature as required for the material. This is realised by heating the die ring outer surface with a 3 kW ceramic band heater and additionally with two heating metal blocks placed on top of the ring, equipped with heating cartridges with a total power of 1.5 kW. The punch is heated by a 0.6 kW band heater wrapped around the punch holder. The heating elements are shown in Figure 2.13.

2.2.4 Controls

The nature of I-ECAP requires synchronisation of the punch travel and the material feeding. A data acquisition system is also needed to gather information about various aspects of the process. All elements of both systems are shown schematically in Figure 2.14. A Zwick control system, called Cubus, has been used to control a servo valve (SV) of the hydraulic actuator (HA). This enables manual control of the actuator to set up an experiment and automatic control of its reciprocating movement during the experiment. Since the automatic movement of the actuator can be controlled as a function of either displacement or force, the Cubus box is connected to an LVDT1 inside the hydraulic cylinder and a load cell (LC1) located at the end of its stem. A National Instruments data acquisition and control system (NI) acquires information about the real movement of the punch (P) from LVDT2 attached to the punch holder. It also

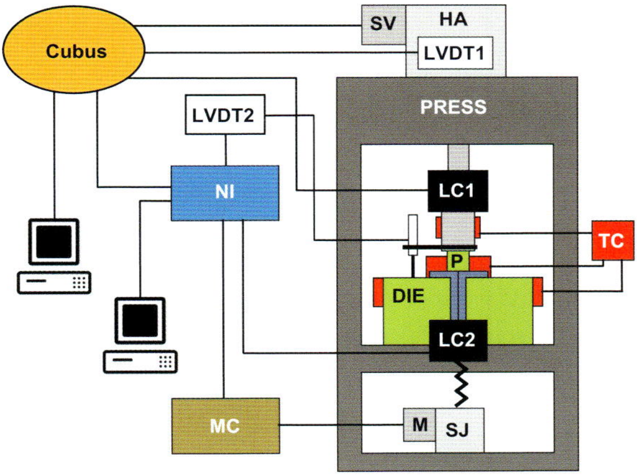

HA	Hydraulic Actuator
SV	Servo Valve
LVDT	Displacement Transducer
NI	National Instruments Modules
MC	Motor Controller
M	Electric Motor
SJ	Screw Jack
LC	Load Cell
P	Punch
TC	Temperature Controller

Figure 2.14 Control and data acquisition system

collects data on the screw jack (SJ) force measured by a miniature load cell (LC2) put between the screw jack and the pusher. The punch force signal is transmitted from Cubus to NI for logging. The output signal generated by NI controls a motor controller (MC) connected to an electric motor (M) driving the screw jack. The control of the screw jack is based on user-defined start and stop points for the motor, which are triggered by the punch displacement signal measured by LVDT2. In this way, the movement between the punch and the pusher is synchronised so that the pusher moves upwards when the punch travels away from the billets and it stops when the punch travels down. There is also another control loop, which triggers the stop command for the motor when the screw jack force measured by LC2 exceeds a predefined level. The NI software application used for realising all these functions is LabView. A separate system controls temperature in which heaters and thermocouples are connected to a temperature controller (TC). The required temperatures are set on the front panel of this controller.

2.2.5 Results

Despite convincing FE results confirming the similarity of I-ECAP and ECAP, the ultimate proof can only be achieved through experimental validation. The experimental rig described above was used to process Al1070 (Rosochowski *et al.*, 2008), Al5083, Cu99E, magnesium alloys AZ31B (Gzyl *et al.*, 2013) and Elektron 43, CP Ti Grade 4 and IF steel. The I-ECAP parameters used in those experiments included the type (sine wave), frequency (0.5–1 Hz) and peak-to-peak amplitude (1.5–2 mm) of the punch command control signal, feeding stroke of the screw jack (0.2–1 mm) and die temperature (20–350°C).

Theoretically, the waveform can be of any shape, which can be generated by a control system, but not having information on any of the possible shapes being better, it was decided to use the simplest sine wave form. It should be noted that the accuracy of control over punch movement depends on the location of a control displacement transducer whose signal is compared to a command signal. Ideally, the control displacement transducer should be as close as possible to the punch. If that is not the case, the punch will follow a different path depending on the elastic deflection of the tool and machine elements between the punch and the control displacement transducer. This can lead to a substantial departure of the punch path from the required one, especially for processes with high force. From the point of view of productivity, increasing punch frequency is recommended. But whether this is possible depends on the flow rate of oil pumped by a hydraulic power pack, the process force and the punch amplitude: low flow rate, high force and-amplitude reduce that frequency. Additionally, some materials such as magnesium and titanium cannot be processed at high strain rate, which limits the frequency of the punch. The punch peak-to-peak amplitude was deliberately chosen to be higher than the feeding stroke to make synchronisation of forming and feeding easier. A feeding stroke of 0.2 mm is very conservative. It can be increased to as much as 10% of the billet thickness (1 mm for 10 mm thick billet) without any adverse effect on the uniformity of strain distribution along the billet. This is related to the fact that the width of the deformation zone is more than 50% of the billet thickness (Figure 2.9(c)) so the consecutive deformations zones overlap for smaller feeding strokes. This has been illustrated (Rosochowski and Olejnik, 2007) as shown in Figure 2.15. The die temperature depends on the material being processed and

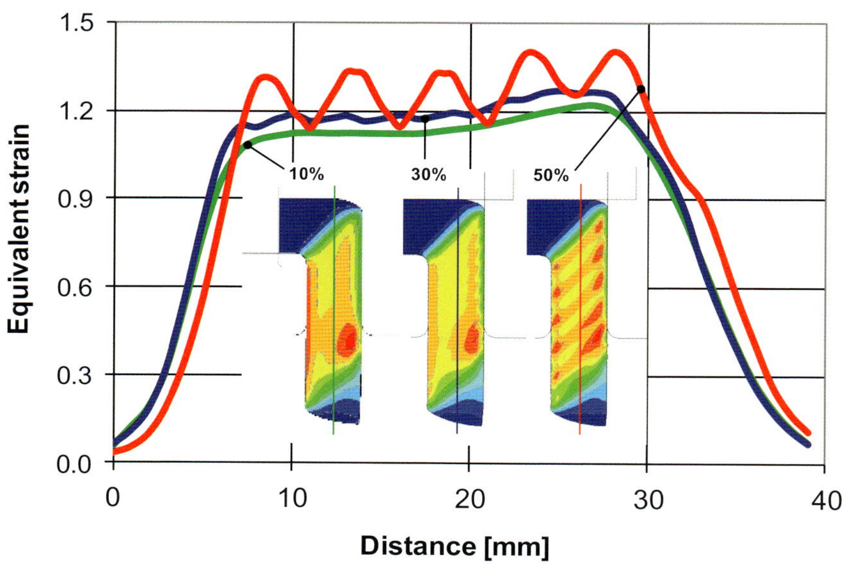

Figure 2.15 Strain distribution along billet for feeding stroke of 10%, 30% and 50% of billet thickness

should be sufficiently high to avoid fracture, but on the other hand as low as possible to promote grain size reduction.

The 100–190 mm long billets were machined to fit the input channel. For the first pass, one edge of the billet's face was chamfered to match the punch spike. Further passes did not require this procedure to be repeated because the shape of the billet's end produced in previous passes and billet rotation between passes created a natural chamfer. Before applying any lubricant, the billet surface was prepared by producing a conversion coating, for example, calcium aluminate on aluminium or by sand blasting. In most cases the billets were lubricated with a molybdenum disulphide (MoS_2) paste. Sometimes proprietary lubricants in the form of a graphite particle suspension in a binder and solvent system were used. While MoS_2 was applied by brush other lubricants could be sprayed and dried. The lubricated billets were inserted into a die that was preheated if necessary. In this case, the die was closed with a preheated punch and left for 10–15 minutes for the tools and the billets to reach a stable temperature. The temperature was measured by a thermocouple inserted in the die, close to the output channel.

During the I-ECAP process, the reciprocating punch gradually deforms the material (Figure 2.16) which is fed incrementally by the screw jack. The process

Figure 2.16 Snapshots of I-ECAP carried out at room temperature

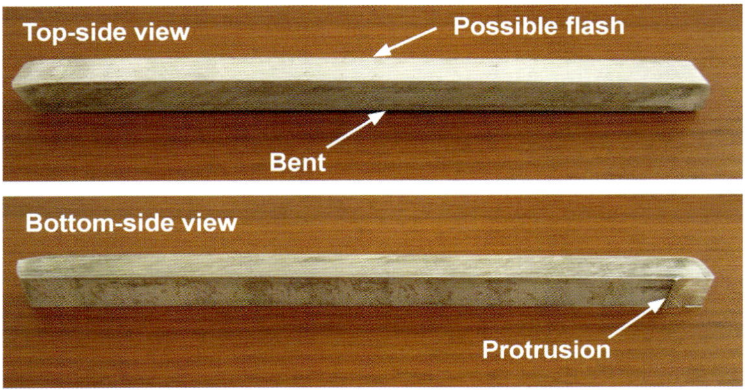

Figure 2.17 Al1070 billet subjected to first pass of I-ECAP at room temperature

ends when a manually operated limit block is inserted into the feeding system, which causes a steep increment of the feeding force which, in turn, triggers stopping of the motor. The size of the block is chosen so that material processing will end when the pusher reaches the bottom surface of the output channel. After that, the reciprocating movement of the punch is stopped and the punch is withdrawn to its uppermost position. Next, the pusher is slightly lowered, which enables the limit block to be removed, and it is then driven upwards to eject the billets from the output channel.

Figure 2.17 shows an Al1070 billet made using this procedure. The billet is slightly bent so it has to be straightened before the next pass. There is also a small protrusion near the rear end of the billet, which results from the material occupying the space between the rounded corner at the crossing of the input and output channels and the pusher in its final position. This protrusion, as well as any flash on the upper surface of the billet, is machined off. Due to the elastic deflection and thermal expansion of tools as well as billet springback and thermal contraction on cooling, the billets have different cross-sectional dimensions from the theoretical ones based on the geometry of unloaded tools. Additionally in I-ECAP, the bottom position of the punch is a little difficult to control, which may affect the thickness of the billet. All these phenomena may lead to oversized billets, which do not fit the input channel in the next pass. This requires some additional machining of billets to reduce the dimensions of their cross-section. Alternatively, for billets subjected to rotation by 90° (so-called route B_C), the width of the output channels can be appropriately reduced and the bottom position of the punch lowered to enable a repeated insertion of the billets in the input channel without machining.

Punch displacement and forces on the punch and the screw jack recorded during the process give an insight into the process. The analysis of Figure 2.18 (Rosochowski *et al.*, 2008), which is a snapshot of the punch and screw jack forces generated during the sixth pass of I-ECAP of Al1070, reveals that during

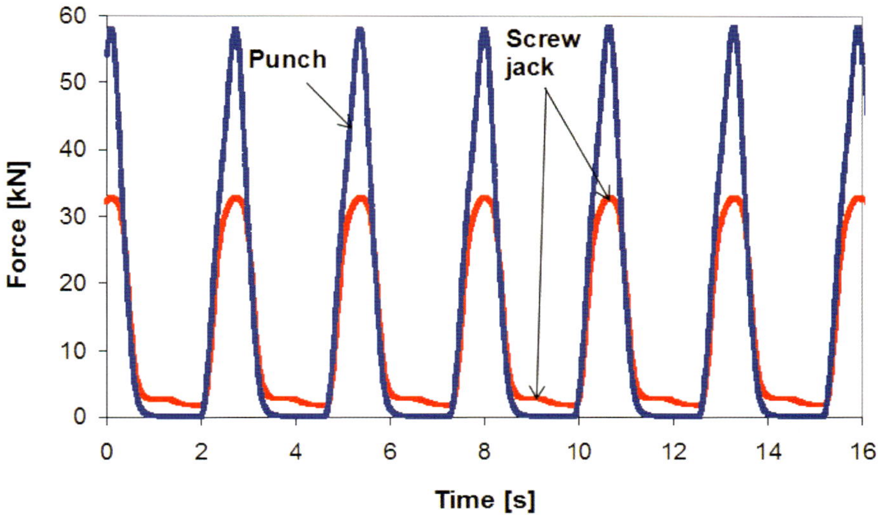

Figure 2.18 Forces on punch and screw jack recorded during sixth pass of I-ECAP of Al1070

the deformation stage of the process only part of the punch force (approximately 60%) is transferred to the screw jack, the rest of it is carried by the die. Another observation is that during the feeding stage the force on the screw jack is minimal because it does not exceed 5% of the punch force.

Routinely, especially for new materials, the micro-structure is established before and after I-ECAP. Figure 2.19 shows the initial structure of Al1070 used in the experiments, which reveals large grains, approximately 300 μm in size.

Figure 2.19 Initial micro-structure of Al1070 subjected to I-ECAP

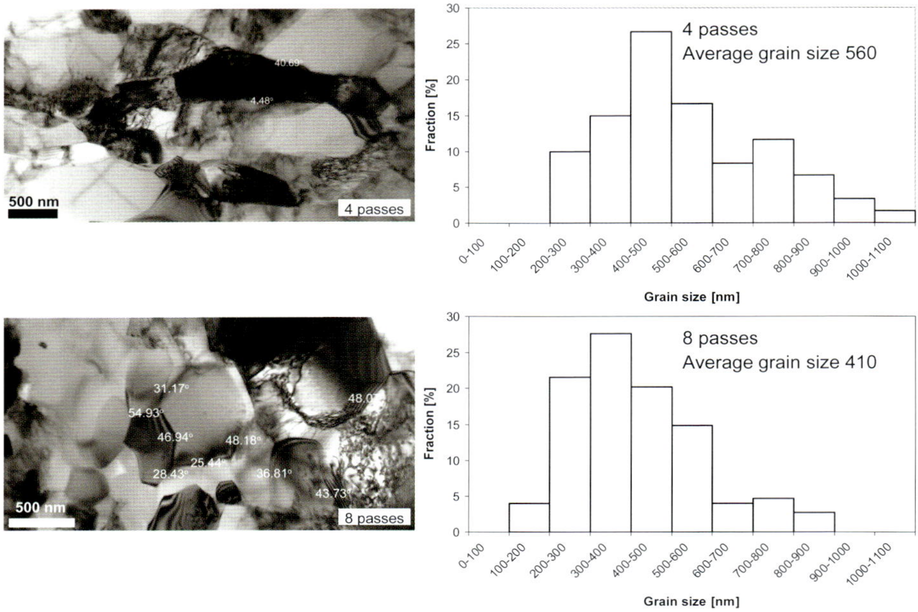

Figure 2.20 TEM images and grain size distribution in Al1070 subjected to four and eight passes of I-ECAP

Figure 2.20 illustrates changes in the micro-structure of the material which occurred as a result of four and eight passes of I-ECAP. The transmission electron microscopy (TEM) technique also enables us to measure grain boundary misorientation angles, which gradually grow indicating evolution from subgrains to grains. The grain/subgrain size distribution is similar for four passes and eight passes of I-ECAP, however, the mean grain/subgrain size is reduced from 560 nm to 410 nm, respectively.

Tensile or compressive tests are often performed on the samples in their initial state and after I-ECAP. For example, pulling 2.5 mm diameter and 7.5 mm long samples at 0.5 mm/min at room temperature resulted in the force–displacement curves shown in Figure 2.21. The initial yield strength of the material has more than tripled while its tensile strength has more than doubled; the ultimate tensile strength, reached 180 MPa. Typically, the total elongation has been reduced by approximately 40% while uniform elongation is nearly non-existent (Rosochowski *et al.*, 2008).

2.3 I-ECAP of plates

Compared to bars, plates have a much larger volume so SPD of plates would increase productivity. The advantage of I-ECAP over other processes is that it can be used for SPD of bars as well as plates. The width of the plates exceeds their thickness, however, this does not necessarily mean that friction in the

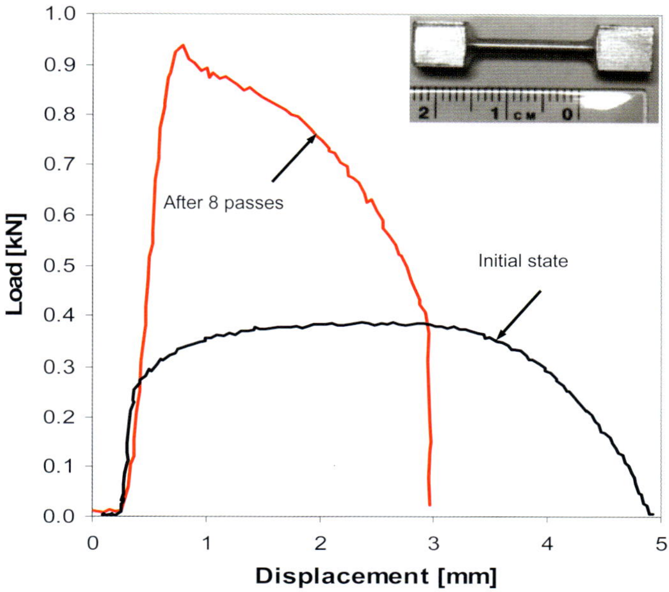

Figure 2.21 Tensile behaviour of Al1070 in initial state and after eight passes of I-ECAP

width direction can prevent material flow in this direction (plane strain condition) as required in ECAP/I-ECAP. Figure 2.22 taken from Olejnik *et al.* (2008) illustrates this point by comparing 3D FE simulation results for two cases: I-ECAP of a plate with the material flow restricted in the width direction and the same process without restricting the flow. The case with the restricted flow is very similar to the frequently published 2D FE results obtained for classical ECAP. However, it is worth noting that the 3D FE simulation reveals some minor strain changes for different cross-sections (middle plane vs. back view) due to friction between the plate and the side wall of the output channel. Despite some friction in the width direction, the latter case results in the material flowing relatively freely in this direction, which leads to a large distortion of the plate's shape and to a highly non-uniform strain distribution in the width direction. Apparently, much wider plates would have to be processed to enable I-ECAP to be used without mechanically restricted material flow in the width direction.

The first attempt to process a plate using I-ECAP took place in 2007 (Olejnik *et al.*, 2008). Figure 2.23 shows the design of the experimental rig that was used. Die 1 and wedge 2 are fixed to a base plate with bolts. The plate is clamped between die 1 and holder 3. Punch 4 is driven down and left by punch block 5. Both punch block 5 and holder 3 are driven in a controlled way by a moving element of the press (not shown in Figure 2.23). The rig was designed to work on any press, with the base plate being fixed to a table and the moving element

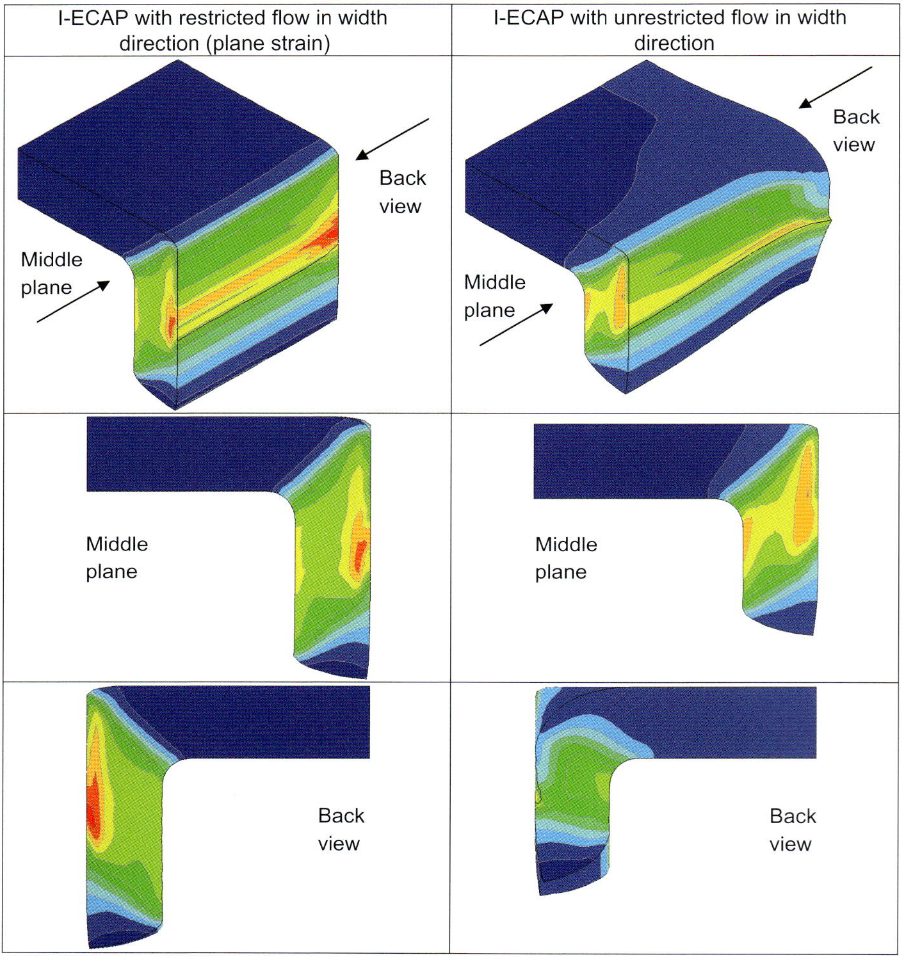

Figure 2.22 Equivalent strain distribution for plate processed by I-ECAP with restricted (plane strain) and unrestricted material flow in the width direction (courtesy of Olejnik *et al.*, 2008)

attached to a ram of the press. Feeding of the plate by 0.5 mm was manual, after lifting the moving element. A 4 mm × 30 mm × 100 mm Al1070 plate was processed in this way up to eight times (Figure 2.23). Despite pretty basic design and operation of the rig, the average grain size after eight passes of I-ECAP was reduced to approximately 500 nm (Olejnik *et al.*, 2008).

A later attempt was more complex because it involved building a dedicated machine for I-ECAP, with two channel turns in order to reduce the number of passes and increase productivity. The process of two-turn ECAP was first proposed by Shchukin and Segal (1976). The idea has since been explored under

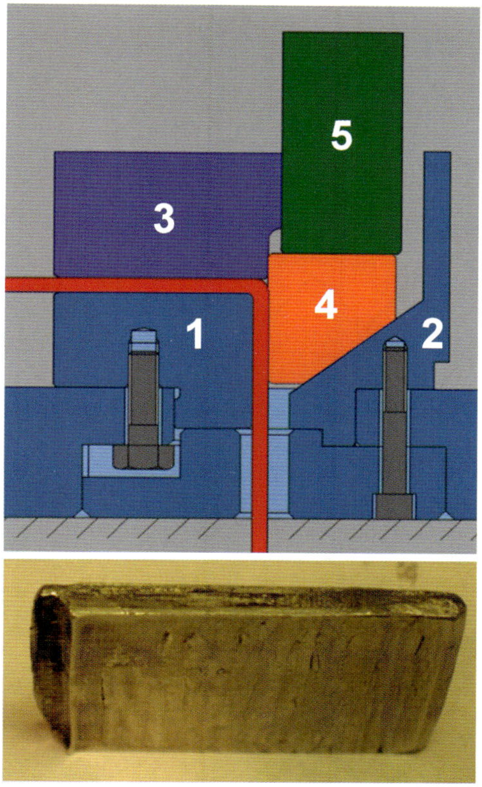

Figure 2.23 Schematics of experimental rig for I-ECAP of plates and processed plate after eight passes

different names, for example, S-type ECAP (Liu *et al.*, 1998), two-turn ECAP (Rosochowski and Olejnik, 2002) and ECAP in parallel channels (Raab *et al.*, 2005). The concept of I-ECAP originated from the case of two-turn channel (Figure 2.2), which was further analysed by Rosochowski and Olejnik (2008). Figure 2.24 explains this idea based on two fixed dies A and B and punch C. The billet is fed incrementally and synchronically with the reciprocating punch C. The billet material is subjected to simple shear twice, as indicated by the dashed zones at the first and the second turn of the channel. At the second turn, the shear is reversed, which is equivalent to route C (180° rotation about workpiece axis) in the classical ECAP process.

This point is also illustrated in Figure 2.25 where the initial square FE mesh distorts to parallelograms at the first turn and then recovers to a more or less square shape at the second turn. The uniform material flow, with well-defined zones of shear deformation, proves the feasibility of the process. However, it should be emphasised that these good results have been obtained for an optimised tool geometry and process parameters. Choosing other conditions may

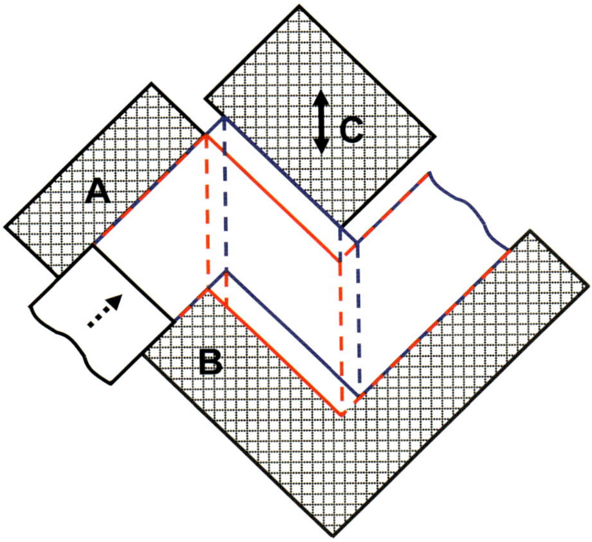

Figure 2.24 Schematics of two-turn I-ECAP

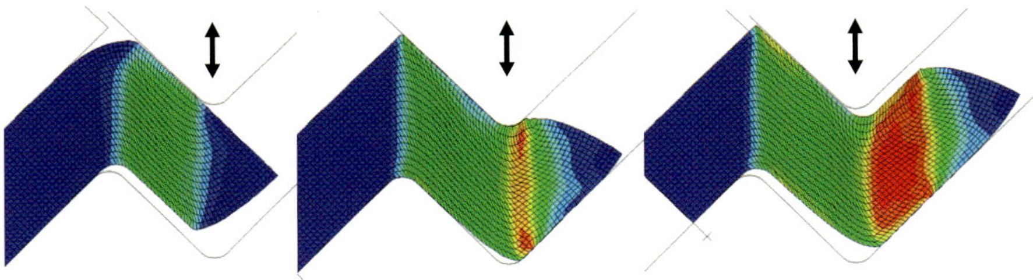

Figure 2.25 Material flow during I-ECAP at second turn of channel, showing maximal equivalent strain 2.37

lead to poor results. For example, if the offset of the output channel is 10 mm (which is the thickness of the workpiece in the simulation) rather than 15 mm, used in Figure 2.25, the result will be a poor channel filling and a non-uniform strain distribution (Figure 2.26(a)). Another entirely defective material flow can be observed for an excessive feeding stroke of 5 mm in Figure 2.26(b).

The design specification for the new machine assumed the cross-section of the processed plate to be 10 mm × 50 mm and the length to be unlimited (for practical reasons that length was 1000 mm). The orientation of the workpiece material in Figures 2.24–2.26 assumed that a classical, vertical press was used so that the punch, attached to a press ram, was moving in the vertical direction. However, for longer plates this could create a problem so it was decided to feed the plate horizontally, then turn it down vertically and finally turn it again to

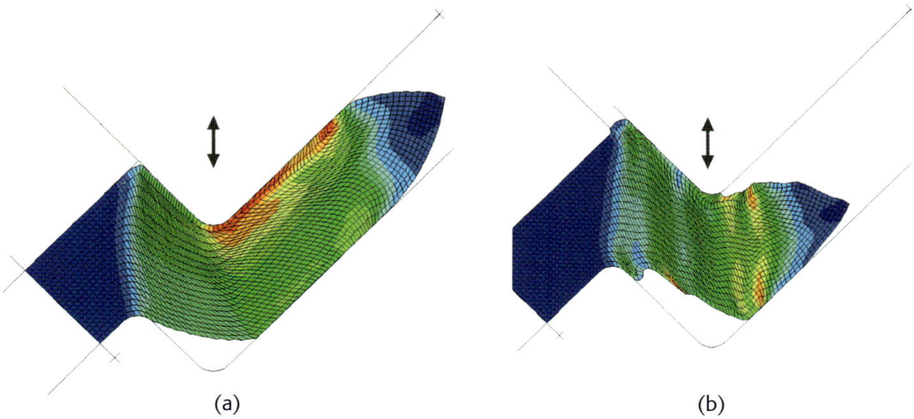

Figure 2.26 I-ECAP with: (a) small channel offset; (b) excessive feeding stroke, maximal equivalent strain 2.8

leave die in the horizontal position. The angles of these turns were both 90° but they could have been increased if necessary. The offset of the input and the output channel was 15 mm and the radii were 2 mm for the punch nose and the die at the first turn of the channel and 4 mm for the die at the second turn of the channel. To deform the material in the same way at both channel turns, the punch was to move at an angle of 45°. Another difference, compared to the I-ECAP of square billets described in Section 2.2, was the replacement of a screw jack driven pusher with an incremental feeder capable of feeding infinitely long plates. Since such a feeder would not be capable of bearing the process force during the deformation stage, a clamp was adopted to close the die and provide the required support for the plate due to friction. At the end of each cycle, the clamp would be released to enable the plate to be fed in. The die was supported by a bolster, which could be moved into position by sliding between two side plates of the machine's frame. The same frame supported the punch, the clamp and the feeder. All these systems, together with a table, are shown in Figure 2.27.

Following this design, the machine was built to prove the concept of the process and to test its industrial feasibility (Figure 2.28(a)). The punch and the clamp were driven by two hydraulic actuators with a capacity of 800 kN and 1000 kN, respectively. The feeder also used two actuators, a vertical one for clamping the plate and a horizontal one to move it forward. Thus, the machine required four hydraulic actuators, which were supplied with oil by four hydraulic pumps fed from a single hydraulic tank (Figure 2.28(b)). Synchronising the work of four actuators required several displacement and hydraulic pressure sensors, which provided information for a PLC unit located in a control cabinet (Figure 2.28(c)). The cabinet pulpit served as a machine–user interface (Figure 2.28(d)). An additional PC-based data acquisition system

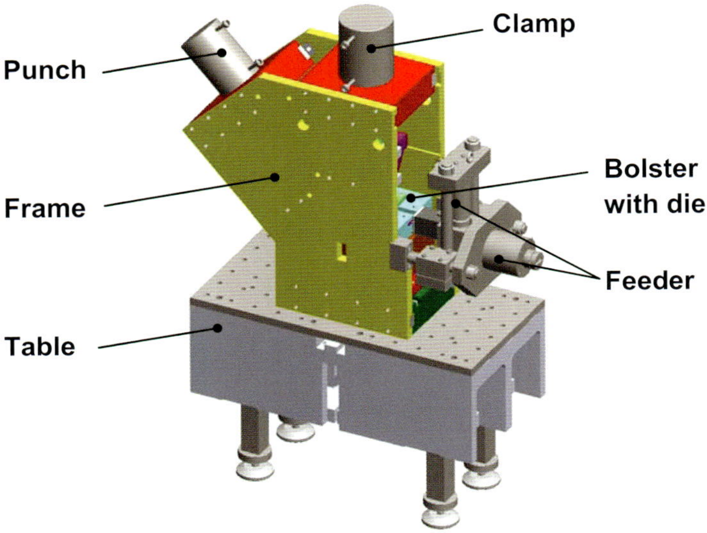

Figure 2.27 Main system of I-ECAP horizontal machine for plates

Figure 2.28 I-ECAP horizontal machine: (a) general view; (b) hydraulic power pack; (c) control cabinet; (d) pulpit

was used to record all relevant data with high resolution. The machine was intended for room temperature as well as for elevated temperatures (up to 300°C) to enable a variety of metals to be processed. The bolster system was equipped with a number of electric heaters and thermocouples positioned at important locations in the bolster and the die. A separate multi-channel system enabled precise control of temperature. To avoid overheating of the machine, there was a closed-circuit water-cooling system incorporating cooling channels and a chiller. There was also a lubricating system, which supplied lubricant to the material deformation zone.

The material used for the initial trials was Al1050 plate (10 mm × 50 mm × 1000 mm) conversion coated with calcium aluminate and lubricated with dry soap. The process was carried out at room temperature using a 1.6 mm feeding stroke and a low cycle frequency of approximately 0.2 Hz. Figure 2.29(a) illustrates the plate being fed to the machine while Figure 2.29(b) shows the plate partly processed and moved forward for inspection of the deformation zone (Figure 2.29(c)). The ultra-fine grained (UFG) structure after the first pass of the process revealed by scanning transmission electron microscopy (STEM) confirmed the feasibility of the process. Figure 2.30 shows this structure in the side plane of the plate, with an average grain/subgrain size of approximately 1 μm.

The results of the initial trials at room temperature have shown the feasibility of building a semi-industrial machine capable of refining grain structure in

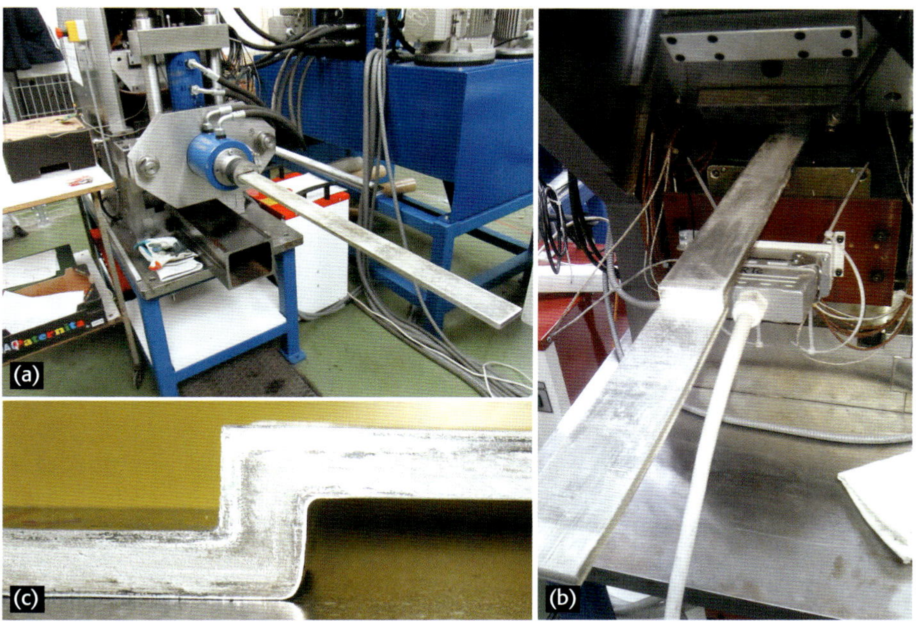

Figure 2.29 I-ECAP horizontal machine for plates in action: (a) feeding of plate; (b) receiving processed plate; (c) deformation zone

Figure 2.30 STEM image of Al1050 structure after first pass of two-turn I-ECAP (side plane)

thick continuous plates with a 10 mm × 50 mm cross-section. For the first time, such a process was used for two turns of the channel doubling the value of the strain achieved in one pass and increasing productivity.

2.4 I-ECAP of sheets

Many manufacturing processes use metal sheet blanks to form metal components. Some of these processes, as well as the manufactured components, might benefit from using UFG sheets. A good example would be using UFG magnesium, aluminium or titanium sheets for low temperature and/or high strain rate superplastic forming. One possible option to obtain UFG sheets is to produce bulk UFG billets by an SPD process and roll them into sheets (Valiakhmetov *et al.*, 2010). Another approach is based on direct conversion of coarse-grained sheet metal into a UFG equivalent by SPD. For example, in accumulative roll bonding (ARB) discussed in Chapter 7, two sheets are stacked together and warm-rolled to 50% of their thickness so that they become bonded. After cutting the resulting sheet into two pieces, the whole sequence is repeated until a UFG structure is created (Saito *et al.*, 1998). There have been attempts to use non-incremental ECAP as an SPD method for sheets, where sheets were pushed into the ECAP-like deformation zone by friction between the sheet and some rotating rolls, drums, etc. (Saito *et al.*, 1999; Lee *et al.*, 2002). However, using just a friction force to feed sheet metal continuously is always a challenge. This is why an increased angle (for example 120°) must be used between the input and output channels in order to reduce the feeding force. In this chapter, a low feeding force I-ECAP-based method will be explored to show that it is useful not only for producing UFG bars and plates, but also for sheets (Rosochowski *et al.*, 2010).

The machine used for I-ECAP of sheets was the one used earlier for I-ECAP of two bars described in Section 2.2. Since the bars were replaced with 2 mm × 50 mm × 190 mm sheet metal strips, some changes to the tools were required. The pusher was extended by adding a feeder (Figure 2.31). It served three functions. Its lower part had a step on each side, which enabled the strips to be fed upwards and prevented their backward movement. It also acted as a movable die, which travelled together with the sheet material and, therefore, did not introduce friction. Finally, it was guiding the punch by entering a gap in the punch. Since the gap had a limited length, the feeder was divided into three segments, of which the top two were pushed out as soon as they appeared above the die. Each face of the punch had a half-spike (too small to be seen in Figure 2.31), to improve the material flow. The die was divided into four segments, which were kept together by dowel pins and bolts. Two

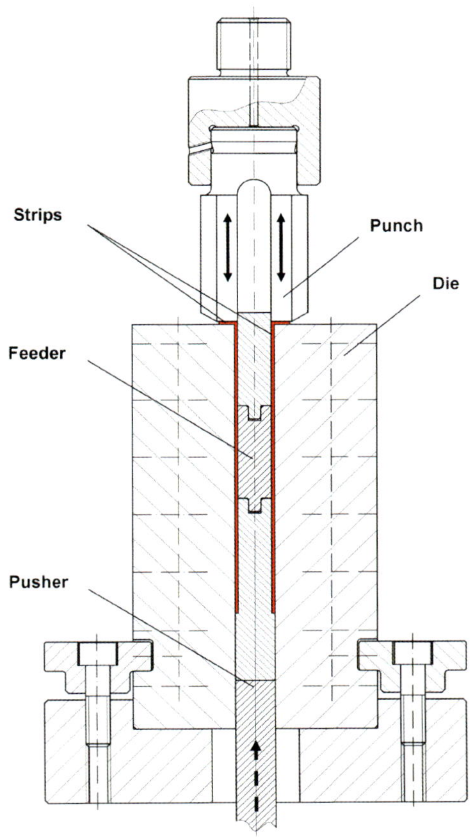

Figure 2.31 Tool configuration used in I-ECAP of sheet metal strips

of these segments were taller to provide walls for the horizontal channel (to secure the plain strain condition) and guide the punch.

The initial trials involved pure Al1070 strips, so they were performed at room temperature. The strips were conversion coated with calcium aluminate and lubricated with MoS_2. The feeding stroke was 0.2 mm or 10% of the sheet thickness. The punch amplitude was 1 mm and its frequency 0.5 Hz. Figure 2.32(a) shows a later phase of this process, with the second segment of the feeder already removed. The same figure presents the pure Al strips obtained in the first pass of the process.

To enable processing of a less ductile aluminium alloy Al–Mg with some addition of Sc, a die heating system was incorporated (Figure 2.32(b)). It consisted of a 1.5 kW band heater mounted around the die near its top and two aluminium heating blocks, each one equipped with a 0.25 kW heating cartridge, placed on top of the die. The location of heaters followed the idea of heating the tools near the deformation zone rather than along the entire feeding channel. The tool temperature was measured with a 0.5 mm diameter thermocouple inserted in a hole near the deformation zone. The temperature was kept at 200±2°C. Two aluminium alloy strips, lubricated in the same way as pure aluminium, were inserted into a die and heated by conduction. The strips were deformed using the same feeding stroke, punch amplitude and frequency as was the case with pure aluminium. The strips were subjected to six passes of I-ECAP, with 180° rotation between consecutive passes (route C). Figure 2.32(b) presents the final strips obtained in this way. Some strips were not perfect geometrically because they tended to bend in the plane of sheet. This was attributed to a non-perfect geometry of the press and the lack of an independent tool guiding system. The material structure obtained by electron backscattering diffraction

(a) (b)

Figure 2.32 I-ECAP of aluminium strips at: (a) room temperature; (b) at 200°C

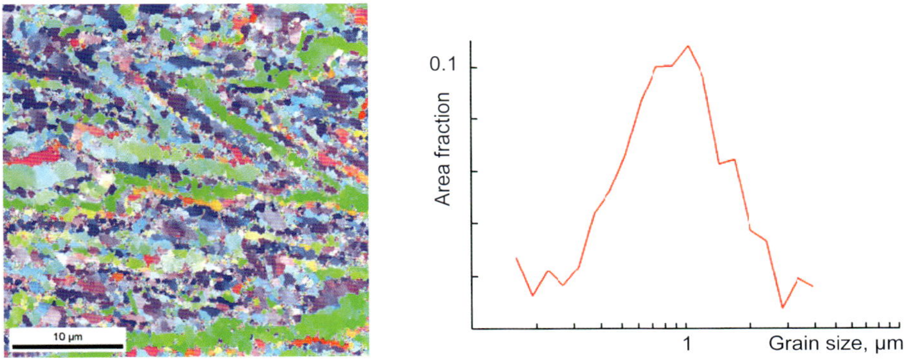

Figure 2.33 EBSD image of aluminium alloy Al–Mg–Sc subjected to six passes of I-ECAP at 200°C and its grain size distribution

(EBSD) indicated a well-developed UFG structure but there were also a few string-like grains with low angle grain boundaries (Figure 2.33). The grain size distribution shown in the same figure, indicated that the mean grain size was 1 μm.

After proving the principle of I-ECAP to be useful for processing sheets at the University of Strathclyde, Glasgow, UK, a new machine was built at Warsaw University of Technology, Poland. It was different from its predecessor in that only a single sheet was I-ECAPed at room temperature. The sheet was incrementally fed upwards and forced into a horizontal channel (channel angle 90°) by a reciprocating punch guided by two columns tilted by 30° from the vertical direction. The punch was pushed by a lever driven by a screw mechanism of the machine. Its return movement was forced by springs. The die was made of a stationary part and a horizontally guided clamping part. However, its clamping function was not utilised in the experiment. Instead, a screw jack was used to drive a pusher and provide incremental feeding of the sheet through a closed die. Figure 2.34 presents the tool configuration and the machine built.

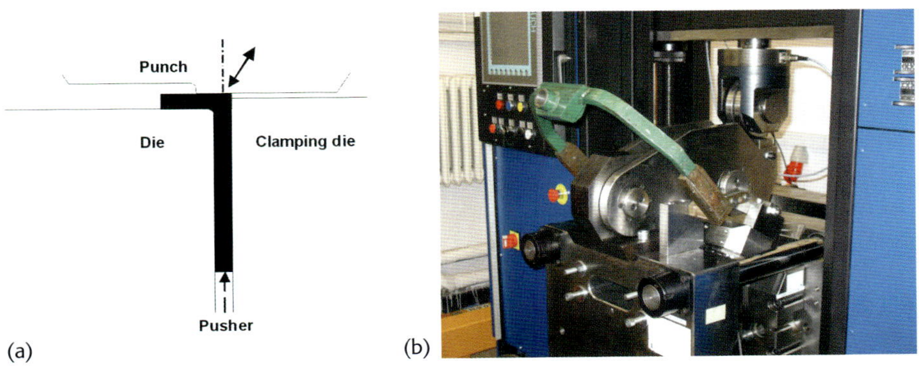

Figure 2.34 (a) Tool configuration; (b) photograph of I-ECAP machine for a single sheet

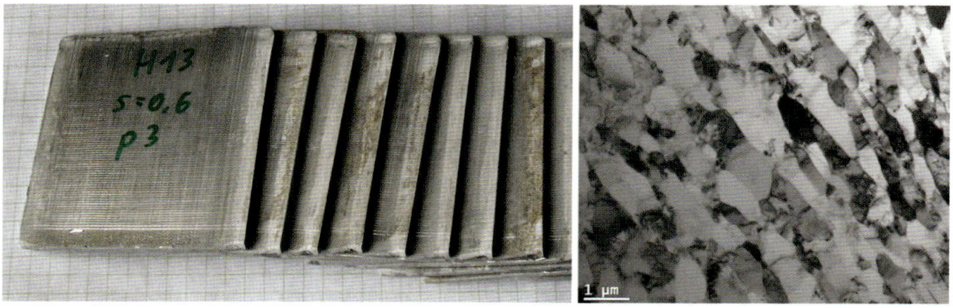

(a) (b)

Figure 2.35 Al1050 strips: (a) subjected to three passes of I-ECAP at room temperature; (b) TEM image of their structure in side plane

The thickness of the sheet strip was 3 mm, its width 62 mm and the maximum length was 105 mm. However, to enable 90° rotation of the strip about its normal between passes, the length of the strip eventually used in the experiment was 62 mm. For a punch amplitude of 1 mm, the feeding stroke was 0.6 mm or 20% of the thickness of the strip. Figure 2.35 shows the Al1050 strips subjected to three passes of I-ECAP. The resulting structure confirmed the feasibility of the method.

2.5 ECAP/I-ECAP with converging billets

The classical ECAP process is simple in terms of the tooling and machines used. However, it suffers from friction present in the die channels, which increases the process force and the tool contact pressure, limits the length of billets processed and causes material pickup. The concept of movable die walls, in both the input and the output channels of the die, was proposed in order to reduce friction (Segal, 2004). A movable wall representing the bottom part of the output channel was realised in practice in industrial ECAP of large plates for sputtering targets (Ferrasse *et al.*, 2008). But even if it is only one movable die wall, it leads to a more complex and expensive ECAP device. Also, even if friction between a billet and the bottom part of the output channel is eliminated, it does not disappear completely. It is simply moved to a new location between the stationary and movable elements of the device. Thus friction reduction in the output channel remains a challenge.

A new solution to the problem of friction in the output channel is based on the idea of ECAP with converging billets (Rosochowski *et al.*, 2013). In its basic configuration (Figure 2.36), the new ECAP system uses two substantially equal square or rectangular input channels converging into a single output channel, which is twice as wide as the input channels so that it can accept two converging billets. The contact surface between converging billets plays the same role as a movable bottom wall in the output channel of classical ECAP. It reduces friction and the process force. This effect is achieved without using a complex

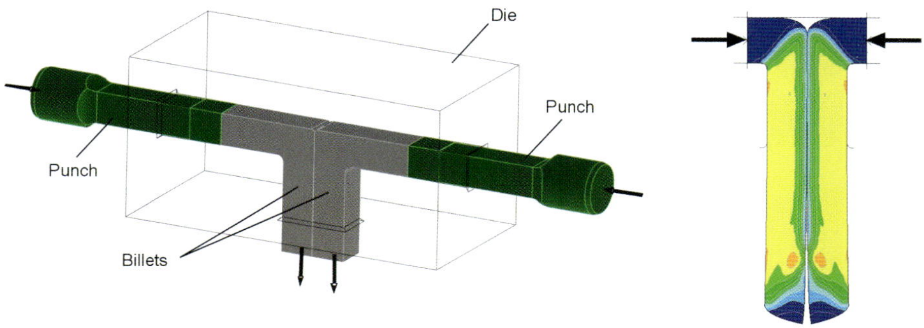

Figure 2.36 Process configuration and equivalent strain distribution for ECAP with converging billets

die with movable parts. Instead, two punches are used to push two billets synchronously from the opposite sides. The system doubles productivity compared to the case of processing a single billet. The strain produced is very similar to that resulting from other ECAP configurations (Figure 2.36).

The concept was tested experimentally on a machine originally built at AGH University of Science and Technology, Krakow, Poland (Richert, 2010) for cyclic extrusion–compression (CEC), which is another SPD process described in this book (see Chapter 5). The CEC machine was equipped with two hydraulic double action horizontal actuators opposing each other. The CEC work cycle involved fixing the die in the horizontal direction with external actuators and then using internal actuators to move the extrusion punch at a controlled speed while the compression punch only provided back pressure. Using the CEC machine for ECAP with converging billets also involved fixing the die in the horizontal direction, but required a different type of punch control to allow the punches to move in the opposite directions at the same speed. Another feature of the original CEC machine was a horizontally split die, whose halves were kept together during the process by a vertical hydraulic actuator. On completion of the process, the actuator lifted the upper half of the die to enable the billet to be removed. In ECAP with converging billets, the billets could leave the die using a lateral hole opposite the vertical actuator so that the die did not have to be split in the horizontal plane. Thus, the clamping action of the vertical actuator was only used to keep the die in the correct position in the vertical direction. Figure 2.37 shows the central part of the CEC machine used and a die set with superimposed arrows indicating the flow of the material.

The design of the die set is shown in Figure 2.38. The elements interacting directly with the actuators were the pushers and the guides. The role of pushers driven by the internal actuators was to push punches and limit their stroke. The role of guides driven by the external actuators was to guide pushers and transmit a horizontal fixing force. The die in the middle was made of five segments. The two short segments on one side of the die created a gap for the billet

Figure 2.37 CEC machine used for ECAP with converging billets (arrows illustrate flow of material)

exit and one long segment on the opposite side of the die featured a profiled spike to facilitate material flow from the horizontal channels into the die exit. The remaining two segments served as side walls for the channel defined by the above three segments. The outer rings were forced on the assembly of five segments to create an interference fit and prestress the die.

The 8 mm × 8 mm × 46 mm Al1070 billets were machined, conversion coated with calcium aluminate and lubricated with MoS_2 before they were placed in

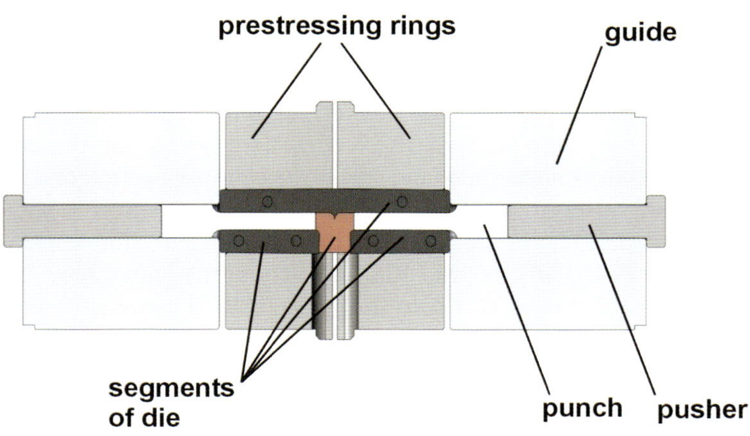

Figure 2.38 Die set used for ECAP with converging billets

Figure 2.39 Al1070 processed by one pass of ECAP with converging billets

the die channels. The die was then clamped in the horizontal direction by the guides using the outer horizontal actuators and clamped in the vertical direction via the prestressing rings using the vertical actuator. This was followed by the synchronised movement of the inner horizontal actuators which, through the pushers and the punches, caused material flow from both horizontal channels to the die exit in the middle. All actuators were withdrawn once the limit position had been reached. This enabled a new pair of billets to be inserted and the whole cycle repeated. This caused the previous billets to be pushed out of the die. This resulted in a fairly symmetrical flow of the material, which produced similar billets in each processed pair (Figure 2.39).

Since the purpose of this experiment was to prove the concept, only one pass of ECAP with converging billets was performed. This resulted in the Vickers hardness HV0.1 of the material increasing from approximately 32 to 52. The structures revealed by STEM in the three planes, *X* perpendicular to the billet axis, *Y* parallel to the billet side and, *Z* perpendicular to *X* and *Y* indicated a significant reduction in the average grain/subgrain size from the initial value of approximately 300 μm to 0.5–2 μm depending on the plane (Figure 2.40). This has proved that the process is suitable for SPD.

Figure 2.40 Grain structure of Al1070 processed by one pass of ECAP with converging billets

ECAP with converging billets suffers from the same problem as any other batch ECAP process: the length of the billets is limited. However, it is possible to remedy this by using tool kinematics, which makes the process incremental like I-ECAP (Rosochowski and Olejnik, 2013). As shown in Figure 2.41(a), in the first half of the process cycle, all four dies move outwards while the billets are held in position by stationery punches. This outward movement of the dies does not produce plastic deformation. In the second half of the process cycle (Figure 2.41(b)), the dies and punches move inwards together at the same speed causing plastic deformation of the billets. Since all tools and billets move together, no friction is involved. If continuous billets were to be processed, the punches would have to disappear and the dies would have to have a freedom of movement in the direction normal to the input channels to act as clamps.

Potentially, ECAP/I-ECAP with converging billets can suffer from an unstable flow of the material if there is a lack of symmetry in the dimensions or properties

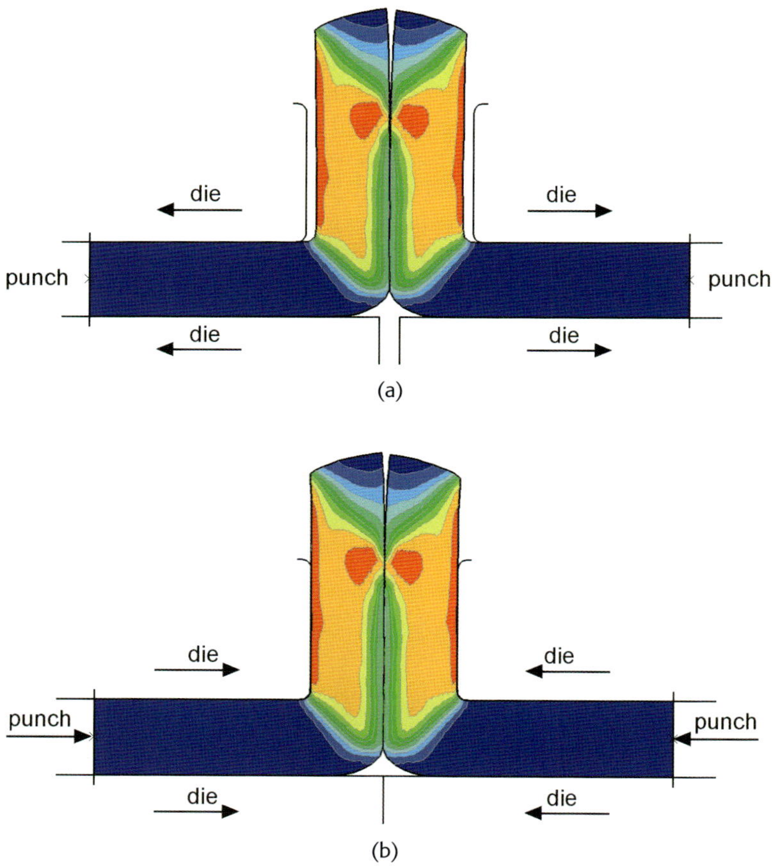

Figure 2.41 Snapshots of FE simulation illustrating principle of I-ECAP with converging billets

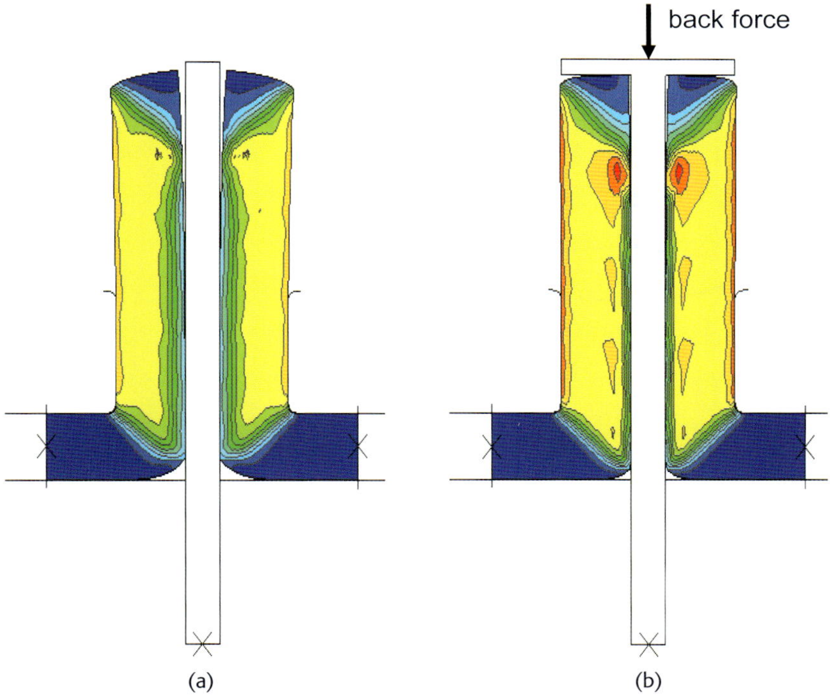

Figure 2.42 Equivalent strain in I-CAP with converging billets with: (a) separating tool; (b) separating tool with back force

of the billets. This can be rectified by using a movable separating tool driven by friction (Figure 2.42(a)). The tool can either be long enough to move together with the two billets all the way, or much shorter to move just the distance required in one cycle and then return to its initial position. A laboratory device is currently being developed to prove this concept. Another possible change is to apply a back force to the separating tool in order to obtain more uniform strain distribution and prevent fracture of brittle materials (Figure 2.42(b)).

2.6 I-ECAP of tubes

Attempts have been made to apply SPD to rings and tubes in order to obtain a UFG structure. The earliest one, shown in Figure 2.43(a), was high pressure torsion (HPT) of rings with conical faces proposed by Erbel (1979). The matching conical faces of the punches were radially toothed to facilitate the application of torque. This enabled pressure applied by punches to be substantially reduced compared with pressure normally required for the HPT of discs. Traditional equipment, normally used for HPT of discs, has also been successfully used for SPD of aluminium rings (Harai *et al.*, 2008). Figure 2.43(b) represents traditional ECAP, which was used to process short blind titanium tubes filled with

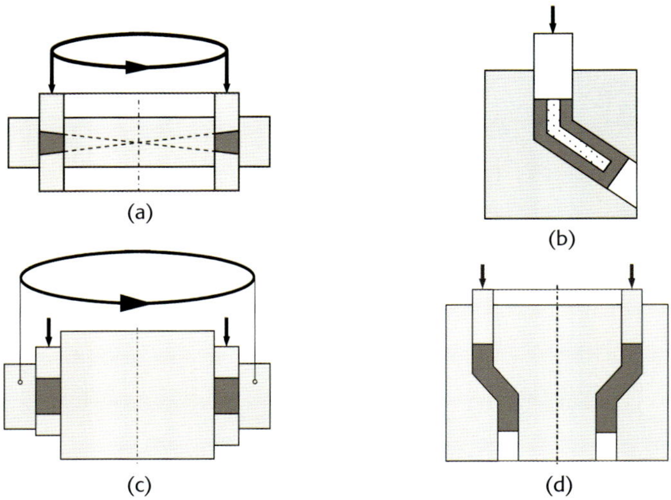

Figure 2.43 SPD processes for rings and tubes: (a) HPT; (b) ECAP; (c) RS/HPTT; (d) two-turn ECAP with changing tube diameter

sand (Nagasekhar *et al.*, 2006). However, the angle of the channel was 150°, which resulted in a low value of strain and only minimal reduction in grain size. Another method was proposed independently by Wang and Shan (2008) for short but rather thick-walled rings, and for thin-walled tubes (Tóth *et al.*, 2009). It has been called rotation shear (RS) and high pressure tube twisting (HPTT), respectively (Figure 2.43(c)). The method is based on fixing the inner surface of the ring/tube and rotating its outer surface so that a shear mode of deformation is developed across the thickness of the ring/tube. The method requires high pressure to be generated on the inner and outer surface of the ring/tube to avoid slippage between the material processed and both tools, a fixed mandrel and a rotating ring-tool. This can lead to problems caused by, for example, a non-uniform contact pressure, especially for longer tubes. Yet another process suitable for direct SPD of tubes has been proposed by Kim *et al.* (2008). It is similar to the process of two-turn ECAP (Rosochowski and Olejnik, 2002). The only difference is that, in the case of tubes, the process implies a change in tube diameter unless it is repeated to reverse that change (Figure 2.43(d)).

The author of this chapter carried out an FE simulation of the process proposed by Kim *et al.* (2008) in March 2006. At the same time, an incremental version of this process was simulated but all these results were published much later (Rosochowski and Olejnik, 2011). Figure 2.44 shows the model used and equivalent strain distribution at the end of classical ECAP. The strain distribution is not very different from that known from the plane strain version of the process. The channel radii used in this case were optimised from the point of view of strain uniformity. The maximum strain is slightly higher because of the

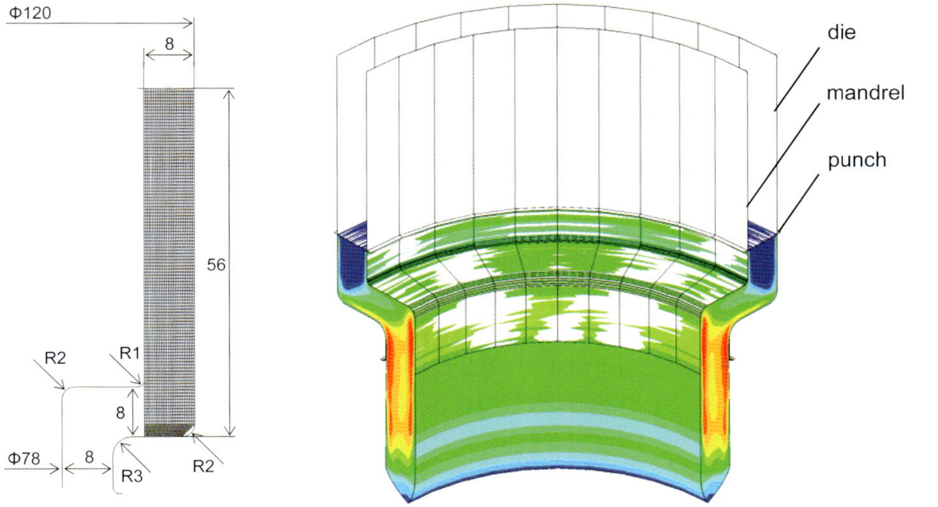

Figure 2.44 Model and equivalent plastic strain obtained in FE simulation of two-turn ECAP of tube

additional circumferential strain caused by the reduction of the tube diameter. There are typical end effects, where the material is less deformed.

The incremental version of the process involved a tubular punch/feeder incrementally feeding a processed tube and either a reciprocating mandrel or a reciprocating die. As can be seen in Figure 2.45(a), the option with reciprocating

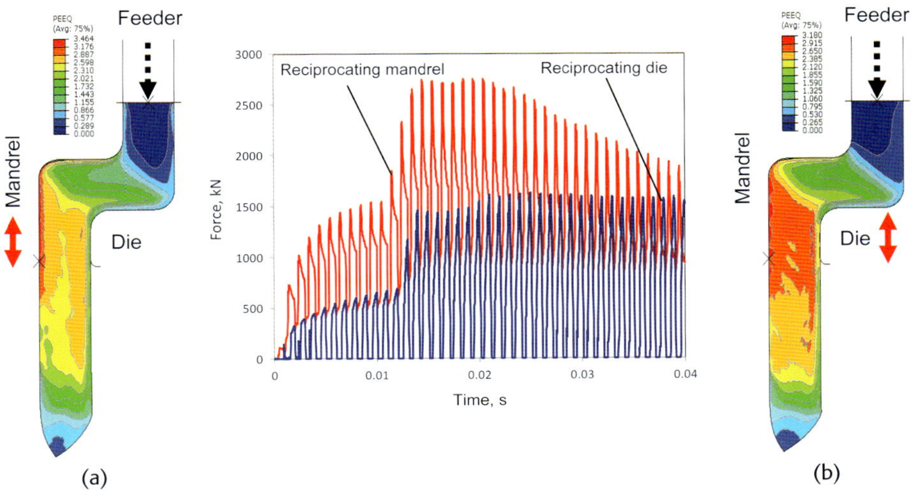

Figure 2.45 Equivalent plastic strain and punch/feeder force obtained in FE simulation of incremental ECAP of tube with: (a) reciprocating mandrel; (b) reciprocating die

mandrel gives a less uniform strain distribution and a higher punch/feeder force compared to the case with reciprocating die (Figure 2.45(b)). This appeared to be related to friction. Without friction, a similar maximal punch/feeder force of approximately 1000 kN was obtained for both cases. Thus, the option used for further analysis was one with a reciprocating die. The radii of the die and the mandrel were optimised to obtain the most uniform strain distribution. The best results were obtained for all the radii equal to 3 mm.

Figure 2.46 compares the equivalent strain distribution in classical ECAP of tube (Figure 2.46(a)) with the I-ECAP of a tube realised with a reciprocating die (Figure 2.46(b)). It can clearly be seen that more uniform strain distribution across the tube wall can be produced by I-ECAP than by ECAP. Figure 2.46(c) gives additional information about the strain distribution across the tube wall, in the middle section of the tube, far from end effects (path A–A). It shows that the I-ECAP strain is higher at the inner surface of the tube (3 vs. 1.9) and in the middle of the wall (2.7 vs. 2.4). Another feature worth mentioning is a smaller wall thickness reduction in I-ECAP, which can be described by a smaller gap between the tube and the die exit. This gap was reduced from approximately 0.2 mm for ECAP to 0.05 mm for I-ECAP.

The SPD of tubes presents a serious problem in terms of the force required because the volume of the material processed is similar to the case for wide plates (width of plate comparable with circumference of tube). A possibility of force reduction by replacing ECAP with I-ECAP could be an important factor, making SPD of tubes more feasible. Figure 2.47 illustrates this point by showing the force history for the die and the punch/feeder for both processes. The die and punch forces during ECAP (slightly different because of the friction force

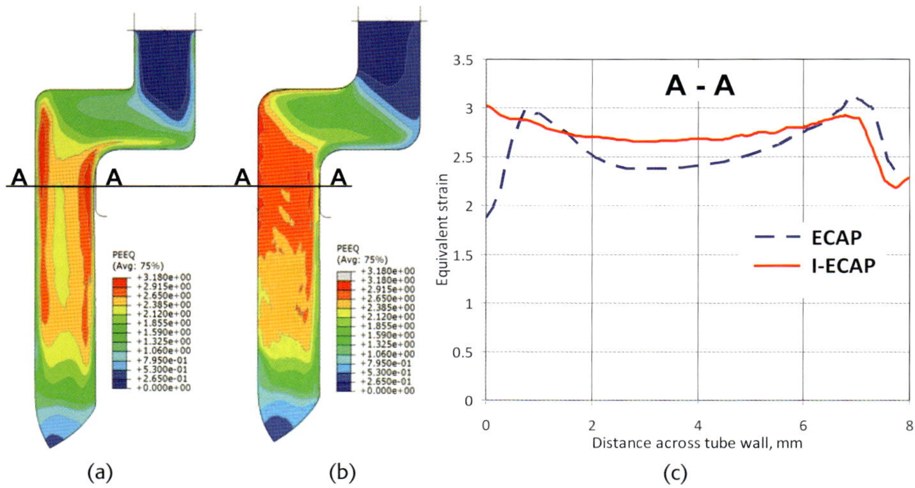

Figure 2.46 Equivalent strain distribution obtained in FE simulation of tube processing: (a) using classical ECAP; (b) using I-ECAP; (c) comparison of both for path AA

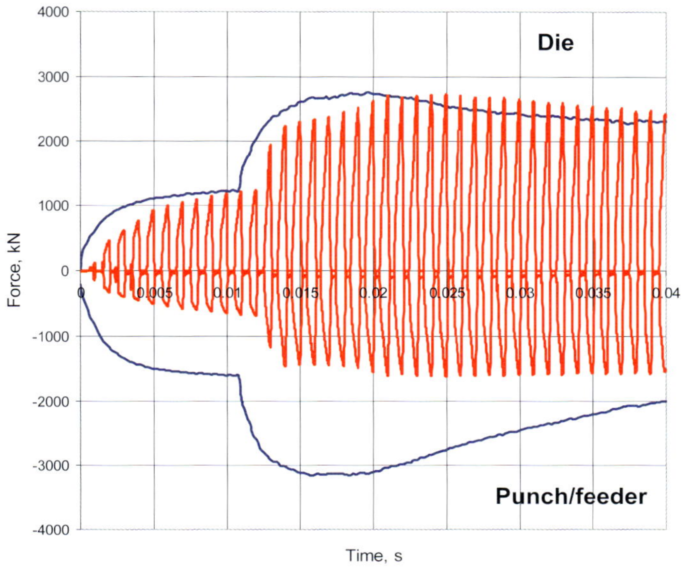

Figure 2.47 Forces acting on die and punch/feeder during ECAP (smooth curves) and I-ECAP (cyclic curves)

on the mandrel) initially increased to about 1500 kN and then jumped to about 3000 kN, when the material took the second turn of the channel. After that, both forces gradually decreased to about 2000 kN due to a reduction in friction caused by the decreasing tube length. The die force in I-ECAP was similar to that in ECAP; it was shifted along the time axis due to a slightly different starting point for both simulations. The force required to feed the tube was close to zero while the force supporting the tube during the deformation part of the cycle was not more than 1630 kN, which indicated about 48% force reduction compared to the ECAP force of 3150 kN. The punch/feeder force for I-ECAP was constant after the tube had passed the second turn of the channel.

I-ECAP of tubes has not yet been realised in practice. However, FE simulation provided an interesting insight into the process and enabled its comparison with classical ECAP. It seems that a better option would be I-ECAP with a reciprocating die. Unlike the process used by Kim *et al.* (2008), it was decided to use a more demanding 90° turning channel, which increases the achievable strain in one pass of the process. It is hoped that one pass, which produces a strain of approximately 2.7, might be enough for some materials to refine their grain structure and improve properties. Another pass would reduce productivity, increase cost and further reduce the diameter of the tube. Increasing the diameter of the tube to its initial value in the second pass is also possible. But it might be problematic in practice, because it may generate circumferential tension in the material which has reduced ductility after the first pass.

It seems that ECAP of tubes has a better chance of becoming an industrial process than ring/tube processing using HPT or RS/HPTT because of a lower force and tool contact pressure as well as simpler process kinematics, which does not require rotating tools. I-ECAP has more complex kinematics than ECAP. However, it enables a substantial reduction in the feeding force, which should enable longer tubes to be processed.

2.7 Incremental angular splitting

Occasionally, I-ECAP with two parallel billets in the input channel (as described in Section 2.2) may lead to an unstable flow of the material in this channel. This manifests itself as one of the billets undergoing local bulging while its neighbour becoming locally thinner. When such a bulge eventually reaches the reciprocating punch, the sharp spike of the punch can shave the bulge off or split the thicker billet. This defect was the inspiration for a new SPD process, known as incremental angular spitting (I-AS). The idea was first published by Rosochowski *et al.* (2012) and then further tested (Rosochowski *et al.*, 2013).

Figure 2.48 illustrates the principle of the new process. Comparing this figure with Figure 2.5, one can only see one difference. Instead of two billets, there is a single billet in the input channel, which has a rectangular cross-section. The output channels have a cross-section equal to half of the cross-section of the input channel. The punch face with a spike plays the role of a splitting tool, which divides the billet into two equal billets and forces them into the output channels. It has to be emphasised that the splitting process is plastic in nature (no fracture) and therefore similar to plastic orthogonal cutting. One can say that it creates a bridge between cutting and ECAP. However, in order to be able to process long billets, an incremental version of ECAP is considered.

Because of its origin, I-AS could easily be performed using the tools (Figure 2.11) and the machine (Figure 2.12) used earlier for I-ECAP. Thus, rather than starting from an FE simulation as usual, an experiment was carried out first. For comparison purposes, it was decided to combine I-ECAP and I-AS in one experiment. This was achieved by designing an Al1070 billet, which

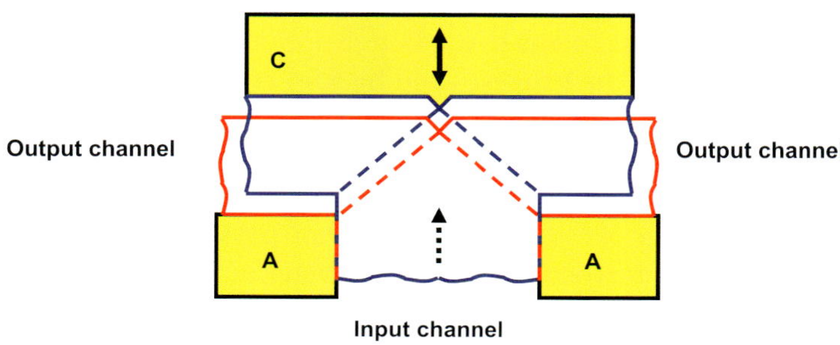

Figure 2.48 Schematics of I-AS of a single billet

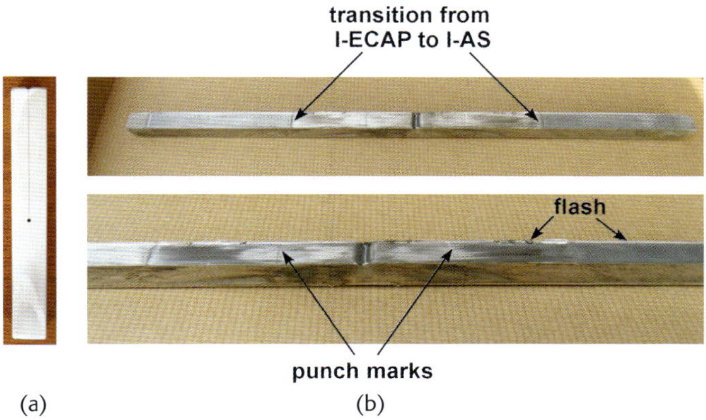

Figure 2.49 Split billet: (a) before; (b) after I-ECAP followed by I-AS

had a rectangular cross-section of 10 mm × 20 mm and was pre-cut along approximately half its length by wire electrical discharge machining. The punch movement followed a sine wave with frequency of 1 Hz and the peak-to-peak amplitude 1.6 mm; the feeding stroke was 0.2 mm. The billet was lubricated with MoS_2 applied to a conversion coat of calcium aluminate created earlier. The pre-cut gap was not lubricated because pre-cutting took place after conversion coating. Additionally, the pre-cut gap was not wide enough to apply the lubricant. The process was carried out at room temperature.

Since the billet was orientated in the input channel in the way shown in Figure 2.49(a), the first process to occur was I-ECAP. After reaching the pre-cut end hole, the punch started splitting the material plastically. I-AS continued until the last portion of billet was fed. Figure 2.49(b) shows the result of this process. The transition from I-ECAP to I-AS is clearly visible as the pre-cut end hole left marks. The quality of the top surface produced by I-ECAP was better than that produced by I-AS despite no lubrication in the gap. The flash caused by a clearance between the punch and the die was more pronounced in the case of I-AS. The punch marks could have been avoided if better control had been exercised over the feeding force causing the process to stop.

The punch force and the feeder force were recorded during the process using the load cells LC1 and LC2, respectively, as shown in Figure 2.14. The results are presented in Figure 2.50. The punch force varied between its maximum value and zero in each cycle. The feeder force varied between a top envelope, when the feeder was stationary and supporting the billet subjected to plastic deformation and a bottom envelope, when the feeder was feeding the billet as a rigid body. A gap in the recorded force history indicates those few cycles, during which the punch was going through the pre-cut end hole. The analysis of the recorded forces does not reveal any substantial difference in their nature or value resulting from the process changing from I-ECAP to I-AS. The gradual changes that are visible are similar to those recorded for I-ECAP alone and

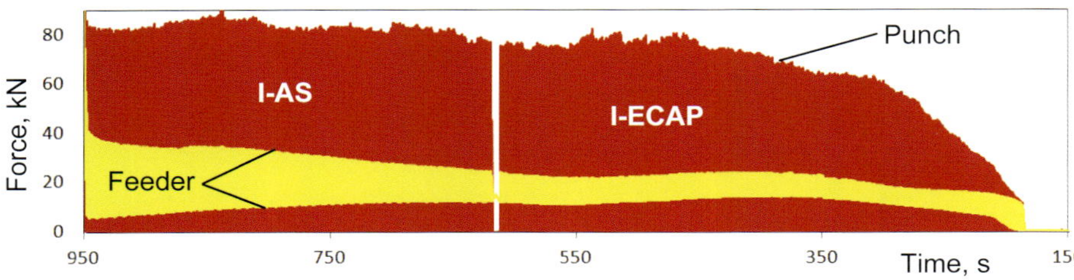

Figure 2.50 Force on punch and feeder during I-ECAP followed by I-AS (time axis has been reversed)

result from a decreasing friction force in the input channel as the billet length in this channel decreases.

To characterise material changes resulting from one pass of I-ECAP/I-AS, the Vickers hardness HV0.1 was measured along the deformed billet as well as at five locations in the transverse direction. The results of these measurements are shown in Figure 2.51. The longitudinal distribution of hardness

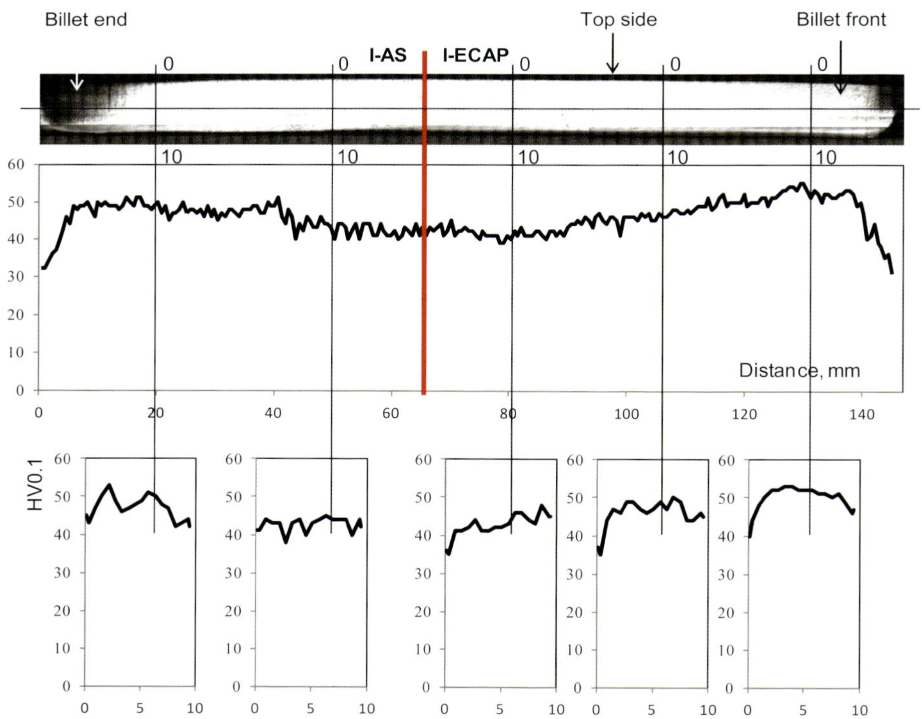

Figure 2.51 Vickers hardness HV0.1 measured along billet subjected to I-ECAP and I-AS and at five locations in transverse direction

is non-uniform, with the lowest value approaching the original hardness of HV32 at both ends, which results from the lack of plastic deformation there. The maximum hardness achieved is approximately HV55 for the I-ECAP side and HV50 for the I-AS side. In the middle part of the billet the hardness drops to about HV40. There is no sudden change of hardness at the point of transition from I-ECAP to I-AS. However, the distribution of hardness in the transverse direction is different for the three locations on the I-ECAP side and the two locations on the I-AS side. While the I-ECAPed billet shows a lower value of hardness on the punch side (<HV40 at the transverse coordinate 0), the I-ASed billet displays more uniform distribution in the transverse direction.

FE simulation of I-AS was attempted using Deform 2D rather than Abaqus because of its ability to automatically remesh the heavily distorted finite elements expected to appear in a process compared to cutting. The material model and friction conditions were the same as in other FE simulations presented in this chapter. The billet geometry and most process parameters were taken from the experiment, only the punch amplitude was reduced to shorten computational time. The mesh of free-form quadrilateral elements was automatically generated but was refined in a predefined area below the punch. The same area was used to apply remeshing during the simulation. The simulation results, in terms of equivalent strain, are presented in Figure 2.52(a). The strain distribution in the I-ECAPed part of the billet is similar to the distribution found in I-ECAP of two billets (Figure 2.7), with the characteristic area of lower strain on the punch side. However, as soon as I-AS starts, the same area undergoes much higher deformation reaching the maximum strain of 2. It was suspected that the geometry of the punch face could change the strain distribution in I-AS. The most radical change would be removing the spike and leaving the punch face flat. Figure 2.52(b) displays the strain distribution in such a case, with the maximum value of strain reaching 8 on the punch side of the billet.

I-AS appears to be similar to I-ECAP and, therefore, to classical ECAP in terms of the force required, material flow and the resulting hardness in the bulk of the billet, except the top (punch) side of the billet where strain and hardness are more uniform. Looking at the benefits of I-AS, it increases the choice of billet shapes by giving an option to plastically split a billet into two billets. I-AS

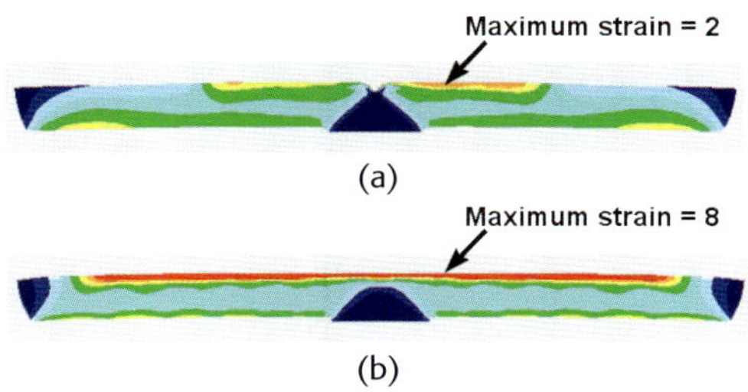

Figure 2.52 Equivalent strain in billet subjected to: (a) I-ECAP followed by I-AS; (b) I-AS only with flat punch

generates a higher strain in the vicinity of the split surface, which leads to more uniform hardness distribution in the transverse direction compared to I-ECAP and ECAP. In the special case of I-AS with a flat punch, a very high strain and possibly high hardness could be achieved in this region. In most cases, when uniform billets are desirable, this will be inappropriate. However, for gradient materials, surface strengthening resulting from I-AS with a flat punch may open up new possibilities.

2.8 Summary

Among many SPD processes for continuous billets, I-ECAP is a unique process belonging to a large group of incremental metal-forming processes. Kinematically, it is more demanding because it requires two actuators with synchronised movements but this enables separation of the deformation and the feeding stage of the process, and results in a reduction of the feeding force to a very low value. The different process configurations presented in this chapter show the versatility of I-ECAP, which can be used for the SPD of bars, plates, sheets and tubes. It can process a single billet, parallel billets in the input channel and converging billets. It can also be used to split billets.

There are still many challenges and unanswered questions regarding further development of the I-ECAP technology. Some of the process configurations described, for example I-ECAP of tubes, have yet to be tried in practice. The clamping/unclamping version of the feeding mechanism should allow really continuous billets to be processed. The limits of the punch/feeder frequency and the feeding increment, which affect process productivity, have not yet been tested. There is also the issue of scaling up the processed billets. Materials that are more difficult to form, such as titanium alloys or steels, will test the capability of the process in terms of forces, tool stresses and temperature. Friction/lubrication remains a very important practical concern. Finally, new developments such as I-AS with a flat punch or ultrasonically assisted ECAP/I-ECAP (proposed in the I-ECAP patent) will sustain enthusiasm about the possibilities of the I-ECAP process that are still open to us.

References

Erbel, S. (1979) 'Mechanical properties and structure of extremely strainhardened copper', *Metals Technology*, 6: 482–486.

Ferrasse, S., Segal, V. M., Alford, F., Kardokus, J. and Strothers, S. (2008) 'Scale up and application of equal-channel angular extrusion for the electronics and aerospace industries', *Materials Science and Engineering A*, 493: 130–140.

Groche, P., Fritsche, D., Tekkaya, E. A., Allwood, J. M., Hirt, G. and Neugebauer, R. (2007) 'Incremental bulk metal forming', *Annals of the CIRP*, 56: 635–656.

Gzyl, M., Rosochowski, A., Yakushina, E., Wood, P. and Olejnik, L. (2013) 'Route effects in I-ECAP of AZ31B magnesium alloy', *Key Engineering Materials*, 554: 876–884.

Harai, Y., Ito, Y. and Horita, Z. (2008) 'High-pressure torsion using ring specimens', *Scripta Materialia*, 58: 469–472.

Kim, K. J., Lee, S. U., Yang, D. Y. and Yoon, J. W. (2008) 'Development of a tube-ECAE process to improve the mechanical properties of a metal tube', Paper presented at *9th ICTP, International Conference on Technology of Plasticity*, held 7–11 September 2008, Gyeongju, Korea. (CD-ROM ISBN 978-89-5708-152-5, 2295–2299).

Lee, J. C., Seok, H. K., Suh, J. Y., Han, J. H. and Chung, Y. H. (2002) 'Structural evolution of a strip-cast Al alloy sheet processed by continuous equal-channel angular pressing', *Metallurgical and Materials Transactions A*, 33: 665–673.

Liu, Z. Y., Liang, G. X., Wang, E. D. and Wang, Z. R. (1998) 'The effect of cumulative large plastic strain on the structure and properties of a Cu-Zn alloy', *Materials Science and Engineering A*, 242: 137–140.

Nagasekhar, A. V., Chakkingal, U. and Venugopal, P. (2006) 'Candidature of equal channel angular pressing for processing of tubular commercial purity titanium', *Journal of Materials Processing Technology*, 173: 53–60.

Olejnik, L., Rosochowski, A. and Richert, M. (2008) 'Incremental ECAP of plates', *Materials Science Forum*, 584: 108–113.

Raab, G. I. (2005) 'Plastic flow at equal channel angular pressing in parallel channels', *Materials Science and Engineering A*, 410(11): 230–233.

Richert, J. (2010) 'Strain–stress conditions of shear band formation during CEC processing on a new machine with control back-pressure', *Archives of Metallurgy and Materials*, 55: 391–408.

Rosochowski, A. (2012) 'Severe plastic deformation of metals', European Patent EP 1861211.

Rosochowski, A. (2014) 'Severe plastic deformation of metals', US Patent 8631673.

Rosochowski, A. and Olejnik, L. (2002) 'Numerical and physical modelling of plastic deformation in 2–turn equal channel angular extrusion', *Journal of Materials Processing Technology*, 125(6): 309–316.

Rosochowski, A. and Olejnik, L. (2007) 'FEM simulation of incremental shear' in *Proceedings of the 10th International Conference on Material Forming, Esaform 2007*, held 18–20 April 2007, Zaragoza, Spain, E. Cuoet and F. Chinesta (eds), pp. 653–658. American Institute of Physics Publishing, Melville, NY.

Rosochowski, A. and Olejnik, L. (2008) 'Finite element analysis of two-turn incremental ECAP', *International Journal of Material Forming*, 1(Suppl. 1): 483–486.

Rosochowski, A. and Olejnik, L. (2011) 'Incremental ECAP of tubular components-FE simulation' in *Proceedings of the 14th International Conference on Material Forming, Esaform 2011*, held 27–29 April 2011, Belfast, UK, G. Menary (ed.), pp. 517–522. American Institute of Physics Publishing, Melville, NY.

Rosochowski, A. and Olejnik, L. (2013) 'Incremental ECAP with converging billets', *Key Engineering Materials*, 554: 869–875.

Rosochowski, A., Olejnik, L., Gagne, J., Ladeveze, N. and Rosochowska, M. (2006) 'Compression behaviour of UFG aluminium' in *Proceedings of the 9th International Conference on Material Forming, Esaform 2006*, held 26–28 April 2006, Glasgow, UK, N. Juster and A. Rosochowski (eds), pp. 543–546. Publishing House Akapit, Krakow, Poland.

Rosochowski, A., Olejnik, L. and Richert, M. (2008) 'Double-billet incremental ECAP', *Materials Science Forum*, 584(6): 139–144.

Rosochowski, A., Olejnik, L., Richert, J., Rosochowska, M. and Richert, M. (2013) 'Equal channel angular pressing with converging billets-experiment', *Materials Science and Engineering A*, 560: 358–364.

Rosochowski, A., Rosochowska, M. and Olejnik, L. (2012) 'New SPD process of incremental angular splitting', *Key Engineering Materials*, 504: 569–574.

Rosochowski, A., Rosochowska, M. and Olejnik, L. (2013) 'Severe plastic deformation by incremental angular splitting', *Journal of Materials Science*, 48: 4557–4562.

Rosochowski, A., Rosochowska, M., Olejnik, L. and Verlinden, B. (2010) 'Incremental equal channel angular pressing of sheets', *Steel Research International*, 81: 470–473.

Saito, Y., Tsuji, N., Utsunomiya, H., Sakai, T. and Hong, R. G. (1998) 'Ultra-fine grained bulk aluminium produced by accumulative roll–bonding (ARB) process', *Scripta Materialia*, 39: 1221–1227.

Saito, Y., Utsunomiya H and Suzuki H (1999), 'Proposal of novel continuous high straining process - development of conshearing process' in *Proceedings of 6th ICTP, Advanced Technology of Plasticity 1999*, held 19–24 September 1999, Erlangen, Germany, M. Geiger (ed.), Vol. III, pp. 2459–2464. Berlin: Springer.

Segal, V. and Segal, L. (1997) 'Method of and apparatus for processing tungsten heavy alloys for kinetic energy penetrators', US Patent 5600989.

Segal, V. M. (2004) 'Engineering and commercialization of equal channel angular extrusion (ECAE)', *Materials Science and Engineering A*, 386: 269–276.

Segal, V. M. (2010) 'Mechanics of continuous equal-channel angular extrusion', *Journal of Materials Processing Technology*, 210: 542–549.

Shchukin, V. Ya. and Segal, V. M. (1976) 'Device for strengthening materials by pressure', USSR Patent SU 515968.

Tóth, L. S., Arzaghi, M., Fundenberger, J. J., Beausir, B., Bouazizc, O. and Arruffat-Massion, R. (2009) 'Severe plastic deformation of metals by high-pressure tube twisting', *Scripta Materialia*, 60: 175–177.

Valiakhmetov, O. R., Galeyev, R. M., Ivan'ko, V. A., Imayev, R. M., Inozemtsev, A. A., Koksharov, N. L., Kruglov, A. A., Lutfullin, R. Ya., Mulyukov, R. R., Nazarov, A. A., Safiullin, R. V. and Kharin, S. A. (2010) 'The use of nanostructured materials and nanotechnologies for the elaboration of hollow structures', *Nanotechnologies in Russia*, 5: 108–122.

Wang, M. and Shan, A. (2008) 'Severe plastic deformation introduced by rotation shear', *Journal of Materials Processing Technology*, 202: 549–552.

Tooling for ECAP

CHAPTER 3

Lech Olejnik

Equal channel angular pressing (ECAP) is the most popular severe plastic deformation (SPD) technique and is widely used for batch production of ultra-fine grained (UFG) and even nanocrystalline bulk metallic materials. ECAP is often used to investigate the relationship between structural changes and physical properties in a number of severely deformed alloy systems. It is also considered to be the best candidate for many practical applications. However, it remains a laboratory technique which is used to prepare small samples. Segal, the founder of the ECAP method, wrote (Segal, 2004)

> ... after more than 30 years since the process was first introduced there is only one report on the process commercialisation.

He meant an application of ECAP for the electronics and aerospace industries (Ferrasse *et al.*, 2008).

Indeed, little technical progress has been made regarding ECAP. The method is usually employed in its classical form (i.e. as it was patented (Segal, 1997) and later applied to grain refinement by Segal *et al.* (1981)). This chapter addresses the issues of tool design, scale up and productivity of ECAP. Its main aim is to share the tool engineering and technological developments of ECAP with the metallurgy and materials processing research community.

3.1 General recommendations for process design

In terms of metal working, ECAP can be classified as side extrusion through a die cavity which consists of two straight channels inclined to each other. In contrast to other bulk metal-forming processes, the extruded material flows down through the channel without any change in its cross-sectional dimensions. Assuming a channel inclination angle of 90°, an equivalent plastic strain $\varepsilon = 1.15$ can be achieved in one pass of the process. This strain value is not sufficient for successful grain refinement for most metallic materials. In order to accumulate the required strain, the workpiece has to be processed up to eight times in the same die.

The ECAP-based microstructure refinement process can be developed by taking the rules from cold or warm bulk metal-forming technology into account. When a bulk metal-forming process is being designed, a number of technology areas are usually considered such as: material characterisation, the forming sequence, configuration of tooling, tool design and die surface modification, billet preparation, lubrication, heat treatment and allied processes. Design guidelines, for example those available for cold extrusion (International Cold Forging Group, 1991), can be used for process planning. Using selected examples, we will show how to plan a multi-stage ECAP process taking into account nomenclature and experience from bulk metal-forming practice.

In process planning, the following three major areas should be addressed

- planning the sequence of operations with associated metalworking equipment
- coordinating the allied processes of surface and heat treatment
- integrating the required workpiece handling and press automation

In fact, any batch SPD process is a multi-stage metal-forming process with the same main operation repeated until the total value of equivalent strain is achieved. Cold or warm solid forward extrusion processes could be considered to be reference operations for any ECAP process. In particular, productivity and material savings are high in cold extrusion due to good surface quality and tight dimensional and shape error tolerances. Due to limited tool load and the limited forming ability of different materials, extrusion processes can also be carried out at elevated temperatures. Lower flow stress during warm extrusion allows for a decreased inclination angle of the ECAP channel, thereby increasing the strain in one pass and reducing the number of passes needed for the required grain refinement and improvement of properties. A combination of warm and cold extrusion processes can also be used for certain materials. The geometry of the billet and extruded part should be determined when planning the process. Processing and dimensional specification and tolerances should be studied taking the limitations of ECAP into account. Process optimisation and basic types of die, including choice of ECAP route with minimum number of passes and channel configuration, should be considered. Die designers must also consider the problems of die deflection, workpiece spring back, die wear, and scratches on the workpiece or even its failure. There is also a need to specify the necessary press equipment, ancillaries as well as the special means required in automated and safe mass production, like monitoring and supervision systems. These factors can be important for the productivity of any SPD process. The range of process parameters, like the required pressing speed or processing temperature for particular metallic materials, is a matter of the necessary process window, hence, it is beyond the scope of this chapter. Productivity should be discussed in terms of the organisation of the production within a single ECAP operation as well as production planning of the whole multi-stage microstructure refining process.

Cost is very important when planning and running cold and warm extrusion processes. The most economical processing can be determined for several alternative production methods based on their overall costs. The cost should include: materials, tooling, direct and indirect labour, and overheads. In the case of cold bulk metal-forming the material and tooling costs are always the main expenses (Clapp, 1966). Their impact depends on the scale of the production. According to Sieber (1966), the cost of tooling can be several times higher than the cost of material for low-scale production and decreases substantially with the increasing number of formed parts, whereas the material cost tends to remain unchanged. This tendency is shown in Figure 3.1. Therefore, in the case of mass production, much attention must be paid to proper utilisation of material and proper tool design to promote long tool life. These issues will be further discussed in Section 3.4 for batch production of UFG metals using ECAP.

Following recommendations for extrusion processes, a cost-effective multistage microstructure refining process would involve one ECAP die used many times. Because the ECAP workpiece does not change its cross-section in a single pass it can be ejected, being pushed out of the channel by the following billet with the same, i.e. unchanged cross-section. Pressing is carried out with rigid tools, possibly at room temperature. The permissible length of the billet or slug depends to a considerable extent on the maximum load-carrying capacity of the punch. If the deformability of the workpiece is limited, ECAP could be performed at an elevated temperature. In this case, the billet/slug should be preheated externally and the required temperature of the die should be controlled via a proper tool heating–cooling system. Billets should be properly lubricated taking into account their hardness, extension of surface and processing temperature. The lubrication layer should be refreshed after each pass of ECAP. The ECAP process is known for a substantial distortion of billet ends. Tipped ends have to be levelled as they can disturb the work of the punch. Sharp edges also have to be made blunt before performing the next operation. In cold extrusion, a maximum stress up to 2500 MPa has to be taken into account when

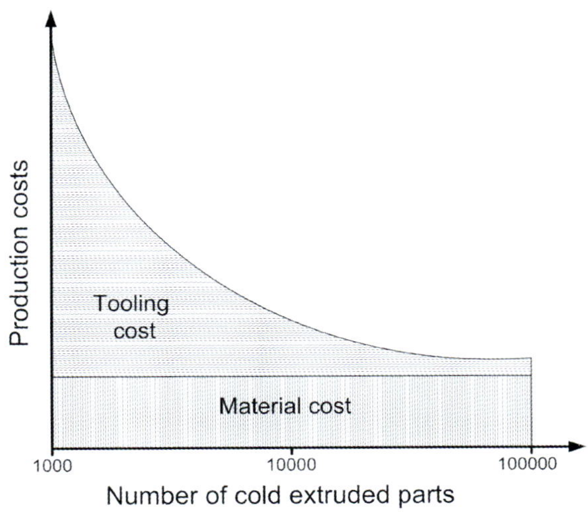

Figure 3.1 Tooling and material costs for cold extrusion

designing the ECAP tools. Tools (i.e. punch and die) should be guided properly with reference to each other. A properly prestressed die should be used and the punch should be wider at its end. Accurate manufacture of tooling is an important factor in the performance of tools.

A typical sequence of manufacturing operations and allied processes for classical ECAP would be as follows

1. lubrication with an extreme-pressure lubricant
2. ECAP through a die channel
3. cutting off a tip at slug end
4. slight reduction of slug cross-section
5. deburring

One cycle of the above operations is shown in Figure 3.2. This single cycle (frequently called a single pass in SPD) has to be repeated many times until the required grain size is achieved. The so-called ECAPed items are always a product of a series of operations which usually involves 4–12 such processing cycles.

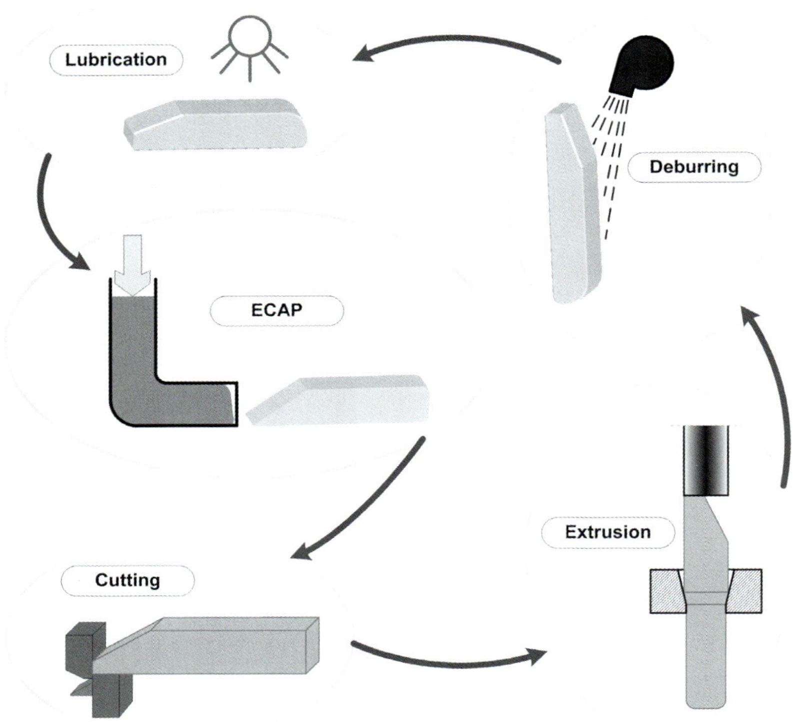

Figure 3.2 Cycle of structure refinement process with classical ECAP as main operation

It should be noted that the above-mentioned general sequence of manufacturing operations recommended for ECAP performed using pressworking tools can be substantially amended, depending on the configuration of the ECAP channel.

3.2 Configuration of ECAP channel

A range of geometric factors affect the configuration of the channel. These will include: the inclination angle between straight segments of the channel, shape and size of the channel cross-section, number of turns along the channel length, length of each straight segment, etc.

3.2.1 Inclination angle

3.2.1.1 Tool loading effects

The pressure required in ECAP depends on the inclination angle of the angular channel (i.e. alignment of straight segments of the channel). This is because the value of the strain achieved in a single pass is a function of the inclination angle of the channel (Segal, 2003). During ECAP a metal billet or slug is plastically deformed through direct application of the compressive punch force. The punch causes the metal to flow within a channel, but not out from it, so all the billet/slug surfaces are confined. When the metal is being moved along two intersecting straight channels it undergoes plastic deformation via simple shear. The shearing process is usually modelled to develop on a plane formed by the intersection of the two channels. From the point of view of tool loading, it is necessary to determine the average value of the stress developed during plastic deformation.

For plane-strain deformation, slip-line theory can be applied to obtain stresses in the plastically deformed material. According to Segal (1995) the punch pressure q_p can be assessed using the following equation

$$\frac{q_p}{2k} = ctg\,(\varphi/2) \tag{3.1}$$

Where k is the material shear yield stress and φ is the inclination angle of the channel.

The maximal punch pressure required in ECAP work depends not only on the yield point of the material after pressing but also on the amounts of friction, depending on the length l of the channel. The extrusion pressure developed by sticking friction when a rectangular slug with the thickness a and length l is pressed under a plane-strain condition through a channel with parallel walls has been analysed by Bednarski (1983). The punch pressure q_t can be calculated by

$$\frac{q_t}{2k} = \exp[\mu \cdot \left(\frac{l}{a}\right) - 1] \tag{3.2}$$

Where μ is the Coulomb friction coefficient.

Using Equations (3.1) and (3.2) one can assess the total punch pressure, by taking into consideration the deformation and friction forces that are required. However, there is another important factor to consider. Punches should be made as short as possible, because the buckling limit of the axial stress varies with the length-to-diameter ratio. Assuming that a Cr tool steel is used for the punch and Al alloy EN AW-5083 is pressed through the classical L-shaped angular channel with punch pressure $q/2k = 2$, the ratio l/a is limited to 7. Taking into account data from the International Cold Forging Group (1970) for this pressing conditions, it is possible to produce 50–100 thousand extruded items with this punch. The limit value of the l/a ratio may be slightly increased if the friction on the channel walls is substantially reduced. Recommendations for suitable lubrication for ECAP work can be found in Section 3.3.3 of this chapter.

It is obvious from Equation (3.1) that by increasing the inclination angle φ the shearing force can be reduced. Therefore, using such an angular channel longer slugs can be pressed with a punch which has limited loading capacity. The application of an obtuse angle φ can be beneficial for the ductility of ECAP processed material. UFG alloys follow the general trend of decreasing ductility with increasing strength (Zhao and Liao, 2010). Poor ductility seems to be a real obstacle to applications of structural bulk nanostructured materials. ECAP with an increased angle φ can help with this issue. Experiments for two sets of tooling with channel passages orientated at 90° and 120° have been reported (Rosochowski *et al.* 2006). The results for Al1070 were compared in terms of the process force, billet end effects, mechanical properties and microstructure of the material. For a 90° channel, the equivalent plastic strain produced due to one turn of the channel is $\varepsilon = 1.15$ while for a 120° channel, $\varepsilon = 0.67$. For both tool sets, ECAP processing was carried out at room temperature and at a constant speed of 1 mm/s. Round billets of about 10 mm diameter were lubricated with PTFE tape and lanolin paste. In order to estimate the process force for 90° and 120° channels, the same ECAP pass and the same friction length in both cases were assumed. Thus, comparing the amount of plastic work per unit volume of the material (unit plastic work = equivalent stress × equivalent strain) and friction work for both cases, one could expect a 40% reduction in force for the 120° channel. The measured forces, for the initial stage of the process, are shown in Figure 3.3. The force in the stationary phase for the 90° channel can be compared with the maximum force for the 120° channel after a punch stroke greater than the diameter of the slug was achieved. This is because of a different geometry of the intersection zone for the 90° and the 120° channels and substantial chamfering of the billet edge at the front end. The force reduction from 14.7 kN to 8 kN is about 45%, which is not far from the estimated value.

After the appropriate number of passes, the maximum strain achieved through the 90° channel and through the 120° channel was approximately 10. After completing ECAP processing with route B_c, mechanical properties of the

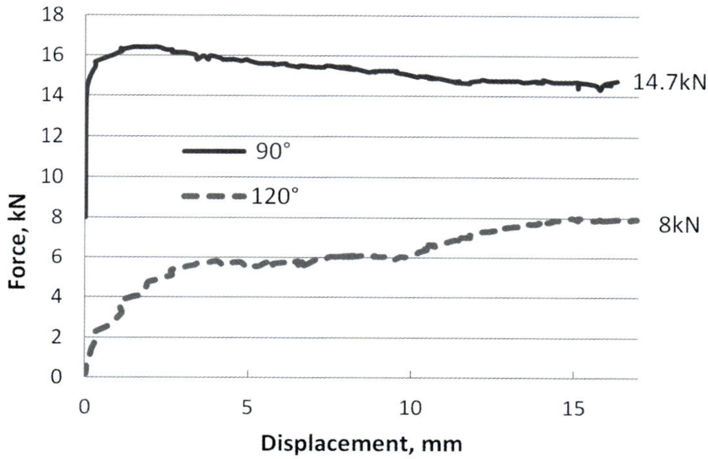

Figure 3.3 Forces vs. punch displacement recorded during first pass of ECAP through 90° and 120° channels

slugs of Al1070 were investigated in a tensile test carried out at room temperature. The sample diameter was 2.5 mm and the gauge length was five times the diameter. Results of ultimate tensile strength R_m and maximal equivalent strain to fracture ε_f are shown in Figure 3.4. Figure 3.4(a) indicates that by changing the channel angle from 90° to 120° the strength is reduced by about 20%. But at the same time the ductility is considerably improved, as is shown in Figure 3.4(b). These effects are related to the material's microstructure which, for the same strain, is similar but not identical after both processes. Grain sizes for ECAP through the 120° channel are slightly larger and characterised by lower misorientation angles compared to the 90° channel case.

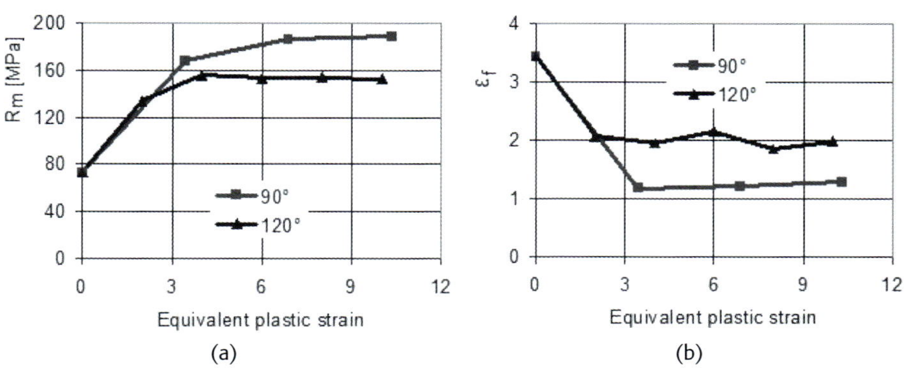

Figure 3.4 (a) Tensile strength R_m; (b) fracture strain ε_f for Al1070 subjected to ECAP (route B_c) through 90° and 120° channels

Strength and ductility are two of the most important mechanical properties of structural materials. Unfortunately, these two feature fundamental inverse proportionality. The strength–ductility trade-off has been investigated quantitatively for Al alloys (Kapoor *et al.*, 2010). As the slugs being pressed through a channel become stronger and difficult to extrude, it is sometimes convenient to perform one or two ECAP passes with an increased angle of inclination. One can expect that with the increased number of ECAP operations ductility will increase again. This is because increasing the equivalent strain of pre-deformation increases the number of high-angle grain boundaries, which has an important influence on ductility. This implies that an increased number of passes are needed for ECAP work, which makes this SPD process inefficient. However, there is an immediate antidote to this: a multi-turn angular channel can be used (see Section 3.2.3).

3.2.1.2 Geometrical effects

Simple shear, which occurs when a unit material square passes the channel's joint plane, converts this square into a parallelepiped. On the macroscopic level, it results in heavy distortion of the ends of the slug. The mechanism of end distortion under low friction conditions is explained in Figure 3.5. It is assumed that the friction on the channel wall is the same as between cooperating slugs. Side views of the slugs are shown on a black background of the channel. A billet with square cross-section is pushed by a punch into a horizontal exit channel to become a slug. It is then pushed further by a new square billet. The front slug, when gradually leaving the channel, affects the front end of the billet. When pressing one slug after another the distortion of the slug ends can assume different forms depending on the friction conditions in the channel, as shown in Figure 3.6. Each slug was pushed out of the output channel by the following slug. Slugs were made of the same material and all pressing parameters except friction were kept unchanged. For ease of presentation, the friction conditions are described qualitatively by a Coulomb coefficient μ.

Much more complicated distortion can be seen when the jointly pressed slugs have different mechanical properties. The problem has been demonstrated by

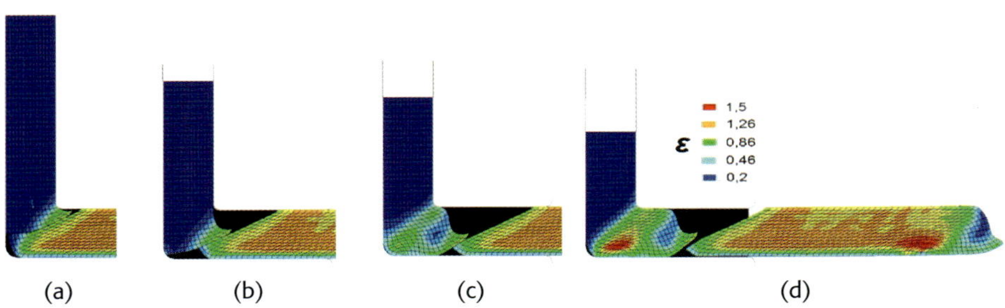

Figure 3.5 Successive stages (a)–(d) of sharp end formation at front and rear ends of slug during classical ECAP

Figure 3.6 Shapes of initially perpendicular billets pressed under different friction conditions through 1 × 90° ECAP channel according to procedure shown in Figure 3.5; μ = friction coefficient

using two Al alloys: soft Al1070 having a hardness of 26.5 HB2.5/31.25 ($5D^2$) and hard Al5083 with a hardness of 76.8 HB5/250 ($10D^2$). The result for different slug combinations of 'hard-pressed by soft' (Figure 3.7(a) upper view and Figure 3.7(b) side view) and 'soft-pressed by hard' (Figure 3.7(c) upper view and Figure 3.7(d) side view) are shown in Figure 3.7. The presented effect can be summarised as: the softer the slug that is pushed out, the more distorted and longer its sheared end will be as can be seen in Figure 3.7(e) for Al1070. It can be particularly dangerous when a soft slug is pushed through the channel by a hard one. An extremely sharp end (shown in Figure 3.7(d) in side view) can tear off and stay on the bottom wall of the channel.

Pressing a stack of slugs through an ECAP channel is a complex metal-forming process. The simple shear mode of plastic deformation and the

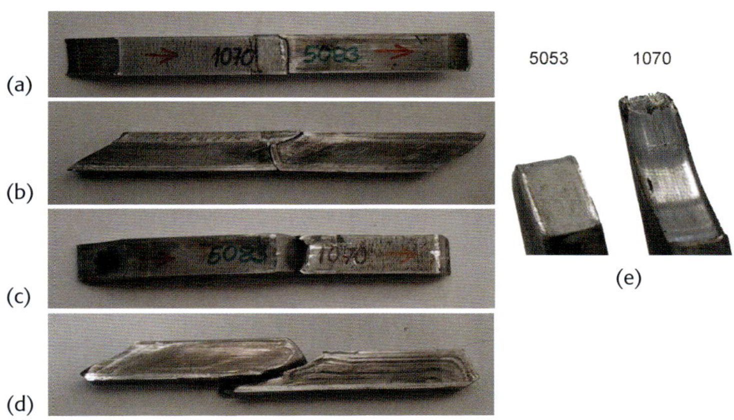

Figure 3.7 Shapes of initially perpendicular billets made of different materials: (a) and (c) upper views; (b) and (d) side views; (e) distorted rear ends

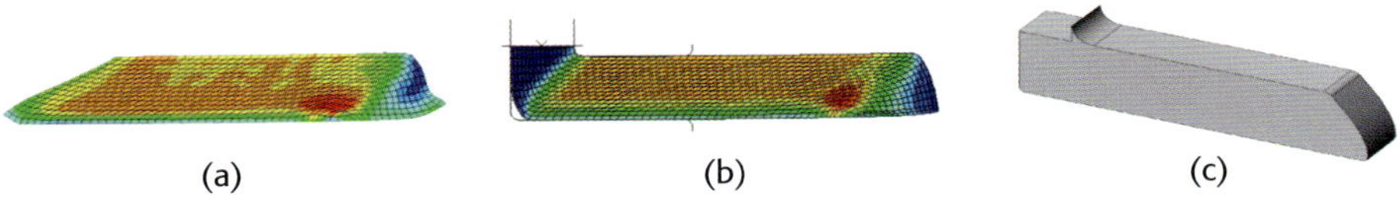

Figure 3.8 Typical end shape defects of slugs: (a) jointly pressed; (b) individually pressed; (c) pressed through angular channel

interaction of consecutive slugs causes strain distribution, which is especially inhomogeneous at the slug ends. This feature is illustrated in Figure 3.8 by the strain distribution in the longitudinal cross-section of a slug pressed through a classical ECAP channel. Roughly, a length which is equal to thickness of the slug has to be separated from the front end of the slug. At the rear end, the cross-section does not remain square along a distance which is again equal to the thickness. Hence, the volume of homogeneously deformed material is substantially limited.

Distortion of the slug ends is obvious when the slugs are jointly pressed through an ECAP channel. When pressed individually, the slug has to be removed from the output channel after the punch stroke has been completed. The shape defect disappears, however, material remains unprocessed at the end of each slug (Figure 3.8(b)). When the punch stops before reaching the upper wall of the output channel (Figure 3.8(b)) or goes forward interacting with the internal radius at the intersection plane of the channels (Figure 3.8(c)), the ejection of the slug is problematic. Certainly, it leads to a more complex and expensive ECAP device. Thus, jointly pressed slugs seem to be the only option for industrial ECAP press-working with a reasonable tool life. It should be noted that, if the cross-section of the channel is round, it is almost impossible to remove the slug.

However, the alignment (inclination) angle between straight segments of the channel can also affect the geometry of the slug ends. Increasing the inclination angle reduces the distortion of the slug ends for both the square and round cross-sections. Using the 110° (Figure 3.9) or 120° (Figure 3.10) channel rounds out the ends of the slugs as compared to the 90° channel.

Figure 3.9 (a) Square billet; (b) slugs pressed through 90° channel; (c) slugs pressed through 110° channel

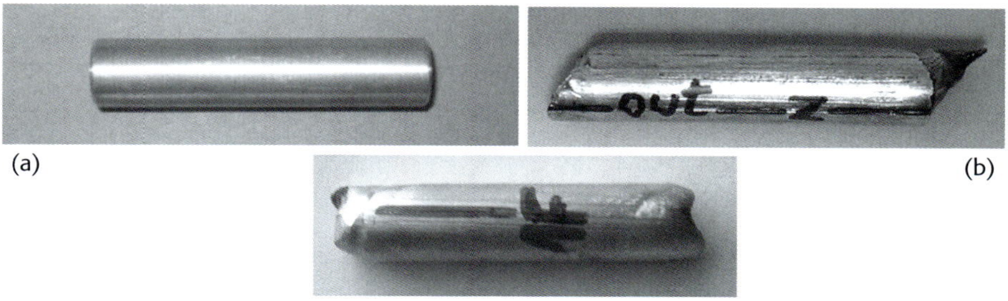

(a) (b) (c)

Figure 3.10 (a) Round billet; (b) slug pressed through 90° channel; (c) slug pressed through 120° channel

The problem discussed here can be avoided when using a die with two-turn, S-shape ECAP channel (Rosochowski and Olejnik 2011), which will be explained in Section 3.2.3.1.1 of this chapter.

3.2.2 Cross-section

3.2.2.1 Shape of cross-section

Both rectangular and circular cross-section channels can be used in ECAP. However, the forming process of ECAP with a near square-section channel is much easier to interpret because it is a typical plane-strain problem. Pressing through a round section channel should be considered as a 3D deformation problem.

Uniform deformation of the workpiece during ECAP is crucial for the homogeneity of grain refinement in bulk materials. The strain distribution is almost the same for both cross-sectional shapes of ECAP channel, because the channel angle has a dominating influence on the effective strains. However, there is another issue to address: the deformation route.

Since a single ECAP pass has to be repeated a number of times, the shearing characteristics may be changed by rotating the slug about its axis between consecutive passes. Segal (1996) has defined a number of ECAP deformation routes, which are shown schematically in Figure 3.11. Their significance for microstructure development in consecutive pressings was first shown by Iwahashi *et al.* (1998) for high purity Al. As shown in Figure 3.11, the end shape features of square or rectangular slugs can be beneficial when choosing the proper position in which to insert the slug into a square channel. In contrast, serious problems may be encountered with maintaining the correct orientation of cylindrical slugs in successive passes. Thus, a round cross-section requires much more attention in industrial applications.

The shape of the cross-section of the billet presents some restrictions regarding tool design, especially for the die. The geometry of the die channel should take into account the shape and dimensions of the product. Sharp corners in the die channel must be rounded by generous radii, otherwise rapid failure will

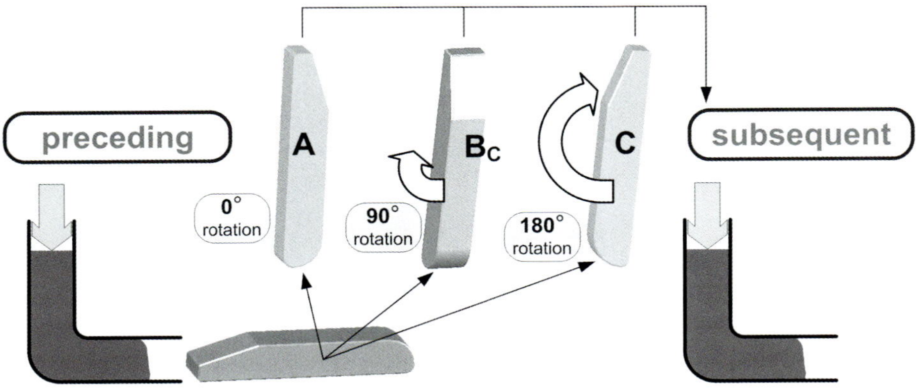

Figure 3.11 Basic ECAP routes applied to multi-pass processing of slug with square cross-section

result from fatigue cracks. From this point of view, round cross-sectional channels can have the advantage of minimising the propagation of fatigue cracks from sharp corners.

The above-mentioned general recommendation will be revised taking into account different types of die design. Some of the possibilities will be discussed in Section 3.4 where examples of ECAP die sets are provided.

3.2.2.2 Size of cross-section

Dies for ECAP can assume different configurations according to design preferences, operational requirements or simply their destination (laboratory vs. industrial application). Despite the shape of the cross-section, billets with a cross-section size of about 10 mm are used in the laboratory to study mechanisms that cause grain refinement during SPD. Scale up of laboratory ECAP to manufacture industrial size components has been attempted for round bars with a diameter of up to 40 mm (Horita *et al.*, 2001) and for 12.5 mm, 50 mm and 100 mm square cross-section bars (Srinivasan *et al.*, 2006). The effect of scaling-up on the mechanical properties and UFG microstructure has been examined for Al alloys. The results showed it was possible to scale up the ECAP process.

Although there are no absolute limits on the cross-sectional dimensions of the ECAP channel, some predictions can be made regarding a practical maximum billet size. Limiting values for a cross-sectional area of square billets are given in Table 3.1. A maximum press capacity of 5000 kN has been assumed for these calculations. Because of possible difficulties with billet handling, the maximum weight of the billet was limited to 50 kg. The length-to-diameter ratio was assumed to be 7, which is a rather large value for punches made of a high-speed steel used in cold forward extrusion and assuming the limit of axial stress of 1000 MPa (which equals the buckling stress). Because of possible material

Table 3.1 Guidance values for possible billet dimensions in L-shaped normal ECAP (square cross-section bars)

Deformability at room temperature	Material designation	Cross-section (mm)	ECAP temperature condition
Excellent	Al99.5 (EN AW-1050)	140	Cold ECAP processing
	Cu-ETP (C11000, CW004A)	85	Cold ECAP processing
Good	Fe (Armco Iron)	70	Warm ECAP processing
	AlMg1 (EN AW-5005)	115	Cold ECAP processing
Medium	Ferritic pearlitic steel (AISI 1025)	70	Warm ECAP processing
	AlMg3 (EN AW-5754)	85	Warm ECAP processing
	Stainless steel (316L)	70	Warm ECAP processing
Difficult to deform	Medium carbon steel (AISI 1045)	70	Hot ECAP processing
	Ti	70	Hot ECAP processing

fracture and/or tool failure, some materials have to be pressed at elevated temperatures up to 200°C (warm processing) or roughly 400°C (hot processing) in order to reduce the flow stress.

3.2.2.3 *Variations in cross-section*

If a metal is plastically deformed it will spring back after the process to a final strain which depends on the elastic properties of the material being deformed and the yield strength achieved during its deformation. Thus, a slug leaving the output ECAP channel is likely to be slightly larger than the dimensions of the channel. This behaviour creates some processing difficulties, as billets and slugs have to be inserted into a die channel with a certain operational clearance. This applies to the slug's re-insertion into an input channel for the next ECAP pass and implies a reduction in the cross-section of the slug before its re-insertion.

In order to reduce the slug's dimensions by the clearance required for its insertion, an ECAP process and a forward extrusion process could be carried out sequentially in a single die. Decreasing the output channel's cross-sectional dimensions is beneficial for powder compaction (Paydar *et al.*, 2009). However, when a solid material is plastically deformed this is rather impractical, as it substantially increases the load on the tool. However, there is one exception in which this solution could be recommended, as proposed by Segal (2004) for ECAP with route B_C. In order to eliminate the material losses caused by the clearance required for slug insertion, a slightly rectangular cross-section of the billet/slug rather than an exactly square one can be used. It would also involve one dimension of the output channel being slightly reduced compared to the input channel, as shown in Figure 3.12. The idea is that the slug leaving the output channel will be rotated about its axis by 90° before being fed into the input channel again (so-called route B_C of ECAP).

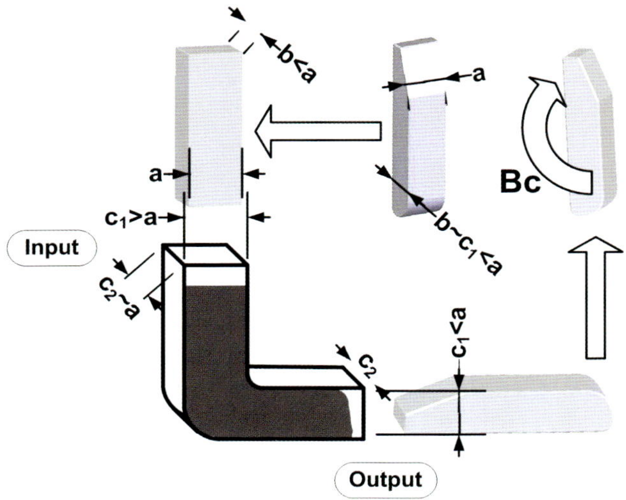

Figure 3.12 Channel dimensions to avoid material waste in conventional ECAP with route B_C

It should be noted that variations in the cross-sectional dimensions of the channel have sometimes been used to optimise the ECAP process. Changes in the deformation state due to two processes combined in one ECAP die can improve the developed UFG structure and lead to a range of new, more efficient, grain refinement methods. A few variations in the shape of the channel in order to combine ECAP with another metal-forming process in a single die are shown in Figure 3.13. A comparative experimental study of Al2124 alloy processed by one pass of conventional ECAP and ECAP in which the thickness of the output channel was substantially smaller than the billet size has been

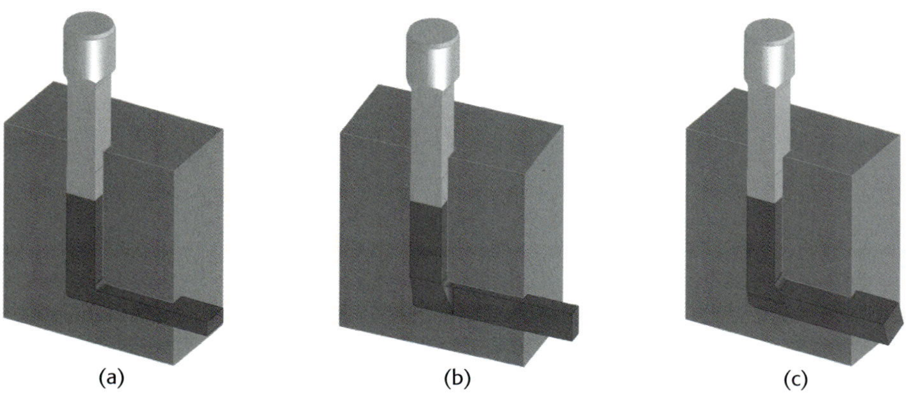

Figure 3.13 ECAP combined with extrusion through: (a) thinner channel; (b) reduced width channel; (c) twisted channel

reported by Toth *et al.* (2009). It was concluded that, although ECAP combined with reduction of thickness (i.e. side extrusion) (Figure 3.13(a)) is not a typical SPD process, it leads to a finer microstructure and a texture rotated increasingly towards the normal direction than is obtained with conventional ECAP. Another modified ECAP which allows a change in cross-section area uses a reduced width of the output channel as shown schematically in Figure 3.13(b). This channel configuration, introduced by Kocisko *et al.* (2009), has only been investigated numerically. It was found that the strain distribution is more homogeneous than for conventional ECAP because the extruded metal flows in two directions: the axial and the lateral ones. The influence of a slight rotation of the output channel about its axis (shown in Figure 3.13(c)) on the microstructure developed in Al was investigated experimentally by Rusz *et al.* (2010). This modification of the channel substantially increases the intensity of the deformation, giving more intensive grain refinement than for conventional ECAP.

It is obvious that additional deformation in the output channel increases tool loading. The application of any of the above-mentioned exit channel alterations should be considered carefully, taking possible benefits into account. The die cavity should always be as simple as possible to promote high tool life and a robust metal-forming operation. Hence, the use of ECAP dies with an altered output channel is not recommended. An alternative solution to the above-mentioned variations in the channel's cross-section, which also aim to avoid any loss of material, is to slightly reduce the cross-section of the slug by using an additional metal-forming operation such as open die extrusion (Olejnik and Rosochowski, 2005).

3.2.3 Number of turns along channel

3.2.3.1 Multi-turn ECAP dies

In ECAP processing, the same slug has to be repeatedly pressed in order to attain the required high strain. Industrial application of such a procedure will take considerable resources and time. One way to increase the effectiveness of ECAP is to construct a die having multiple channel turns. During a single processing pass through such a die, a number of shearing events will occur, equal to the number of channel turns.

3.2.3.1.1 Two turns along channel

The simplest channel configuration capable of this function is based on a two-turn channel. As shown in Figure 3.14, there are three possible ways to design an ECAP die with $2 \times 90^\circ$ channels. In all these cases, the amount of plastic strain in one pass reaches 2.3. This allows the number of ECAP passes to be halved compared to the classical L-shaped channel. Changing the spatial position of the output channel in relation to the input channel makes it possible to press a slug using different ECAP routes as defined in Figure 3.11.

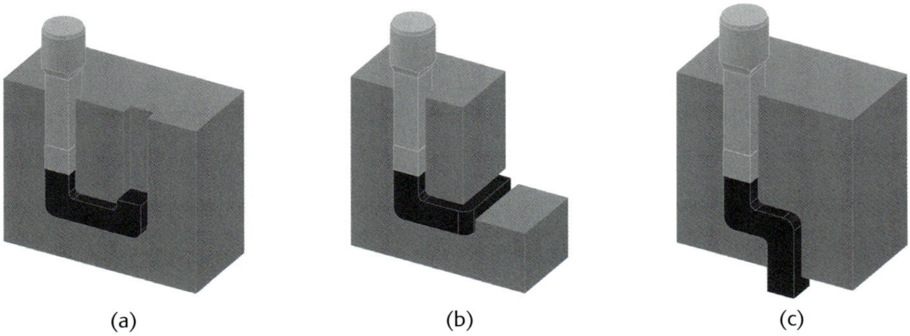

Figure 3.14 Design of ECAP dies with two-turn channel to perform: (a) route A; (b) route B_C; (c) route C

The die shown in Figure 3.14(a) performs route A (no rotation), which is neither very useful for grain refinement nor suitable for pressworking applications. A die with two consecutive channel passages angled at 135° resulting in a side output was constructed by Nakashima *et al.* (1998). It was found that such a die was not capable of producing ultra-fine grains with high-angle grain boundaries in pure Al. The die shown in Figure 3.14(b) is difficult to design, made and use. Material flow through this die is equivalent to route B_C of classical ECAP, in which the slug is rotated about its axis by 90° between consecutive passes. It means that the third passage of the channel is perpendicular to the plane defined by the first two passages, which gives a side output of the slug. This mode of ECAP, incorporating two pressings in a single die, was mentioned very early by Iwahashi *et al.* (1998), however, it was not used in practice until it was adopted by Olejnik and Rosochowski (2005).

Among the three main routes, route C in a single die (Figure 3.14(c)) represents the simplest version of two-turn ECAP to be realised in tooling for pressworking. It seems that this configuration of the ECAP channel was first proposed by Shchukin and Segal (1976). The idea has since been explored under different names by Zuyan and Zhongjin (1999) and Zuyan *et al.* (2000) (S-type ECAP), Rosochowski and Olejnik (2002) (two-turn ECAP), Raab *et al.* (2007) (ECAP in parallel channels) (Raab, 2005; Raab *et al.*, 2007), and Chen *et al.* (2008) (changing channel ECAP). Only few of these proposals have been explored experimentally to produce UFG metals and alloys.

Figure 3.15 shows the case of a two-turn ECAP with both channel turns at 90°, which is the most viable mode of ECAP in a single die for industrialisation of grain refinement. In this case, the amount of plastic strain in one pass reaches 2.3. This allows the number of ECAP passes to be halved compared to the classical L-shaped channel.

The value of u is the main design parameter for this type of ECAP die, influencing both the flow pattern and the strain distribution in the slug. Geometrical details of a two-turn, S-shape channel were optimised by simulation using both

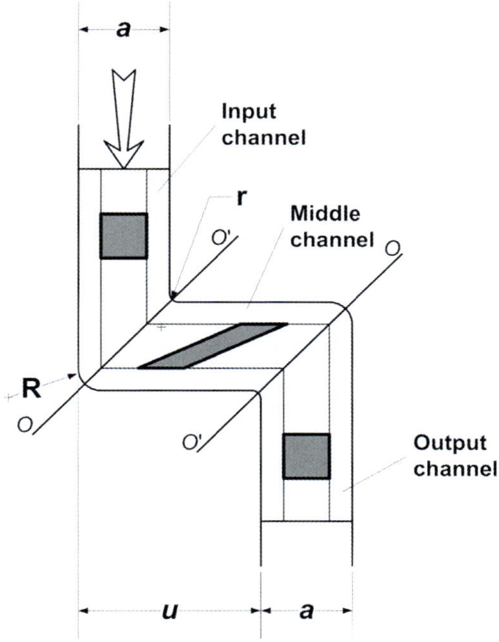

Figure 3.15 Schematics of two-turn ECAP with opposite channel turns at 90°: a is channel size; u is distance between input and output channels; r is inner corner radius; R is outer corner radius; O–O' are consecutive shearing planes

finite element (FE) analysis and physical modelling (Rosochowski and Olejnik, 2002). It has been found that the optimal value of this parameter, leading to the largest strain homogeneity under plane-strain condition is $u = 1.5a$, where a is the size of the square cross-section channel. The deformation pattern makes the structure homogeneous along almost the entire length of the rectangular slug giving a substantial increase in the material utilisation ratio. The internal shaded areas between two flow lines in Figure 3.15 depict the shearing as the slug passes the shearing zones O–O'. Simple shear, which occurs when a unit material square passes the first channel turn, converts this square into a parallelepiped which, when passing the second channel turn, is converted back into a square shape.

For an experimental evaluation, the process of two-turn ECAP (2 × 90°) was applied to small Al1070 billets having a volume of approximately 2.5 cm^3. This amount of material was enough to carry out material testing under both static (Rosochowski *et al.*, 2006) and dynamic conditions (Kokkonen *et al.* (2008). Slugs of exactly the same size were processed using classical ECAP (1 × 90°) for comparison purposes. A batch-type ECAP was performed according to process recommendations formulated by Olejnik and Rosochowski (2005), with an advanced lubrication procedure suitable for bulk cold metal-forming of Al.

Table 3.2 Plastic strain cumulated in ECAP and accompanying processes

Type of ECAP die	Plastic strain due to shear ε_{ECAP}	Additional plastic strain $\varepsilon_{extrusion + upsetting}$	Number of ECAP passes	Total strain imposed $\Sigma\varepsilon$
1 × 90°	1.15	0.2	9	12.0
2 × 90°	2.30	0.2	4	9.9

Route B_C was used for the 1 × 90° channel while, in the case of the 2 × 90° channel, it was a combination of an 'in-die' route C and route B_C (slug was rotated between consecutive passes). The processing sequence, consisting of lubrication, pressing, cutting of ends (only for 1 × 90°), open die extrusion and deburring, was performed repeatedly until the total plastic strain required to develop a UFG microstructure was obtained. Details of straining for both ECAP processes are listed in Table 3.2. The third column includes the cumulative data for material plastic straining due to open die extrusion after ECAP in order to allow insertion of the slug into the die and some upsetting of the slug before ECAP in the input channel.

Figure 3.16(a) shows a collection of slugs obtained by nine consecutive passes of 1 × 90° ECAP, whereas Figure 3.16(b) presents the whole set of slugs produced by four consecutive passes of 2 × 90° ECAP. The difference in the ratio of material utilisation regarding both ECAP configurations can easily be seen. Figure 3.17 shows it more precisely in a plot of the material loss measured after each pass with respect to the cumulative plastic strain and the ECAP process being analysed.

Slugs processed by one-turn and two-turn ECAP were investigated by mechanical and structural testing. The results for Al1070 shown in Figure 3.18 indicate that both processes change the mechanical properties in the way which is typical for SPD processed metals. However, classical ECAP with a pure B_C route is more effective in increasing the strength than the two-turn ECAP

(a) (b)

Figure 3.16 Square cross-section 8 mm × 8 mm slugs produced by ECAP: (a) 1 × 90° classical; (b) 2 × 90° two-turn

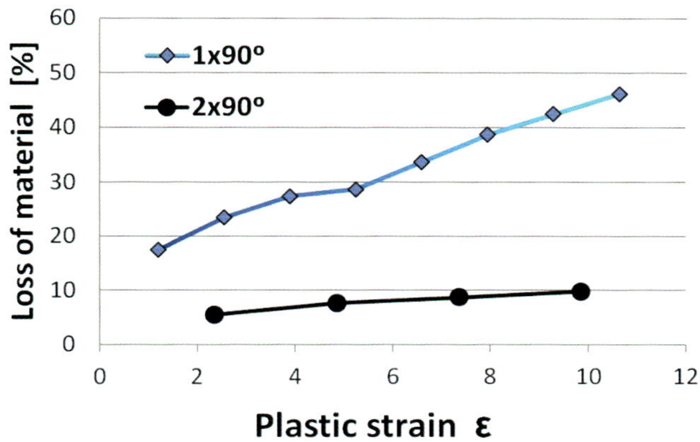

Figure 3.17 Loss of material vs. plastic strain in one-turn and two-turn ECAP processes

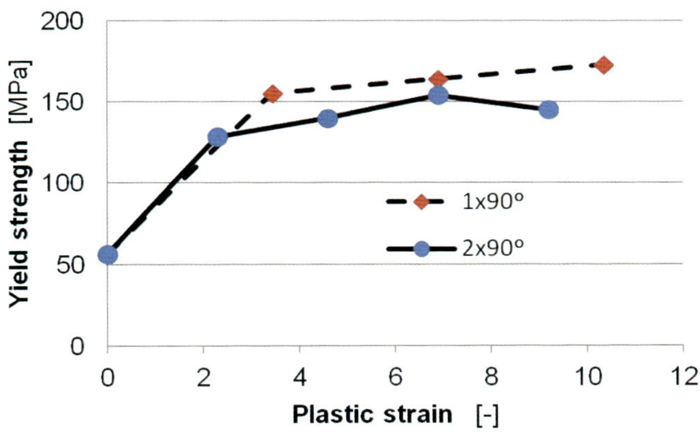

Figure 3.18 Evolution of yield strength in Al1070 subjected to ECAP processing in one-turn (1 × 90°) and two-turn (2 × 90°) ECAP dies

with the combined C + B_C route. Therefore, the material was next investigated by using transmission electron microscopy (TEM) to confirm the existence of a UFG microstructure.

TEM observations shown in Figure 3.19 indicate that both ECAP processes change the microstructure down to the UFG range. The averaged grain size after one-turn ECAP was reduced below 500 nm, whereas a similar reduction was not achieved after two-turn ECAP. However, the grain size distributions shown in Figure 3.20 for both Al slugs look very similar, being relatively wide and having the typical features of a well-developed UFG microstructure. Therefore, both grain refinement processes can be treated as equivalent SPD methods in terms of properties and microstructure. But

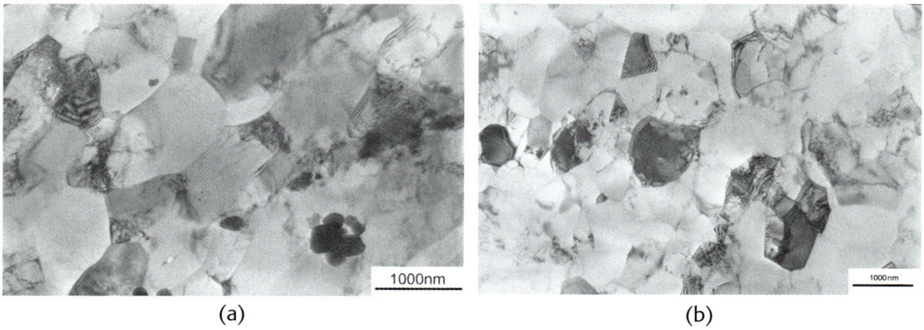

Figure 3.19 Microstructure of Al1070: (a) after nine passes of 1 × 90° ECAP; (b) after four passes of 2 × 90° ECAP

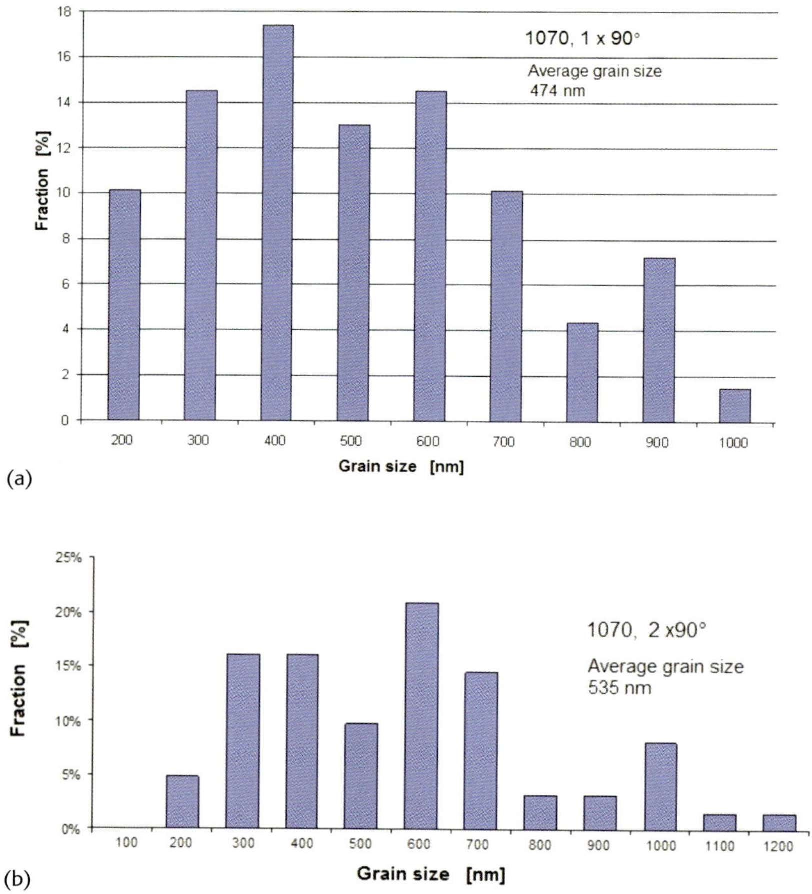

Figure 3.20 Grain size distribution in Al 1070: (a) after nine passes of 1× 90° ECAP; (b) after four passes of 2 × 90° ECAP

differences in productivity are obviously giving priority in industrial applications of two-turn ECAP.

It is worth emphasising that the shape of a slug after ECAP through a two-turn die remains identical to the shape of the initial billet. This is of considerable practical importance, not only because of the increased utilisation of the material, but also the potential for automating the pressworking operation. The multi-pass tool for C route ECAP can be easily constructed as a two-piece die. It also provides the easiest method for increasing the productivity of ECAP by adding another two-turn section to the angular channel. In this way the amount of plastic strain achieved in one pass can be doubled while the unit material parallelogram passing the even number of opposing shearing planes resurrects its initial square shape. However, this approach is rather impractical because the long total length of the angular channel substantially increases the tool loading, makes slug lubrication difficult and hence decreases tool life.

3.2.3.1.2 More turns along channel

A multi-pass C route die which contained a channel having up to five 90° bends has been used for research purposes (Nakashima *et al.*, 2000). The microstructure of a specimen subjected to five passes in a single-turn die and to one pass in a five-turn die was investigated. It was found that the evolution of the microstructure and hardness was identical for both the conventional ECAP die with single shearing plane and for the multi-pass die.

However, it is recommended that, in order to achieve a reasonable tool life, no more than three turns along the ECAP channel be used. The upper part of the Figure 3.21 shows configurations of the ECAP channels, which were proved experimentally in the pressworking applications. Both 8 mm × 8 mm square (Rosochowski *et al.*, 2005) and ø10 mm round (Rosochowski *et al.*, 2006) cross-sections of a three-turn ECAP channel with 90° turns were tested using

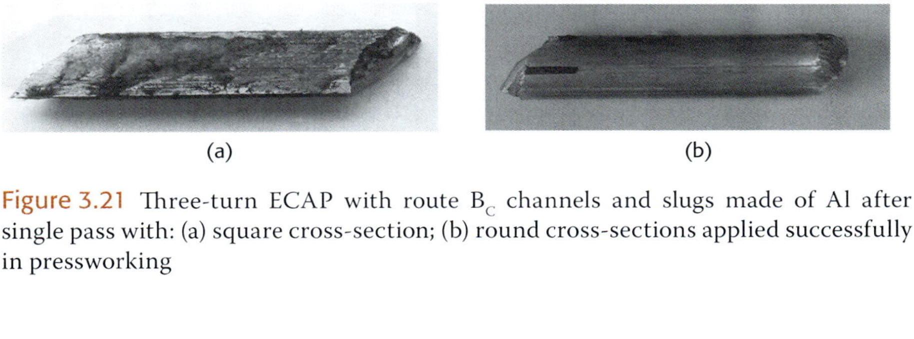

(a) (b)

Figure 3.21 Three-turn ECAP with route B_C channels and slugs made of Al after single pass with: (a) square cross-section; (b) round cross-sections applied successfully in pressworking

Al1070 alloy. Slugs which have been pressed through the die by a following slug are shown in the lower part of Figure 3.21. Their ends are geometrically distorted reflecting the mode and history of deformation during three-turn ECAP with route B_C, this is also known as 3D-ECAP. Since the slugs were in contact with each other inside the die, the end effects (undeformed material at the slug ends) were reduced, except for the leading end of the first slug.

Since the billets, jointly pressed in the 3D-ECAP die, were only slightly longer than the total length of two middle channels (shown in the upper part of Figure 3.21), there was always a slug inside the die which extended over two or even three turns. Figure 3.22(a) shows such a slug pressed halfway and removed from the segmented die. The slug did not show any flash, which can be attributed to the appropriate tool design and manufacture of the die-inserts which compose the 3D-ECAP channel. It can also be seen that the material did not completely fill the channel, despite the much higher forming force which is required for the multi-turn ECAP compared to a classical one-turn L-shape channel. This may be due to the entrapped lubricant.

Altogether three passes have been applied during the process trials with 3D-ECAP dies. To achieve an equivalent strain larger than $3 \times 1.15 = 3.45$, the process was carried out twice more, which resulted in a strain of 6.9 after the second pass and 10.35 after the third pass. The slugs were rotated by 90° between consecutive passes to conform to the in-die rotation taking place during 3D-ECAP. For a round cross-section, a position of the slug at the die exit had to be marked with a line as can be seen in Figure 3.21(b). Since the slugs leaving the die undergo elastic unloading, they no longer fit the input passage of the channel. Thus, despite the shape of the cross-section, slugs for the second and third passes were prepared by reducing their lateral dimensions. Their ends were cut off to provide good contact with the punch and the slug in front.

Since Al1070 is a very ductile material, the forming trials were carried out at room temperature. The forming speed was kept low (about 1 mm/s) to avoid heating the material, which could lead to grain growth. The maximum punch

Figure 3.22 Aluminium slugs pressed halfway using 3D-ECAP channel that has: (a) square cross-section; (b) round cross-section

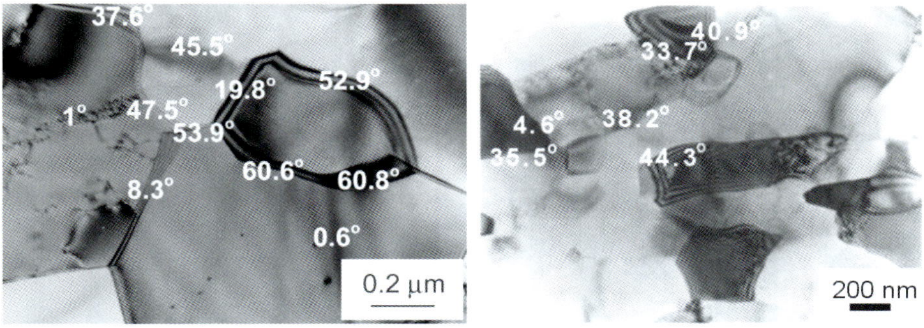

(a) (b)

Figure 3.23 TEM micrographs of Al1070 subjected to severe plastic deformation ($\varepsilon \approx 10.4$) by 3D-ECAP with channel that has: (a) square cross-section; (b) round cross-section

pressure, resulting from the maximum forming force applied, was 1325 MPa for square and 1540 MPa for round cross-sections of the channel, respectively. The increase in tool loading for the round cross-section was due to a slight reduction of the cross-sectional area to achieve additional extrusion in the output channel, as can be seen in Figure 3.22(b).

The average grain size of the initial coarse grain Al1070 was about 200–300μm. TEM was used in order to reveal the UFG grain microstructure after three-turn ECAP with route B_C. As is common for SPD processing, the mechanism of grain refinement is based on the fragmentation of initial grains into smaller domains (cells). Whether these cells are subgrains or fully developed grains, can be judged by revealing their misorientation angles. The technique used for this purpose was based on establishing the Kikuchi diffraction patterns for all cells under consideration, calculating the orientation of each pattern and determining the angle of misorientation between the neighbouring cells. The cells in Figure 3.23 are already fully developed grains because their misorientation angles are mostly above a threshold value of 15°. In the case of a multi-turn ECAP of Al with a substantial value of strain in one pass, some advanced UFG grains were free of dislocations due to possible recovery and/or recrystallisation. These microstructural features are valid despite differences in the cross-sectional shape. Submicrometre range microstructure with an average grain size of about 0.7μm has been achieved after three passes through both the square (Figure 3.23(a)) and the round (Figure 3.23(b)) cross-section of a 3D-ECAP channel.

3.2.3.1.2 Multi-pass ECAP concepts

An alternative way to achieve effective functionality of ECAP is to eliminate the need to remove and reinsert slugs from and into the channel. Several concepts developed to avoid this labour-intensive and time-consuming limitations are shown schematically in Figure 3.24.

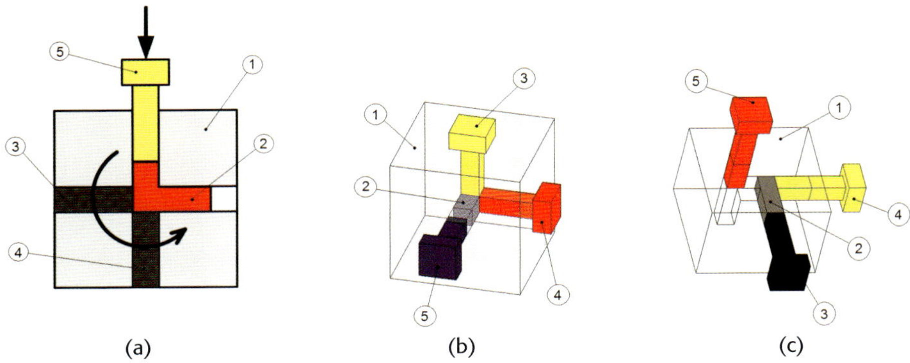

Figure 3.24 Complex ECAP concepts featuring multi-pass ability: (a) rotary die; (b) three-way die; (c) multi-punch facility

The simplest method of multi-pass pressing in a single ECAP tool is to make use of a rotary die (Nishida *et al.*, 2001) as shown in Figure 3.24(a). Body, (1), of the die contains two channels with equal cross-sectional dimensions intersecting at the central point of the die at an angle of 90°. This configuration uses two additional counterpunches, (3) and (4). Punch, (5), running in the upper vertical channel, presses a single slug, (2), into an opened side channel during the first pass. After completing the stroke, the die is rotated by 90°, together with counterpunches, as the arrow in Figure 3.24(a) indicates. The vertical punch, (5), may then press the slug back into the initial input channel. This repetitive ECAP work is very simple. However, it has a drawback. The slug does not leave the channel until the required number of passes has been completed. Therefore, the slug's surface cannot be re-lubricated and any distorted ends cannot be cut off. Thus, the slenderness ratio of the slug should be small and the end effects may lead to inhomogeneity.

An approach illustrated schematically in Figure 3.24(b), has similar disadvantages. The die, (1), has three mutually perpendicular channels converging at the die central point. The slug, (2), can be pressed sequentially by using three punches, (3)–(5), with different combination of routes. According to the original proposal (Akira and Masakazu, 2003), this multi-pass facility was used as a two-way channel for route A completed with a separate feeding channel.

The multi-punch facility (Rosochowski, 2005) shown in Figure 3.24(c) provides more suitable conditions for ECAP processing. The die, (1), has an ordinary 3D-ECAP channel through which each slug, (2), only passes in one direction. Punches, (3)–(5), sequentially press the slug, additionally sealing the channel.

Each of the above-mentioned multi-pass processes requires the construction and manufacture of a complex pressing facility. It involves a die with multiple channels and a number of individually driven punches. The lubrication of the working surfaces is problematic and the end effects may lead to significant inhomogeneities. Such a design is therefore not very suitable for industrialisation.

3.3 Reduction of friction

The key factor in ultimately utilising ECAP technology in industrial applications is to reduce friction. We will now present a modification and modernisation of the die-set design and die work technique as well as establishing new processing routes and regimes.

3.3.1 Movable die parts

During ECAP processing the wall of the output channel can move with the slug to reduce the friction in this channel. According to Segal's earliest proposal (Segal *et al.*, 1995), flat wall slides at the bottom portion of the die can be used for this purpose. Such a tool arrangement is shown schematically in Figure 3.25(a). Figure 3.25(b) illustrates another solution (Stecher and Thomson, 2003) in which a rotatable cylindrical surface having a substantial diameter was used to support the slug exiting the ECAP die.

When designing metal-forming tools, it is common to use some movable tool elements (e.g. stripper, ejector, etc.). However, these elements are applied when the deformation process has been completed. Using any movable part, which operates under loading during the working stroke of the punch, is rather impractical and is not recommended by any guidelines for tool design. Therefore, the modification of an ECAP die by adding movable tool parts does not actually improve the process, because reducing friction in this way leads to more complex and expensive tool design and tends to generate new problems.

3.3.2 Common face pressing

Simultaneous pressing of two rectangular slugs through a die having channels of equal cross-sectional area can effectively reduce friction and, hence, decreases punch pressure. It also eliminates any difficulties associated with lubricating movable tool parts especially when ECAP is performed at elevated

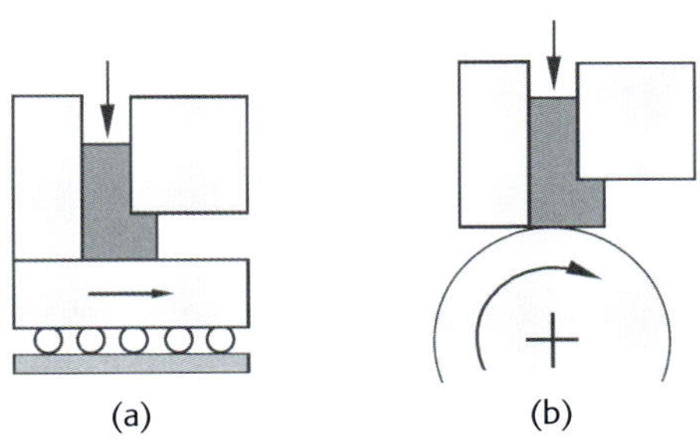

Figure 3.25 Reduction of friction in output channel by using: (a) movable wall; (b) rotatable roll

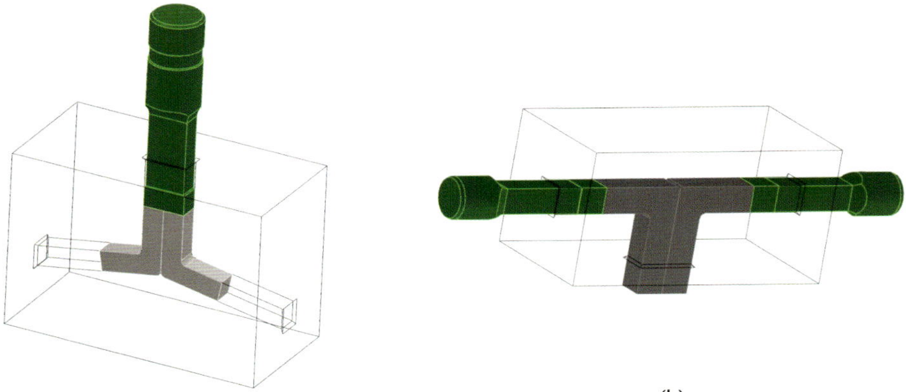

Figure 3.26 Simultaneous pressing of two rectangular slugs through an ECAP die: (a) at entrance of channel; (b) at exit of channel

temperatures. Such pressing, with the sticking faces of two slugs, can be carried out in the input (Segal and Segal, 1997) or in the output channels (Rosochowski and Olejnik, 2011). The first case is illustrated in Figure 3.26(a). Two parallel billets are fed into the input passage of the channel and directed in the two diverging output passages connected to the input channel. The second configuration, shown in Figure 3.26(b), uses two equal square or rectangular input passages converging into a single output passage, which is twice as wide as the leading passages so that it can accept two converging slugs. The contact surface between converging slugs plays the same role as a movable die wall or a rotatable roll discussed in Section 3.3.1. It reduces friction and the process force. This effect is achieved without using a complex die with movable parts. Instead, two punches are used to push two slugs synchronously from the opposite sides. It should be noted that processing two slugs simultaneously results in symmetry of tooling which is beneficial for pressworking applications. The system doubles productivity compared to the case of a classical L-shape ECAP die.

3.3.3 Lubrication

In SPD processes, as in any other metal-forming process, friction directly influences tool loading and wear. Thus, adequate lubrication is required. Different lubrication techniques may be selected depending on the metallic material being processed, SPD operation, and processing conditions. Ordinary die lubricants break down because of high pressure, excessive surface extension and the elevated temperature sometimes required in ECAP operations.

For cold and low-temperature ECAP, a suitable lubrication layer consists of conversion coating which is subsequently treated with a solid lubricant. The lubrication is usually a chemical process consisting of several baths, as shown in Figure 3.27. The lubricating procedures selected for ECAP pressworking

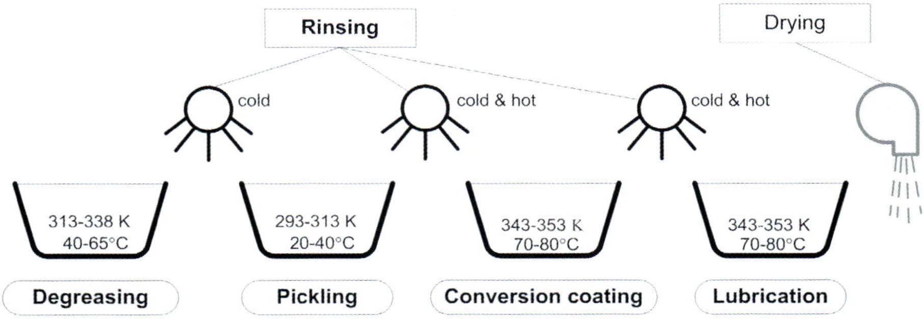

Figure 3.27 Bath sequence for generation of lubricated conversion coatings before ECAP

purposes can usually be based on lubrication which is recommended for heavy metal-forming operations, especially for cold extrusion. Table 3.3 presents conversion coatings which are suitable for different metals which are cold extruded, with increased severity of die operation. The conversion coating acts as a partition which reduces the metal-to-metal pickup and serves as a carrier for the lubricant.

The bonded layers listed in Table 3.3 are subsequently covered with a layer of a suitable lubricant having good adhesion and lubricating properties in the conditions encountered in ECAP work. An aqueous, fatty acid type soap emulsion can be used for carbon and low alloy steels. Substances such as lanolin and combinations of fatty acids and soaps have been used successfully for Al alloys (Rosochowski and Olejnik, 2011). Mg alloys are ECAPed lubricated with dried MoS_2 fat compounds. The lubricants are applied to the billets by either dipping them in the heated fluid lubricant and then air or oven drying, or by tumbling them in a quantity of powder lubricant in a drum. The above-mentioned combination of conversion coating and overlaid lubricant has proved quite successful for ECAP pressworking.

The wear behaviour of ECAP tools depends not only on the appropriate selection of a lubricant for the processed alloy but also on the suitable design of the tool, including hardness and chemical composition. The finish of the contact surfaces is very important. The surface finish of the channel walls is

Table 3.3 Recommended conversion coatings for ECAP pressworking operations

Stock material	Conversion coating
Carbon steels	Zinc phosphate surface layer
Stainless steels	Oxalate surface layer
Aluminium alloys	Calcium aluminate surface layer
Magnesium alloys	Silicated surface layer

determined by the number of pressings to be made. On jobs where the number of pressings is less than 100, a highly polished channel is not required. However, when mass production is considered, it is most advantageous to polish the channel surface since that leads to increased die life. The essential factor to consider is also the selection of a hard coating for the dies. An optimal metal billet–lubrication–tool coating system has to be validated experimentally (e.g. using a ring test (Male and Cockcroft, 1964)).

Despite good lubrication and the finish of the channel surfaces, a little maintenance is sometimes required for ECAP dies. Where billet material has been welded to the channel surfaces, it can be removed by grinding or chemical etching. In this context, composite dies, which can easily be disassembled to give good access to these surfaces, are recommended.

3.4 Die design

The starting point for the design of ECAP dies is certainly the shape of the channel and its cross-section (see Section 3.2.2). However, for reasons of economy, it is worth considering various types of dies for both short-run ECAP and large-scale production. Small-scale production of UFG metals, including research purposes, usually results in a low die cost. Simple dies may be favoured over the high-production dies, because of the lack of special skills required to design, construct and maintain high-production dies. Low-cost, low-production dies are made with techniques which may not be applicable to the making of more costly dies. Following the good die design principles should result in a die life which meets ECAP requirements at different production levels. Three categories with increasing demand for productivity can be distinguished in the development of ECAP dies: bolted dies, split dies and prestressed dies.

3.4.1 Bolted dies

This type of die design involves several die blocks which are held together by the outer plates. The plates are bolted with heavy bolts in order to secure the constant position of the channel-defining blocks, even when a substantial pressure exerted on the channel walls by the processed material.

The simplest designs for a bolted die for classical ECAP are shown in Figure 3.28. Generally, it consists of at least two metal blocks. The required angular channel can be made partly in both die blocks, as shown in Figure 3.28(a). The bolted die design illustrated in Figure 3.28(b), in which the whole profile of the channel has been machined in one block, is much more practical. In this case only the channel block has to be exchanged instead of both blocks when the die fails or there is a need to change it for another die with different channel shape or size. The most useful die design is shown in Figure 3.28(c). It contains additional flat segments located in a die centre between side blocks. These segments serve as the die-inserts which define the ECAP channel. They also release any possible stress concentration along the channel edges.

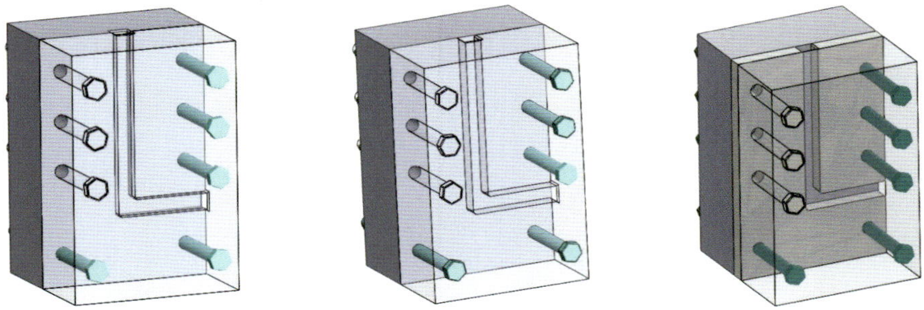

Figure 3.28 Bolted die assembly consisting of: (a) and (b) two blocks; (c) blocks and flat segments

ECAP dies made of the assembled tool steel blocks and segments are mostly used for research purposes (Furukawa *et al.*, 2002; Valiev and Alexandrov, 2002) to produce a small number of billets pressed through a single-turn channel. The bolted die design provides considerable flexibility in terms of channel geometry including billet shape, size and channel angle. The central plate is normally made of two segments. But in one proposal (Purcek *et al.*, 2004) it was made as one tool piece and mounted in an ECAP die. Bolted dies should be constructed to withstand the internal pressure to which they are subjected during the pressing of the billet. High-strength threaded bolts should be used to maintain the geometry of the channel and to avoid generating gaps at the segment/block interfaces into which the material could flow to create so-called flash.

Bolted ECAP dies are usually designed to be mounted on a hydraulic testing machine. The most important consideration is a careful alignment of the tool set before processing begins. The pressing speed is kept constant during the deformation process and rather low speeds are used, usually below 0.5 mm/s.

3.4.2 Split dies

A typical example of universal tooling using a split ECAP die is shown in Figure 3.29. The die-inserts can be identical so that the channel may be formed partially in the upper and lower die-inserts. For ease of manufacture and maintenance it is reasonable to machine the channel in only one die-insert, leaving the second one flat. This is the case of a split-die tooling in Figure 3.29, which is equipped with an interchangeable lower die-insert that creates a square cross-section, L-shape ECAP channel.

The upper and lower die-inserts making a split-die set are aligned and mounted onto the clamping plates. The plates have guide or leader pins and bushings to ensure the correct alignments. Pins are press-fitted into the lower plate whereas bushings are assembled in the upper plate. The number of guide pins and bushings varies from two to four. An accurate slip fit has to be established between the bushings and the pins. Standard or purchased bushing-pin

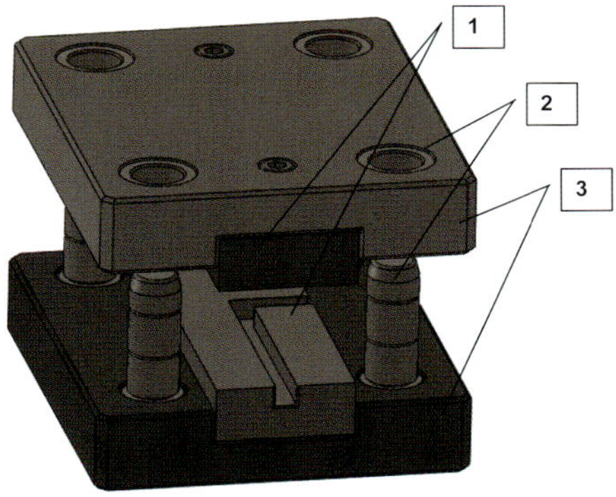

Figure 3.29 ECAP tooling with a split die: (1) die-inserts; (2) guide system; (3) clamping and bolster plates

sets, usually used to construct tools for metal forming or polymer processing, can be applied. The length of the guide pin should be such that the pin does not extend beyond the top surface of the upper plate when the tooling is closed.

Before ECAP, the top plate and the bottom plate come together to close the split die. The billet is then pressed in the ECAP die by a punch which is operated horizontally. This technique requires a horizontal forging machine, which has two degrees-of-freedom, one for clamping the billet in the split die and another one that is used for the motion of the punch.

3.4.3 Prestressed dies

During ECAP the billet is stressed to its elastic limit in compression and simple shear, which results in the lateral pressure acting on the channel walls. High tool pressure causes elastic deflection of the tools but, in the case of a multi-piece die, can also lead to relative movements of the die elements. Some kind of prestressing is required to prevent loss of the channel's cross-sectional shape by excessive elastic deformation or relative movement of die-inserts. This can be achieved with bolts as described in Section 3.4.1, but it can be more effectively achieved by using a shrink fit between the die and a massive prestressing ring so that the die elements are always in compression.

The design of a prestressed ECAP die should satisfy the following general principles. Die-inserts made of heat-treated tool steel are hard and wear resistant. They are supported by a softer, but strong and tough, ring. This type of die is useful for both cold and warm ECAP work. However, at elevated temperatures, the tool steel should also resist thermal softening. Hence, careful selection of tool steels and heat treatment of the die elements is necessary. It is good practice to follow the guidelines for armoured dies for metal forming (International Cold Forging Group, 1992). In this way the design of the die can be based on recommendations which have resulted from years of practical experience.

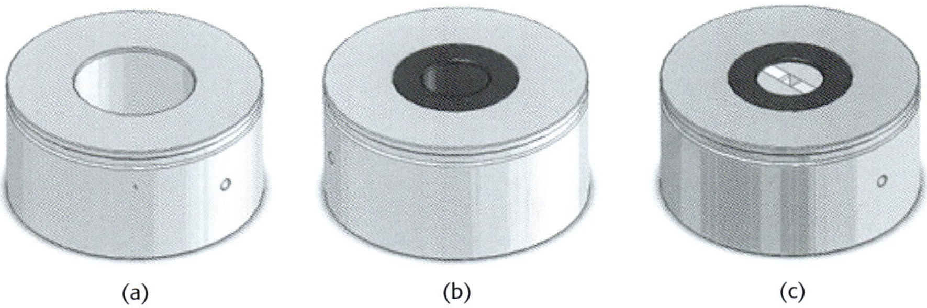

Figure 3.30 Armouring sequence for prestressed ECAP die with two stress rings and segmented die-insert: (a) outer ring; (b) two rings shrunk together; (c) die assembly completed with tapered die-insert

The sequence for assembling a typical prestressed die for ECAP is shown in Figure 3.30. Since the outer ring (Figure 3.30(a)), machined from a hot forged or rolled steel, will be stressed in tension its hardness must not be greater than Rockwell 45. The outer ring is shrunk onto a hardened inner ring. The inner surface of the resulting assembly (Figure 3.30(b)) is ground to obtain a 0.5° to 1° taper (per side) corresponding to the outer taper of the segmented die-insert. The ring assembly is forced over the die-insert to provide a substantial (0.15–0.25% per side) interference fit (Figure 3.30(c)). Such compression preloads the die-insert so that it can better withstand the process-generated stresses, which tend to expand the cross-sectional area of the channel.

This type of die is especially useful for mass production because the design is robust, facilitates interchangeability and reduces die cost through increasing die life. Slugs produced by using these dies range from small cross-sections about 60 mm^2 (Olejnik and Rosochowski, 2005) to almost 700 mm^2 (Olejnik and Rosochowski, 2008). ECAP processed items are the product of a series of pressings, which may combine any type of channel. Both square (Figure 3.21(a)) (Rosochowski *et al.*, 2007) and round (Figure 3.21(b)) (Rosochowski *et al.*, 2006) cross-sectional shapes have been successfully ECAPed.

Any ECAP process can be realised by using a suitable segmented die-insert. A one-turn classical ECAP channel or any type of multi-turn ECAP channel can be defined by the design of the segments which build up the die-insert. However, it is worth remembering that the classical ECAP and B_C route two-turn ECAP (Figure 3.14(b)) dies force metal to flow in a direction which is perpendicular to the direction of the movement of the punch. As a result, a side output in the die-insert and a side orifice in the stress ring assembly are required.

A lateral hole in prestressing rings is necessary to enable the processed slugs to leave the die. The feasibility of making a hole in the ring assembly, which is subjected to tensile stresses due to prestressing, was checked using the FE method (Rosochowski and Olejnik, 2011). Figure 3.31 shows the results of this analysis. At least two stress rings are necessary to keep the tangential stress

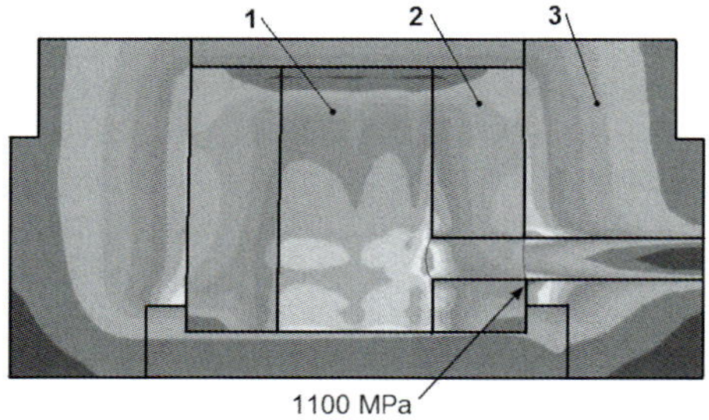

Figure 3.31 FE simulation results of equivalent stress distribution resulting from assembling stress rings with interference close to 0.5%: (1) location of die-insert; (2) inner stress ring; (3) outer stress ring

component compressive at the hole, (1), for die-insert. The internal ring, (2), is shrunk by an interaction with the outer ring, (3), which is pushed over it with an interference. The interference fit should not be greater than 0.5% of the interface diameter. Using this model, the distribution of equivalent stress was calculated in the ring assembly. Careful inspection shows that equivalent stress reaches the maximum value of 1100 MPa. The maximum cyclic equivalent stress that can usually be sustained by tooling is about half the yield strength of the tool material. The yield strength of the common tool steel is normally 2000 MPa. Therefore, in the longer term, fatigue fracture can be expected at the lower edge of the lateral hole in the outer ring. Hence, prestressed dies for any type of ECAP, which requires a lateral hole, should be used with caution.

A die-insert for ECAP should be robustly constructed to withstand repeated loading. Therefore, some geometrical design considerations should be addressed. Radii and fillets should be used in the problematic areas. Otherwise, the high and cyclic load applied to the die during pressing may cause its premature failure. A possible means of reducing that risk is based on splitting the die-insert into segments to achieve a stress-concentration-free design. In this design, all edges of the angular channel are defined by different segments to avoid fracture. Tightly fitted pins may be inserted to keep the segments together, especially before assembly. Finally, the segmented die-insert has to be supported by toughened steel plates to distribute the high local load.

A die-insert with two channel turns (2 × 90°) in a single die, in which the slug can be rotated about its axis by 90° when moving along the channel die (the method was previously explained using Figure 3.14(b)), is shown in Figure 3.32(a). The die was used to press 8 mm × 8 mm × 44 mm Al1070 billets as reported by Rosochowski and Olejnik (2011), where the die-insert design was shown in detail. Application of the internal segments with intricate geometry gave a construction recommended for heavily loaded ECAP dies, in which all edges of the channel are defined by different segments. The die-insert presented in Figure 3.32(b) was much easier to design, make and use. The process realised in this die is equivalent to route C of the classical ECAP and was previously

(a)

(b)

Figure 3.32 Channel layouts and 3D models of segmented die-inserts for two-turn ECAP of square cross-section billets using: (a) 2 × 90° route B_C; (b) 2 × 90° route C

outlined in Figures 3.14(c) and 3.15. Here four segments were also used to achieve a stress-concentration-free tool design. The channel-defining segments are flat, however, a number of tightly fitted pins must be applied to keep segments together in a proper position in the die-insert. This type of channel geometry enables the channel exit to be at the bottom of the die, which is convenient for pressworking. In the first place, it was used to process small 8 mm × 8 mm cross-section slugs to produce UFG Al, mostly for micro-forming (Olejnik *et al.*, 2009) and compression at high strain rates (Kuokkala *et al.*, 2008). Next, the two-turn ECAP process was scaled up (Olejnik and Rosochowski, 2008) to produce a high volume of UFG Al to enable both advanced post-processing (e.g. diamond machining of large optical structures for ultraviolet applications) (Brinksmeier *et al.*, 2011), and extended investigation of dynamic behaviour (e.g. using both tension (Kokkenen *et al.*, 2009) and torsion (Hokka *et al.*, 2012) split Hopkinson pressure bar devices).

Two design proposals for a die-insert with three channel turns to realise route B_C in a single die are presented in Figure 3.33. This type of die triples the

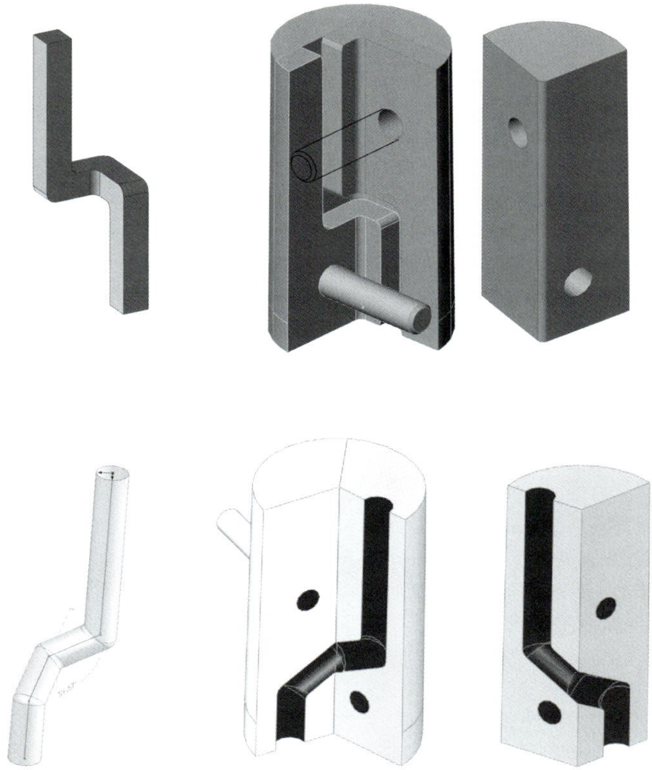
(a)
(b)

Figure 3.33 Channel layouts and 3D models of segmented die-inserts for three-turn ECAP with: (a) route B_C for $3 \times 90°$ square bars; (b) $3 \times 120°$ round bars

strain achievable in the classical one-turn channel while the slug favourably exits the die at its bottom. In the case of slugs with a square cross-section, the die-insert consists of four segments, which are simply fitted together with two pins. The die-insert design shown in Figure 3.33(a) was revealed in a patent application (Rosochowski, 2005) and used to produce UFG Al1070 slugs (Rosochowski *et al.*, 2007) for a variety of technological applications (e.g. electrical discharge machining (Rosochowski *et al.*, 2007)), backward extrusion of miniature components (Rosochowski *et al.*, 2007) and micro-indentation with bulging (Presz and Rosochowski, 2006).

A die-insert for ECAP of round cross-section slugs in a three-turn channel can be constructed using three segments (Figure 3.33(b)) compared to four segments for the square cross-section channel. Half of the channel's longitudinal cross-section is located in two segment of a composite die-insert. It should be noted that there is a strict geometrical relationship between the inclination angle of a channel's axis and the central angles of segments of a composite cylindrical die-insert. For the $3 \times 120°$ angular channel the cylinder can be split into three segments: two segments with a central angle of 125° and a complementary segment with an angle of 110° (right-hand side segment in Figure 3.33(b), which completes the cylindrical body of the die-insert). The selected design results in a slight deviation from the 120° inclination angle of the channel's axis, which is indicated with an angular dimension on a channel profile shown on the left-hand side in Figure 3.33(b). The 3 × 120° die-insert performs three-turn ECAP with a reduced strain in one pass compared to 3 × 90° die-insert, it was therefore used to fabricate round slugs for purposes of comparison (Rosochowski *et al.*, 2006).

A segmented die-insert for a 3 × 90° round channel is much easier to construct because three equal segments of a cylindrical die-insert (i.e. having the central angle of 120°) can be used in this case. This type of die-insert can produce the same type of round slugs as the die shown in Figure 3.33(b). It was successfully used to produce round bars of UFG Al1070 for ultra-precision diamond turning of metal optics. Compared to coarse-grained Al, the surface roughness of diamond tip turned UFG Al was reduced by 20–30%. High structural accuracy and a surface roughness of $Ra = 3$ nm was achieved (Osmer *et al.*, 2007), which verified the applicability of UFG Al for machining optical quality surfaces.

3.5 Punch design

Pressure is applied to the billet by means of a punch, which matches the ECAP channel. The punch must be designed to withstand the process pressure and a tendency to buckling. Any lack of symmetry in the punch loading forces the punch out of alignment. This is why, among other reasons, sharp tips at the ends of the slug have to be cut off before the slug is fed into an input channel.

A shoulder punch is the most efficient type of punch for ECAP operations. Figure 3.34 shows a typical punch for pressing slugs with a round cross-section. The punch could have been designed as a cylinder with a constant diameter close to the diameter of the input channel. The clearance between punch and input channel is calculated to accommodate elastic distortion of the punch and

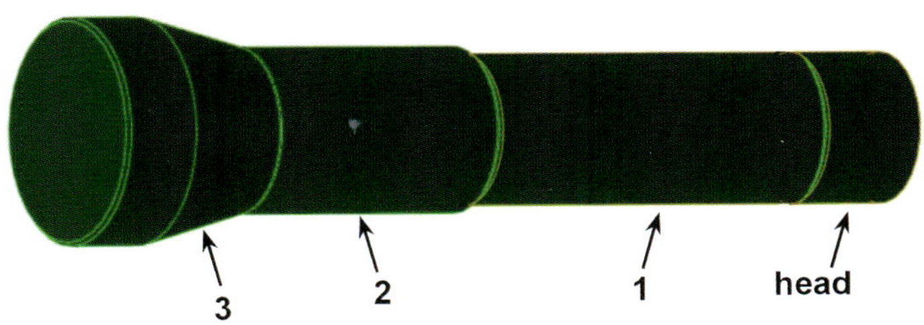

Figure 3.34 Punch for ECAP of round cross-section slugs: (1) relieved part; (2) aligning part; (3) shank

die and still obtain the desired tool fit. With a tight fit, there is a danger of substantial additional loading of the punch during pressing and that it will be more difficult to remove the punch on the up-stroke of the press ram. On the other hand, an excessive clearance enables backward flow of metal between the punch and the input channel. Thus, it is a good practice to limit the length of the punch-fitted section to approximately 60–80% of the channel's input diameter. As a result, the punch for ECAP has a short guiding head followed by a relieved part (detail 1), to remedy any inaccuracy in the alignment of the punch with reference to the channel, and finally changes to an aligning part (detail 2) and a shank.

The punch can have a tapered shank, as shown in Figure 3.34 (detail 3). In such a case, the conical shank surface is used to connect the punch with the upper plate, either by a bushing within a threaded nut or by a tapered ring screwed to the upper plate. A straight cylindrical shank, designed to align with the bushing, can also be used while being restrained by side screws, preferably with a cone point.

In hot ECAP, the shoulder-type punch is usually fixed by a flange to the punch holder or fitted directly into the ram of the press. In general, the punch holder can be adjusted in both the axial and the radial directions. Radial adjustment helps to centre the punch in the channel and is performed by aligning the holder by side screws. In the case of a slug with a square cross-section, it must be possible to rotate the punch in order to correct its angular position with respect to the die.

In cold and warm ECAP, the punch is usually aligned with the channel by a guide system. For pressworking applications, guide or leader pins and bushing are incorporated in commercially available die sets. Standard die sets consist of an upper and a lower plate, on which the punch and die components may be mounted. The plates are guided with reference to each other. An accurate slip fit is established between the lower plate pins and the upper plate bushings to hold the punch and die in alignment.

3.6 Tool materials for die and punch

A variety of tool steels are suitable for ECAP tool components. The selection of the proper tool material for the die and punch depends on the type of tool design chosen for pressing and is governed by the size of the run.

Successful ECAP dies have been made from the following AISI-SAE tool steels: Mo–W high-speed tool steels M2 and M3, high-carbon, high-Cr cold-work tool steels D2 and D6, Cr–Mo hot-work tool steels H11 and H13. These are all-purpose die steels with very good abrasion resistance and good shock resistance at high hardness. Resistance to heat checking is also quite good especially among hot-work and high-speed tool steels. The selection of the tool steel depends partly upon the required die life. High-speed tool steel, preferably made by using powder metallurgy for best homogeneity, is the best choice for die-inserts shrunk in stress rings. The criteria for selecting the material for the armouring rings should follow the recommendations for the assembly of the prestressed die.

The magnitude of force in efficient ECAP pressworking is rather large. Due to the high contact pressure there is a danger of local collapse of the tool surface. To prevent this, the surface hardness of the die components must be approximately Rockwell C60.

Punches should be made of a good grade tool steel, which has been hardened and ground. Tool steels commonly used are AISI-SAE M3 (Mo–W) for any condition and D2 (high-Cr) for cold working. Careful heat treatment is necessary because high working stresses are encountered. Hardness treatment is recommended to Rockwell C60–62.

Dies and punches are mainly produced by machining. Dies can also be manufactured by die sinking electrical discharge machining (EDM) (mostly with graphite and sometimes with Cu electrodes). Wire EDM is also being used, especially to manufacture segmented die-inserts.

Die components for ECAP require a high surface finish especially on their working surfaces. Tool surface finish has an impact on productivity, because material pickup depends not only on the natural adhesion of the processed metal to the steel tool but also on the surface roughness. A high surface finish also reduces the risk of local fracture due to temporary overloading or fatigue.

If a high quality surface coating is to be applied to tools, then it is always recommended that high gloss polishing of the tool surface be done before the coating process. To achieve a surface with an almost mirror finish the preparation process must involve several grinding and diamond polishing steps and these have to be performed in a clean workplace.

3.7 Working conditions

Although room temperature ECAP is sometimes used for grain refinement in pure metals, harder and less ductile metals and alloys have to be processed at elevated temperatures. However, it is desirable that the ECAP temperature does not exceed half of the melting temperature and it should be as low as possible in order to avoid grain growth. From the point of view of ECAP productivity, automatic heating is required for billets. Depending on the material being processed, this can be accomplished by adding either an inductive coil or a resistance heating element to the die chute. In some cases an electric radiant heating can be added to the feeder, making it an integral part of the feeding mechanism. In addition to heating the billets before pressing, the dies must be heated to approximately the same temperature to avoid a chilling effect due to the often slow speed of ECAP processes.

The dies are usually heated electrically. All types of standard heating elements designed to heat solid bodies can be applied. Electric heaters are attached to the dies, usually by clamping them around or otherwise fastening to the external die surfaces. Sometimes flexible tubular heaters can be embedded in irregular surfaces or fitted into milled grooves in a supporting plate placed in the vicinity of the tool elements. Cartridge heaters can also be inserted with an appropriate fit into the input and/or output channel to speed up the initial preheating of the die before starting ECAP.

When a higher range of elevated temperatures is needed, ECAP dies should be wrapped with a thermally insulating material to slow down heat loss and to protect the press. Specifically, insulation between the bolster/plate and the die is required to minimise heat transfer through the press table. Any high-temperature insulating plate, designed for thermo-mechanical applications which feature extremely high resistance to pressure, can be used. In this case a compressive strength of at least 600 MPa at room temperature is required.

Additional cooling around the die assembly is also important when wishing to intensify high-temperature ECAP. It reduces heat transfer to the press systems, which could be adversely affected by the high temperature. Cooling may be accomplished by a coolant passing through a system of channels in the die assembly supporting tool blocks. Therefore, an additional water/oil distribution system is required in the ECAP tooling.

ECAP with back pressure is another means, which can be used to avoid the fracture of slugs made of low ductility materials. It can also improve the uniformity of the strain distribution across the slug body (Lapovok, 2005). However, ECAP with back pressure makes the operation a very severe press-working task, since the flow of the material is entirely confined in the channel closed by two cooperating punches. The counterpunch is designed to slide in the output channel of the die assembly and is actuated by a hydraulic drive with a substantial stroke.

All the above-mentioned features have been applied in an ECAP tooling shown in Figure 3.35. The tooling is designed to perform hot ECAP of demanding

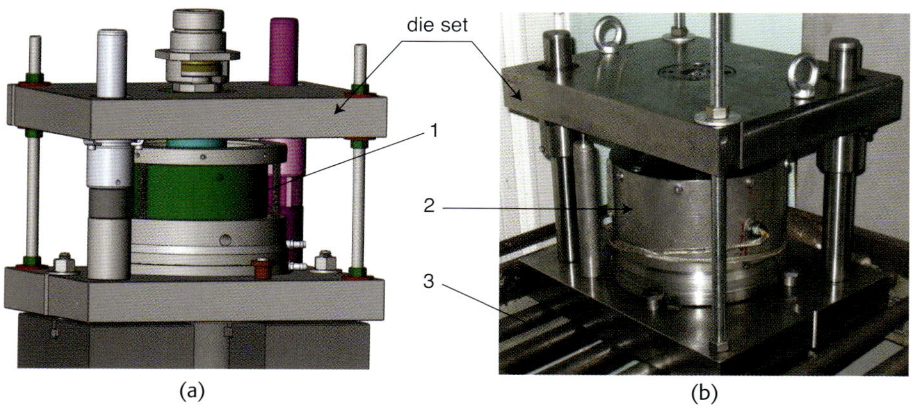

Figure 3.35 ECAP die and punch assembled in standard guide pin die-set: (a) model of ECAP tooling attached to bolster plate; (b) assembled tooling for hot ECAP operation ready to changeover; (1) prestressed die assembly; (2) electric band heater; (3) rolls of changeover table

materials, which require both high temperature and back pressure. It can work on a specially designed double action hydraulic press. The press has an unusual spectrum of ram velocities and its working sequence can be programmed to meet the different requirements of any ECAP process. Figure 3.35 shows the tooling assembled in a standard die-set. The prestressed die assembly, (1), has been equipped with an electric heater, (2), and water-cooling system for hot ECAP operations. The whole die-set was transported from a setting up station and is located near the press on its changeover table, (3).

Hot ECAP combined with back pressure is carried out on a hydraulic press equipped with a special hydraulic cushion system. The cushion can be adjusted in both the axial and radial directions. Axial adjustment helps to set up the counterpunch to operate properly with the ECAP tooling. Radial adjustment is necessary to centre the counterpunch in the output channel.

A 3D model illustrating the final stage of setting up the ECAP tooling is shown in Figure 3.36. Both punch, (1), and counterpunch, (2), are presented in their extreme positions in the 2 × 110° square cross-section channel, (3). There is a

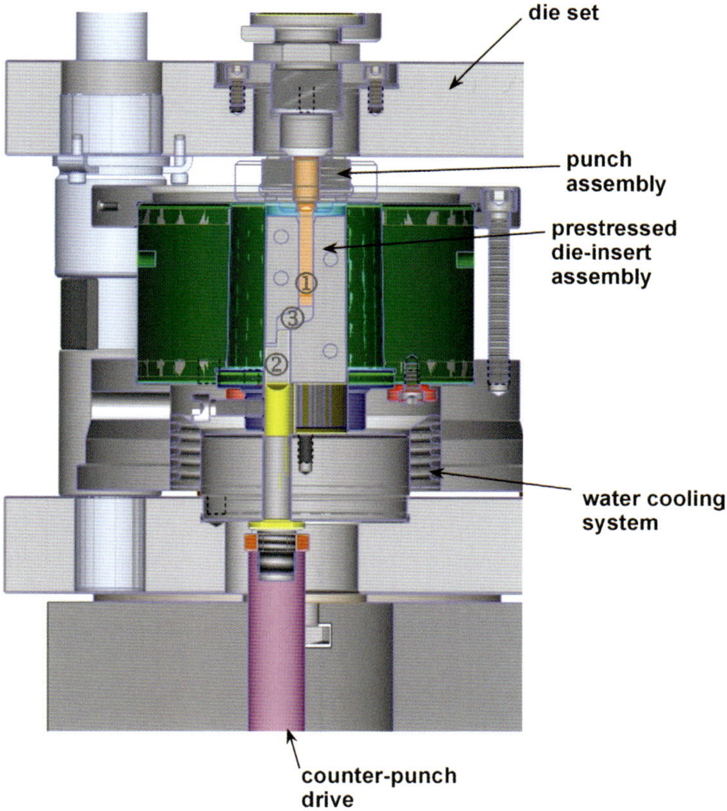

Figure 3.36 Cross-section of tooling for 2 × 110° hot ECAP operation with back pressure

working space, (3), between them for the processed material. The billet, or slug with cut off sharp ends and a reduced cross-section, can be fed into the input channel when the punch, (1), is in its upper position. The subsequent slug, further down the channel may already be supported by the counterpunch, (2). The front part of the counterpunch has been shaped to match the geometry of the output channel. Initially, the main ram of the press descends quickly to insert the punch's head into the input channel. The press ram then slows down before the punch's head approaches the billet/slug and it finally moves with constant velocity to press the slugs. During that stage the preceding and the subsequent slugs are gripped between the punch and counterpunch inside the angular channel. After the upper punch, (1), has completed its pressing function, counterpunch, (2), actuated by a lower hydraulic drive of the press, descends to open the output channel.

3.8 Monitoring of ECAP

In the case of hot ECAP, the automatic heating of the billets and the control of the die temperature are the most important auxiliary systems surrounding the press. Heating can be accomplished with electrical resistance heating elements attached to, or embedded in, the tool body and preferably placed near the tool-forming surfaces to control the temperature in the forming area. Using many independently powered electric heaters to control the heating zones is quite effective in providing reasonably accurate control of the die temperature. A multi-channel electronic temperature controller would normally be used to maintain the forming temperature during ECAP operations, characterised by the cyclical opening and closing of the press. Although the use of many separately powered and controlled heating zones is sometimes cumbersome, it provides, in conjunction with the internal cooling and insulation systems, an effective method for thermal maintenance of hot ECAP dies. It is more difficult to manage the mechanical loading of ECAP tools, especially during pressing at high temperatures.

Under normal conditions, tool loading in ECAP depends upon: the yield point of the processed material (type and temperature of material), the length and cross-sectional area of the billet, the efficiency of lubrication and plastic strain depending on the channel's geometry. Following good ECAP practice, all of these factors should give a constant response in terms of the processing force. However, cooperation of slugs jointly passing through the channel and their interaction with tool walls produce a number of additional effects, which influence the processing force.

The ECAP operation is also described in terms of the time taken to carry it out, which can be controlled by setting the stroke and velocity of the press. The working conditions result in a specific force–displacement relationship, which can be recorded to monitor the ECAP process. The resulting force–displacement curve or diagram can be treated as a process signature for each ECAP operation.

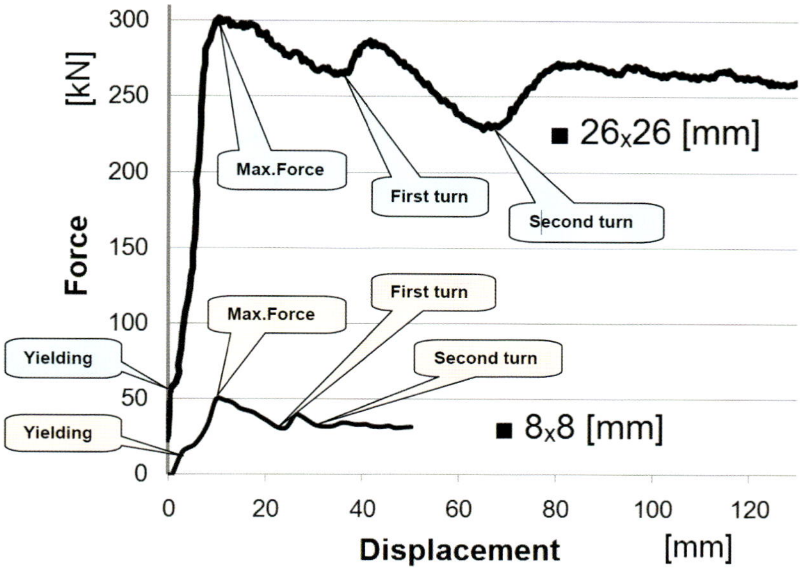

Figure 3.37 Force vs. punch displacement diagram recorded during third pass of two-turn ECAP of Al 1070 billets with different cross-section areas

Figure 3.37 illustrates the history of the force (derived from the oil pressure of the hydraulic press) vs. ram stroke continuously recorded during third pass of 2 × 90° ECAP at room temperature. Billets were made of Al1070 alloy with different cross-sections. The upper curve was recorded for a 26 mm × 26 mm cross-section billet and the lower one represents a smaller billet with a cross-section of 8 mm × 8 mm. Billets had a similar slenderness ratio of about 6 and were treated in the same way during both preparation (e.g. lubrication) and processing (e.g. ram velocity). There are obvious differences between the two curves such as different spans of the *X* and *Y* axes, which result from different lengths and cross-sections, respectively. On the other hand, local minima due to interaction between adjacent slugs jointly passing through consecutive turns of the channel are similar. Thus, the intrinsic features of the process are clearly seen despite the size of the billet. Figure 3.37 shows a typical force vs. ram displacement relationship, which can be recorded during any batch ECAP processing of jointly pressed billets. A more detailed comparison of forces for different billet sizes can be found in Olejnik and Rosochowski (2008) and for different channel configurations in Rosochowski *et al.* (2006).

The above-mentioned force–displacement relationship can be treated as a process signature and utilised in the automated production of UFG metals and alloys using ECAP. For batch ECAP with enhanced productivity not only is auxiliary equipment necessary but a suitable automatic supervision strategy is also needed. Automatic supervision in manufacturing deals with unavoidable

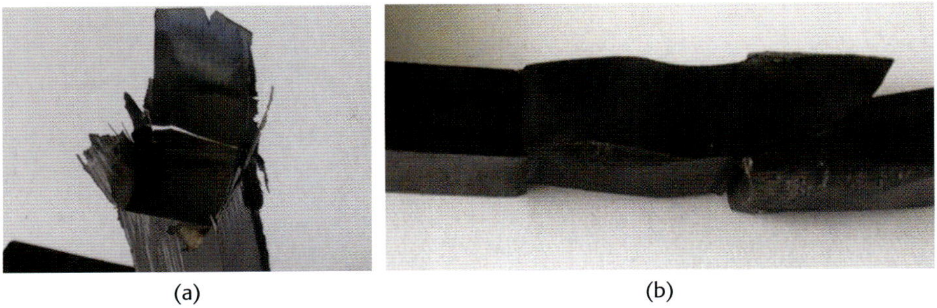

Figure 3.38 Failed slugs produced by classical ECAP: (a) unexpected flash under punch caused by fracture of prestressing ring; (b) collision between slugs in output channel

disturbances during production, thus making possible the intended functioning of the ECAP rig. The force–displacement diagram is recognised as the best way of characterising metal-forming processes performed on a press (Siegert *et al.*, 1990) as has been successfully demonstrated for extrusion (Chodnikiewicz and Olejnik, 1994). Force and displacement signals are easily available from hydraulic presses and can be used to monitor the ECAP process to develop an appropriate action. In the case of a catastrophic failure (Figure 3.38(a)) or a collision (Figure 3.38(b)), the automatic supervision system must interpret and interrupt the pressing process as quickly as possible.

The disturbance shown in Figure 3.38(b) is characteristic of classical ECAP, where slugs leaving the output channel cannot easily drop away from the die by gravity. The explanation for this collision is given in Figure 3.39(a). It is the most frequent disturbance in the classical ECAP operation carried out at elevated temperatures. Multiple billets passing consecutively through a long horizontal output channel are likely to stack since a hot and soft slug pushes

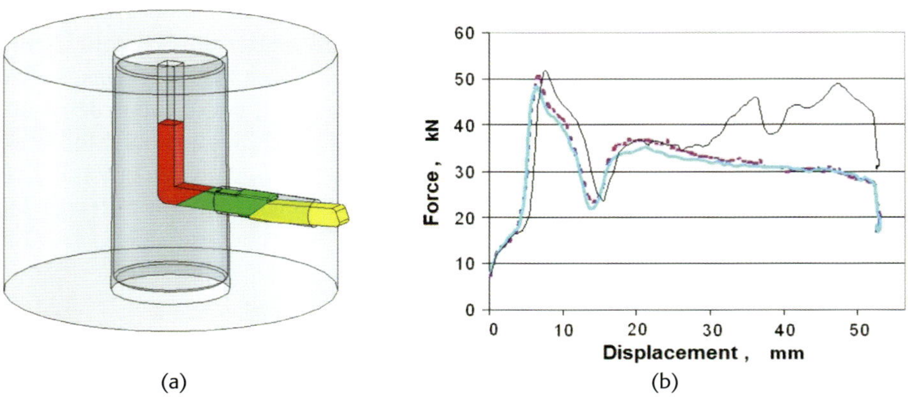

Figure 3.39 (a) Disturbance likely in hot ECAP; (b) force–displacement response

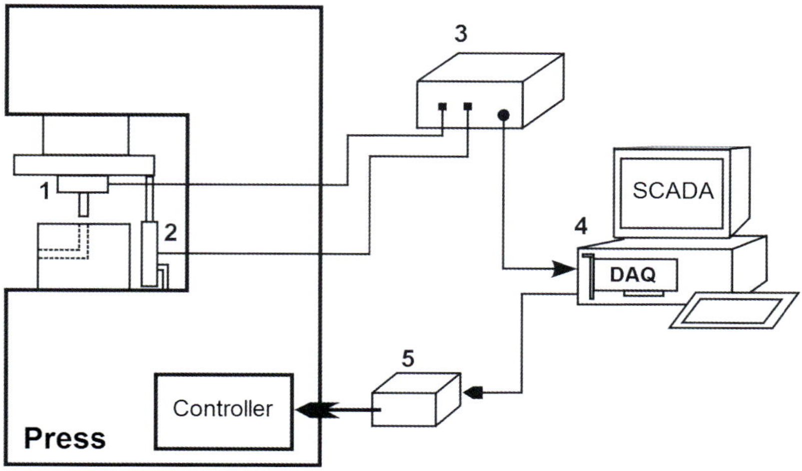

Figure 3.40 Main elements of supervision system for hot ECAP performed on press

a cold and hard one out of the die orifice. Inherent deformation of the leading and rear ends of the slugs which are in contact may block the passage for the following slugs. This causes the punch force to increase. Experimental evidence of such a case, recorded for warm ECAP of 8 mm × 8 mm cross-section slugs made of IF steel, is shown in Figure 3.39(b). Thick lines represent the force–displacement relationship for correct pressing, whereas the thin line indicates jammed slugs.

Careful analysis of force–displacement diagrams enables a suitable control strategy for hot ECAP to be developed. It requires monitoring of the force and displacement of the punch to trigger an alarm at any unwanted increase in force. The monitoring system has to be connected to the control system of a hydraulic press to stop and automatically lift the ram of the press immediately a jam occurs. An example of a simple supervision system for hot ECAP is shown in Figure 3.40. The system uses a load cell, (1), as the primary transducer to detect force during press operation. An LVDT sensor, (2), is also used to provide stroke information to the system. Force and stroke data are processed using a conditioning device, (3), and sent to a data acquisition (DAQ) board, (4), which is hosted by a computer. The computer supports the SCADA[1] operations of the DAQ board, performs display and console functions and communicates the stop and lift commands to the press controller via an interface, (5).

An automatic supervision system having the above-mentioned structure has been tested to work on a hydraulic press accordingly to the different control strategies in order to develop a strategy suitable for the hot working operation of classical ECAP. Figure 3.41 shows an experimental rig developed for supervised ECAP.

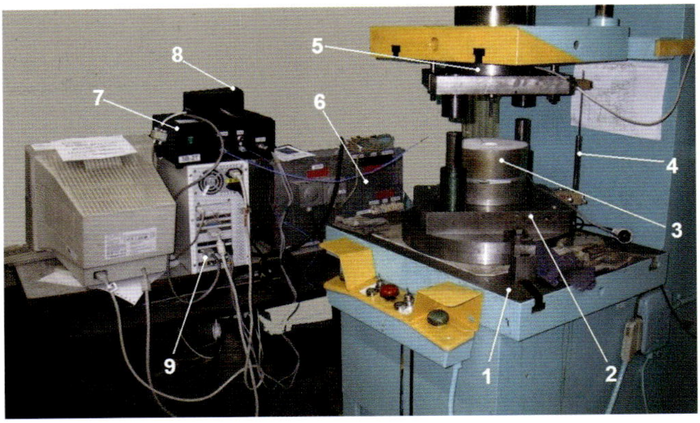

Figure 3.41 Automatic supervision rig for hot ECAP on hydraulic press: (1) press table; (2) die-set; (3) prestressed die assembly with electric heater; (4) displacement transducer; (5) load cell; (6) temperature controller; (7) multi-channel conditioning device; (8) interface with press controller; (9) DAQ board

References

Akira, O. and Masakazu, M. (2003) 'Method and device for manufacturing metallic material of micro structure' (Mitsubishi Heavy Ind. Ltd.). Japanese Patent 019532.

Bednarski, T. (1983) *Teoria procesów obróbki plastycznej, cz.I kształtowanie brył*, WPW Warsaw.

Brinksmeier, E., Riemer, O., Gläbe, R. and Meier, A. (2011) 'Material aspects for the diamond machining of submicron optical structures for UV-application', *International Journal of Nanomanufacturing*, 7: 63–72.

Chen, D.-Ch., Wang, J.-C., Pan, J.-S., Jhang, Ch.-F. and Tzou, G.-Y. (2008) 'Study of titanium alloy in the changing channel angular extrusion' in *Proceedings of the Ninth International Conference on Technology of Plasticity (ICTP 2008)*, held 7–11 September 2008, Gyeongju, Korea, pp. 1537–1542, Hanrimwon Publishing, Seoul.

Chodnikiewicz, K. and Olejnik L. (1994) 'Automatic supervision in metal forming' in *Automatic Supervision in Manufacturing*, M. Szafarczyk (ed.), pp. 121–138, Springer, London.

Clapp, B. P. (1966) 'A consideration of the economics of cold extrusion', *Sheet Metal Industry*, 43(465): 6–17.

Ferrasse, S., Segal, V. M., Alford, F., Kardokus, J. and Strothers, S. (2008) 'Scale up and application of equal-channel angular extrusion for the electronics and aerospace industries', *Materials Science and Engineering A*, 493: 130–140.

Furukawa, M., Horita, Z., Nemoto, M. and Langdon, T. G. (2002) 'The use of severe plastic deformation for microstructural control', *Materials Science and Engineering A*, 324: 82–89.

Hokka, M., Kokkonen, J., Seidt, J., Matrka, Th., Gilat, A. and Kuokkala, V.-T. (2012) 'High strain rate torsion properties of ultrafine grained aluminum', *Experimental Mechanics*, 52: 195–203.

Horita, Z., Fujinami, T. and Langdon, T. G. (2001) 'The potential for scaling ECAP: effect of sample size on grain refinement and mechanical properties', *Materials Science and Engineering A*, 318: 34–41.

International Cold Forging Group (1970) 'Production of steel parts by cold forging. ICFG document DOC.1/77', also in German: 'Kaltfliesspressen von Stählen und NE-Metallen'. VDI-Richtlinien VDI 3138, VDI-Handbuch Betriebstechnik, VDI-Verlag, Düsseldorf, Germany.

International Cold Forging Group (1991) 'General recommendations for design, manufacture and operational aspects of cold extrusion tools for steel components. ICFG document DOC.6/82', *Wire*, 41: 422–431.

International Cold Forging Group (1992) 'Calculation methods for cold forging tools. ICFG document DOC.5/82', *Wire*, 42: 141–159.

Iwahashi, Y., Horita, Z., Nemoto, M. and Langdon, T. G. (1998) 'The process of grain refinement in equal-channel angular pressing', *Acta Materialia*, 46: 3317–3331.

Kapoor, R., De, P. S. and Mishra, R. S. (2010) 'An analysis of strength and ductility of ultrafine grained Al alloys'. *Materials Science Forum*, 633: 165–177.

Kocisko, R., Kvackaj, T., Bidulská, J. and Molnárová, M. (2009) 'New geometry of ECAP channel', *Acta Metallurgica Slovaca*, 15: 228–233.

Kokkonen, J., Kuokkala, V.-T., Olejnik, L. and Rosochowski, A. (2008) 'Dynamic behavior of ECAP-processed aluminum at high strain rates at room and sub-zero temperatures' in *Proceedings of the 11th International Congress and Exposition of the Society for Experimental Mechanics (SEM2008)*, held 2–5 June 2008, Orlando, FL, Proceedings paper 164, Society for Experimental Mechanics, Bethel, CT.

Kokkonen, J., Kuokkala, V.-T., Seidt, J. D., Walker, A., Gilat, A., Olejnik, L. and Rosochowski, A. (2009) 'High strain rate deformation analysis of UFG aluminum sheet samples' in *Proceedings of the Annual Conference and Exposition on Experimental and Applied Mechanics (SEM09)*, held 1–4 June 2009, Albuquerque, NM, Proceedings paper No. 318, Society for Experimental Mechanics, Bethel, CT.

Kuokkala, V.-T., Kokkonen,J., Song, B., Chen, W., Olejnik, L. and Rosochowski, A. (2008) 'Dynamic response of SPD processed 1070 aluminum at various temperatures' in *Proceedings of the 18th DYMAT Technical Meeting*, held 10–12 September 2008, Bourges, France, pp. 39–46.

Lapovok, R. Ye. (2005) 'The role of back-pressure in equal channel angular extrusion', *Journal of Materials Science*, 40: 341–346.

Male, A. T. and Cockcroft, M. G.(1964) 'A method for determination of the coefficient of friction of metals under conditions of bulk plastic deformation', *Journal of the Institute of Metals*, 93: 38–46.

Nakashima, K., Horita, Z., Nemoto, M. and Langdon T. G. (1998) 'Influence of channel angle on the development of ultrafine grains in equal-channel angular pressing', *Acta Materialia*, 46: 1589–1599.

Nakashima, K., Horita, Z., Nemoto, M. and Langdon T. G. (2000) 'Development of a multi-pass facility for equal-channel angular pressing to high total strains', *Materials Science and Engineering A*, 281: 82–87.

Nishida, Y., Kume, S. and Imai, T. (2001) 'Large deformation apparatus, the deformation method and the deformed metallic materials' (Agency of Industrial Science and Technology, Japan), US Patent 6209379B1.

Olejnik, L. and Rosochowski, A. (2005) 'Methods of fabricating metals for nanotechnology', *Bulletin of the Polish Academy of Sciences: Technical Sciences*, 53: 413–423.

Olejnik, L. and Rosochowski, A. (2007) 'Nanostructuring-a new task for metal forming' in *Proceedings of the 4th International Conference on Advances in Production*

Engineering (APE'07), held 14–16 June 2007, Warsaw, Poland, L. Dabrowski (ed.), pp. 139–146. Warsaw, Warsaw University of Technology.

Olejnik, L. and Rosochowski, A. (2008) 'Scaled-up ECAP with enhanced productivity', *Steel Research International*, 79 (Special edition, vol. 2), pp. 439–446.

Olejnik, L., Presz, W. and Rosochowski, A. (2009) 'Backward extrusion using micro-blanked aluminium sheet', *International Journal of Material Forming*, 2 (Suppl. 1): 617–620.

Osmer, J., Riemer, O., Brinksmeier, E., Rosochowski, A., Olejnik, L. and Richert, M. (2007) 'Diamond turning of ultrafine grained aluminium alloys' in *Proceedings of the 7th International Conference (euspen 2007)* held 20–24 May 2007, Bremen, Germany, vol. 2, pp. 316–319, euspen, Bedford, UK.

Paydar, M. H., Reihanian, M., Bagherpour, E., Sharifzadeh, M., Zarinejad, M. and Dean, T. A. (2009) 'Equal channel angular pressing–forward extrusion (ECAP–FE) consolidation of Al particles', *Materials and Design* 30: 429–432.

Presz, W. and Rosochowski, A. (2006) 'The influence of grain size on surface quality of microformed components' in *Proceedings of the Ninth International Conference on Material Forming (ESAFORM 2006)*, held 26–28 April 2006, Glasgow, UK, N. Juster and A. Rosochowski (eds), pp. 587–590, Publishing House Akapit, Krakow, Poland.

Purcek, G., Altan, B. S., Miskioglu, I. and Ooi, P. H. (2004) 'Processing of eutectic Zn–5% Al alloy by equal-channel angular pressing', *Journal of Materials Processing Technology*, 148: 279–287.

Raab, G. I.(2005) 'Plastic flow at equal channel angular processing in parallel channels', *Materials Science and Engineering A*, 410(11): 230–233.

Raab, G. I., Botkin, A. V., Raab, A. G. and Valiev, R. Z. (2007) 'New schemes of ECAP processes for producing nanostructured bulk metallic materials' in *Proceedings of the 10th Int. Conference on Material Forming (Esaform 2007)*, held 18–20 April 2007, Zaragoza, Spain, E. Cueto and F. Chinesta (eds), pp. 641–646, American Institute of Physics Publishing, Melville, NY.

Rosochowski, A. (2005) 'Method of treating a metal billet' (University of Strathclyde), International Patent Application WO-05039792 A1.

Rosochowski, A. and Olejnik, L. (2002) 'Numerical and physical modelling of plastic deformation in 2-turn equal channel angular extrusion', *Journal of Materials Processing Technology*, 125(6): 309–316.

Rosochowski, A. and Olejnik, L. (2011) 'Current practice and future opportunities for two-turn ECAP', *Materials Science Forum*, 667(9): 121–126.

Rosochowski, A. and Olejnik, L. (2011) 'Equal channel angular pressing with converging billets-FE simulation' in *Proceedings of the 10th International Conference on Technology of Plasticity, (ICTP 2011)*, held 25–30 September 2011, Aachen, Germany, G. Hirt and A. E. Tekkaya (eds), pp. 235–240, Verlag Stahleisen, Düsseldorf, Germany.

Rosochowski, A., Olejnik, L., Gagne, J., Ladeveze, N. and Rosochowska, M. (2006) 'Compression behaviour of UFG aluminium' in *Proceedings of the Ninth International Conference on Material Forming (ESAFORM 2006)*, held 26–28 April 2006, Glasgow, UK, N. Juster and A. Rosochowski (eds), pp. 543–546, Publishing House Akapit, Krakow, Poland.

Rosochowski, A., Olejnik, L. and Richert, M. (2005) '3D-ECAP of square aluminium billets', *International Journal of Forming Processes*, 8: 637–640.

Rosochowski, A., Olejnik, L. and Richert, M. (2006) 'Channel configuration effects in 3D-ECAP', *Materials Science Forum*, 503(4): 179–184.

Rosochowski, A., Olejnik, L. and Richert, M. (2007) '3D-ECAP of square aluminium billets' in *Advanced Methods in Material Forming*, D. Banabic (ed.), pp. 215–232. Berlin: Springer.

Rosochowski, A., Olejnik, L., Roginski, S. and Richert, M. (2007) 'Micro-EDM of UFG aluminium', *Proceedings of 3rd International Conference on Multi-Material Micro Manufacture (4M 2007)*, held 3–5 October 2007, Borovets, Bulgaria, pp. 203–206, Whittles Publishing, Dunbeath, UK.

Rosochowski, A., Presz, W., Olejnik, L. and Richert, M. (2007) 'Micro-extrusion of ultra-fine grained aluminium', *International Journal of Advanced Manufacturing Technology*, 33: 137–146.

Rusz, S., Cizek, L., Cizek, L., Dobrzanski, L. A., Tylsar, S. and Kedron, J. (2010) 'ECAP methods application on selected non-ferrous metals and alloys', *Archives of Materials Science and Engineering*, 43(6): 69–76.

Segal, V. M. (1977) 'The method of material preparation for subsequent working', USSR Patent 575892.

Segal, V. M. (1995) 'Materials processing by simple shear', *Materials Science and Engineering A*, 197: 157–164.

Segal, V. M. (1996) 'Plastic deformation of crystalline materials', US Patent 5513512.

Segal, V. M. (2003) 'Slip line solutions, deformation mode and loading history during equal channel angular extrusion', *Materials Science and Engineering A*, 345: 36–46.

Segal, V. M. (2004) 'Engineering and commercialization of equal channel angular extrusion (ECAE)', *Materials Science and Engineering A*, 386: 269–276.

Segal, V. M., Goforth, R. E. and Hartwig, K. T.(1995) 'Apparatus and method for deformation processing of metals, ceramics, plastics and other materials', (Texas A&M University System), US Patent 5400633.

Segal, V. M., Reznikov, V. I., Drobyshevski, A. E. and Kopylov, V. I. (1981) 'Plastic working of metals by simple shear', *Russian Metallurgy* (Engl. transl.), 1: 99–105.

Segal, V. and Segal, L. (1997) 'Method of and apparatus for processing tungsten heavy alloys for kinetic energy penetrators', US Patent 5600989.

Shchukin, V. Ya. and Segal, V. M. (1976) 'Device for strengthening materials by pressure', Patent Application SU 515968.

Sieber, K. (1966) 'Beispiele zu den Fertigungskosten des Kaltfliesspressens', *Industrie-Anzeiger*, 88(37): 115–118.

Siegert, K., Klamser, M. and Straube, O. (1990) 'Error identification during deep drawing', *Zeitschrift fuer Wirtschaftliche Fertigung*, 85(8): 416–420.

Srinivasan, R., Cherukuri, B. and Chaudhury, P. K. (2006) 'Scaling up of equal channel angular pressing (ECAP) for the production of forging stock', *Materials Science Forum*, 503: 371–376.

Stecher, G. and Thomson, P. (2003) 'Improved angular channel processing' (Monash University, Australia), International Patent Application WO-027337 A1.

Toth, L.S., Lapovok, R., Hasani, A. and Gu, Ch. (2009) 'Non-equal channel angular pressing of aluminum alloy', *Scripta Materialia*, 61: 1121–1124.

Valiev, R. Z. and Alexandrov, I. V. (2002) 'Development of severe plastic deformation techniques for the fabrication of bulk nanostructured materials', *Annales de Chimie Science des Matériaux*, 27(3): 3–14.

Zhao, Y. and Liao, X. (eds) (2010) 'Ductility of bulk nanostructured materials', *Materials Science Forum*, 633/4: 1–734.

Zuyan, L. and Zhongjin, W. (1999) 'Finite-element analysis of the load of equal-cross-section lateral extrusion', *Journal of Materials Processing Technology*, 94: 193–196.

Zuyan, L., Gang, L. and Wang, Z. R. (2000) 'Finite element simulation of a new deformation type occurring in changing-channel extrusion', *Journal of Materials Processing Technology*, 102: 30–32.

Note

1. SCADA (Supervisory Control And Data Acquisition). This type of application software is used for automatic supervision systems in manufacturing.

High pressure torsion (HPT)

CHAPTER 4

Reinhard Pippan and Anton Hohenwarter

4.1 Introduction

In the field of deformation technology, the stress–strain behaviour of materials at large degrees of deformation is of considerable significance. This information can, for instance, be obtained by torsion tests, which enable much larger degrees of deformation compared to tensile or compression tests. In this test the deformation is applied by shear stresses and, supported through the absence of a hydrostatic tensile stress component, damage evolution in the material is postponed. Nevertheless, the applied deformation strain is radius dependent and thus restricts its attractiveness and applicability in material testing. Another further step towards even higher strains is achieved by combining this torsional test with a compression test. The compressive stress component suppresses damage evolution under shear and the maximum applicable strain before failure increases markedly. The combination of torsion and compression was first intensively investigated by Bridgman (1935, 1943). The basics of this pioneering work were continued by scientists in Russia in the 1980s. It was further developed and is now known as high pressure torsion (HPT) (Langdon, 2009; Valiev *et al.*, 2009). It is currently one of the most popular severe plastic deformation (SPD) techniques. As is typical for all SPD processes the technique is mainly used to produce ultra-fine grained (UFG) or even nanocrystalline (NC) materials. This is achieved by imposing extraordinarily high strains onto materials, inducing a substantial grain refinement. The grain refinement and the connected beneficial effects on the mechanical and physical properties have been the focus of a vast number of papers and several overview articles (Valiev *et al.*, 2000: Zhu and Langdon, 2004; Zhilyaev and Langdon, 2008). One feature, often referred to as a disadvantage, is that the applied strain is linearly dependent on the radius, as in classical torsion tests. In addition, the restricted sample size is frequently regarded as a drawback. This is a result of the technical challenges which are encountered to reach the high compressive stresses. Nevertheless, in the authors' experience, interest in HPT has increased, particularly in the last 5–10 years, reflected by a large number of contributions in the literature and also in the rising number of scientific groups working with this technique. This tendency seems to be driven by a variety of attractive features which make it worthwhile to use this deformation technique for material research on UFG and NC materials. Some interesting benefits are briefly outlined in Sections 4.1.1–4.1.7.

4.1.1 Unlimited strain

The most significant feature and advantage of HPT in comparison with other techniques is the possibility of applying almost infinitely high deformation strains onto a material without causing serious damage to either the specimen or the tool. This is realised by the high hydrostatic pressure component, which prevents the materials from failing. It is also significant that the material is mostly deformed without changes to its outer dimensions. As an example of the significance of this feature, a simplified comparison to rolling can be made: assuming a required true strain of 7, an HPT disc having a typical diameter of 8 mm and a thickness of 0.8 mm reaches this strain at the outer radius after less than half a rotation. To introduce the same strain by rolling, a sheet would have to be rolled from a thickness of 1 m down to approximately 1 mm.

4.1.2 Efficiency of the technique

The process is also very time efficient when very high processing strains are required (e.g. for the synthesis of nano-composites). Regarding the aforementioned HPT disc dimensions, a true strain of 100 can be reached within several minutes. Imposing the same strain for example with equal channel angular pressing (ECAP) would require applying at least 100 passes to the billet! (An ECAP die with a 90° angle gives a strain of about 1 per pass.)

4.1.3 Simplicity of the process

In general, every ordinary testing rig can be used and customised to perform HPT experiments by obeying some general advice given in this chapter. The investments required for the tool and the changes to an existing rig are rather low compared to other investments for synthesis techniques.

4.1.4 Reliability of the technique

This technique is highly reliable despite being quite simple. Failure of the anvils under reasonable handling conditions is unusual. Worn anvils can be re-used simply by re-machining their geometry.

4.1.5 Flexibility of testing parameters

The process is highly flexible with regard to the adjustable processing parameters. The deformation temperature can easily be varied between temperatures of liquid N (−196°C) and about 700°C. The strain rate can be changed by several orders of magnitude and the applied strain is, for the majority of the deformed materials, a question of the operator's time and patience rather than of technical feasibility.

4.1.6 Versatility of high pressure torsion

A main feature of HPT, the high hydrostatic compressive stress component, allows deformation of almost every material, ranging from very soft pure metals to high-strength steels, refractory metals and even some ceramics. HPT is not only exceptional at reducing the grain size of metals and alloys but can also be used for powder consolidation combined with subsequent deformation. It provides a unique possibility to create nano-composites with peculiar properties which cannot be synthesised by any other commercial synthesis processes so far.

4.1.7 Sufficient sample dimensions

Although the sample size is very often small in comparison with other SPD processes such as ECAP, a large variety of investigation methods are possible covering almost every microstructural characterisation technique. In addition, many mechanical properties can be evaluated. Nevertheless, from an engineering point of view there are almost no limitations to upscaling the process to larger specimens if that is desired.

The intention of this chapter is to outline and discuss the main engineering aspects and ideas required to build an HPT tool, highlighting the problems and pitfalls that might occur and to share long-term experiences with interested readers. In order to pursue this aim, a few microstructural features concerning HPT will initially be summarised, then some basic versions of HPT setups will be reviewed before significant design criteria for building a HPT device are discussed. Additionally, advice for implementing some useful attachments, for instance for measuring the torque, are given. Although this chapter contains new, unpublished aspects of HPT engineering, some of its principal ideas and conclusions can also be found in earlier publications dealing with this topic (Pippan *et al.*, 2008; Hohenwarter *et al.*, 2009), which formed the basis for this chapter.

4.2 Some characteristics of HPT microstructures

Although this chapter focuses on the technical aspects of HPT, some basic features and peculiarities of microstructures obtained by HPT should be noted as they are relevant to this chapter. Explanations will only be made for single-phase materials. Figure 4.1 shows various specific stages of the fragmentation of an initially coarse-grained undeformed material in inverse pole figure maps describing the local crystal orientation in a specific colour code.

After a relatively low degree of deformation by SPD standards, strong fluctuations (expressed by local colour changes) of the local crystal orientation in the still visibly coarse-grained metal can be found (Figure 4.1(a)). The newly stored dislocations arrange themselves in dislocation cells, and with ongoing deformation in cell blocks which have a greater difference in orientation from each other. With ongoing deformation, misorientations in the different cell structures increase, their size decreases and the initial grain boundaries are no longer perceptible (Figure 4.1(b)). Gradually, the cell blocks and dislocation cells, typical and well investigated for 'low' deformation strains, convert into regular grains with typical crystalline sizes well below a micron (Figure 4.1(c)).

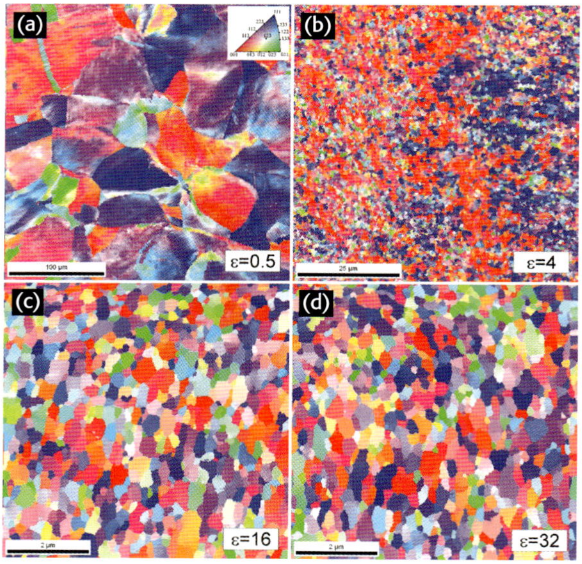

Figure 4.1 Microstructural evolution of Ni subjected to HPT presented at certain stages of fragmentation: (a) $\varepsilon = 0.5$; (b) $\varepsilon = 4$; (c) $\varepsilon = 16$; (d) $\varepsilon = 32$. Colour code in inverse pole figure maps represents crystal orientation of crystal parallel to surface normal. Note use of different scale bars

Typically, after applying strains of 16–32 the grain size stabilises (compare Figures 4.1(c) and (d)). A state often known as 'saturation of grain fragmentation' is reached in this deformation regime (Pippan *et al.*, 2010). An increase in the strain does not significantly change the grain size any further and a sort of equilibrium between dynamic hardening and softening processes is reached. The occurrence of a lower limit of grain size in HPT deformed materials is mainly dependent on the purity of the metal, alloying content, process temperature and strain rate. By applying a sufficiently high number of rotations, the lower limit of grain fragmentation can also be reached at very low disc radii. Hence a homogenisation of the microstructure and the mechanical properties throughout almost the entire disc is possible, despite the fact that the applied strain depends linearly on the radius. An example is shown in Figure 4.2 for pure Fe subjected to 1, 5 and 10 rotations. After only one rotation an almost linear increase in hardness with the radius is present. After five rotations a plateau in the hardness is reached in a large part of the disc. The maximum hardness is reached since the minimum strain for the maximum grain refinement is obtained at a lower radius of the HPT disc. Applying an even greater number of rotations, for example 10, the low hardness area becomes even smaller. Thus, applying an adequate number of rotations may homogenise the mechanical properties of the disc, which is an important pre-condition when mechanical tests are performed on larger specimens. This homogenisation is possible not

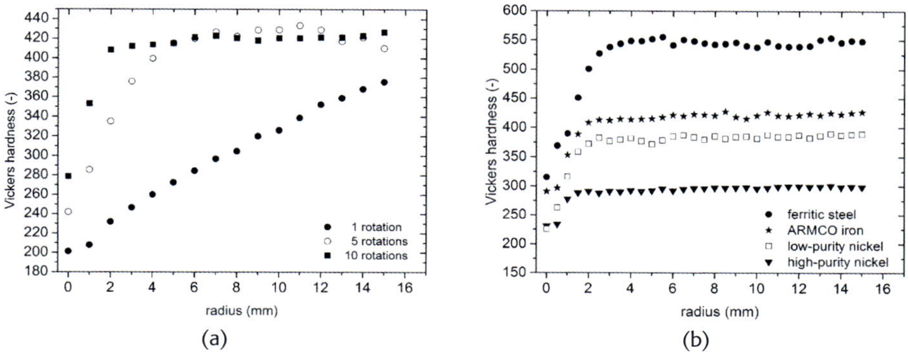

Figure 4.2 Hardness plots along the radius of HPT discs. (a) Hardness evolution in pure Fe for 1, 5 and 10 rotations. Although a well-fulfilled linear dependence of hardness on radius is measured after one rotation, a pronounced plateau after 5 and 10 rotations can be found. (b) Hardness distribution of several metals and alloys subjected to 10 rotations, respectively

only for pure metals, but also for more complex and industrially relevant materials such as the steels shown in Figure 4.2(b).

Naturally, the minimum grain size obtainable with HPT depends on a large number of physical factors. In general, the principal viewing directions are also relevant for studies of grain sizes. A commonly used coordinate system is presented in Figure 4.3. For an HPT disc it includes a tangential direction (TD), an axial direction (AD) and a radial direction (RD). Representative micrographs of HPT-processed pure Fe, taken in the saturation regime and parallel to the major viewing directions are shown in Figure 4.4. In an RD (left-hand side of Figure 4.4), the granular structure exhibits an aligned and elongated shape with a certain aspect ratio. It is a consequence of the monotonically induced shear strain and depends strongly on the homologous deformation temperature. In a TD (middle of Figure 4.4), the grains also exhibit certain elongation. Parallel to

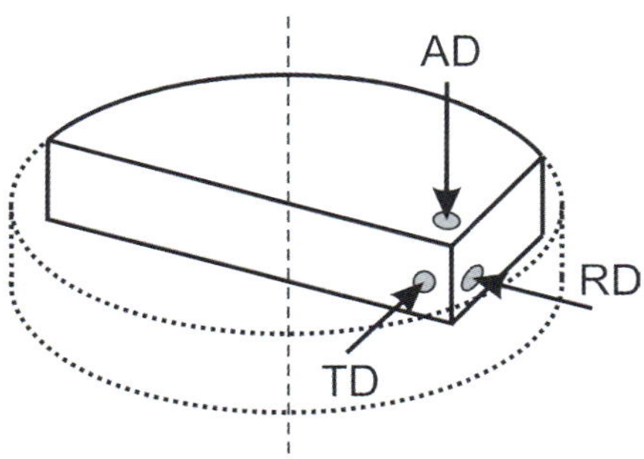

Figure 4.3 Schematic of coordinate system with main viewing directions onto microstructure defined as TD, RD and AD

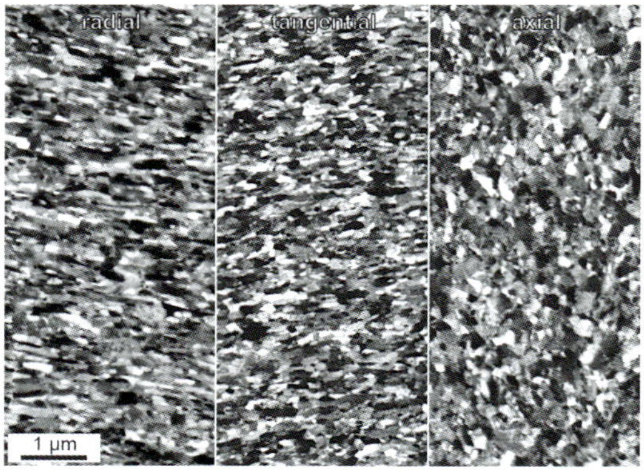

Figure 4.4 Representation of saturation microstructure of pure Fe inspected in three principal viewing directions

an AD (right-hand side of Figure 4.4), the grains exhibit a more or less equiaxial shape and appear to be somewhat larger. To sum up, the grain morphology is lamellar and can also be termed a pancake structure. Similarities in the grain morphology are also found when metals are subjected to ECAP with route A or accumulative roll bonding.

This distinctive alignment of the microstructure can lead to pronounced anisotropy in the mechanical properties (Hohenwarter and Pippan, 2010). Nevertheless, even with HPT the structure can be altered by changing the strain path (i.e. the straining direction and deformation route becomes similar to ECAP route C). In the cyclic version of HPT (Wetscher and Pippan, 2006; Kawasaki and Langdon, 2008) the amplitude of the applied strain and the number of alternating cycles can freely be tuned and can lead to an equiaxial microstructure even for a RD, as long as the strain amplitude remains low. This is shown in Figure 4.5 in a straightforward comparison of the microstructure of Ni subjected to monotonic (Figure 4.5(a)), and cyclic HPT (Figure 4.5(b)).

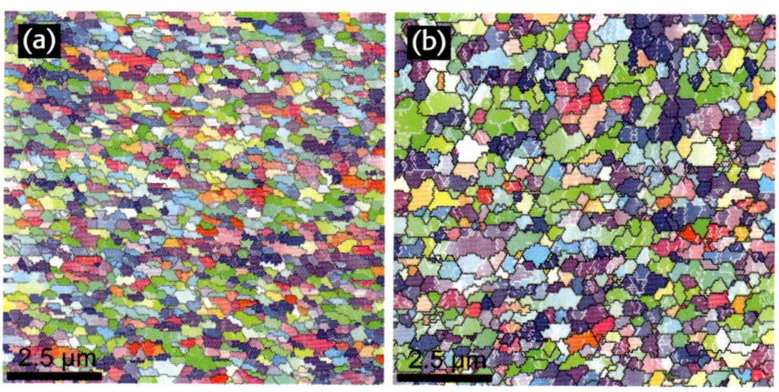

Figure 4.5 Comparison of saturation microstructure of Ni subjected to: (a) monotonic; and (b) cyclic HPT observed in the RD. In both cases the total applied von Mises strain ε_{total} was 64. For cyclic HPT a strain amplitude $\Delta\varepsilon = 4$ was chosen requiring 16 cycles

Additionally, it can be seen that the microstructure of the cyclic HPT is somewhat coarser than for monotonic HPT. Another important feature of cyclic HPT is that it can be more efficient in certain cases (e.g. pearlitic steels) to fragment microstructures.

4.3 Principles of HPT: description of different setups

Although HPT is presented in a vast number of publications, various setups can be found in papers, which at a first glance seem to only be slightly different. Nevertheless, the setup used can have a large impact on the actual processing and thus also on the microstructure which is obtained and the properties of the material that has been processed. Different tool setups which could be generally used are described in Sections 4.3.1, 4.3.2 and 4.3.3 and are shown in Figure 4.6 (Hohenwarter *et al.*, 2009).

4.3.1 The idealised HPT version

In general, in HPT the samples are discs having a distinctive thickness t and diameter d where t is usually very small compared to d, typically 1/10 or even smaller. In the idealised version of HPT the deformation process can be regarded as a simple shear process. Assuming that t remains constant during the experiment the applied shear strain can be calculated as

$$\gamma = \frac{2\pi r}{t} n \tag{4.1}$$

where r is the radius, t is the thickness of the disc and n is the number of revolutions. This simple equation implies that in HPT the shear strain always remains zero in the middle of the disc and increases linearly with increasing r.

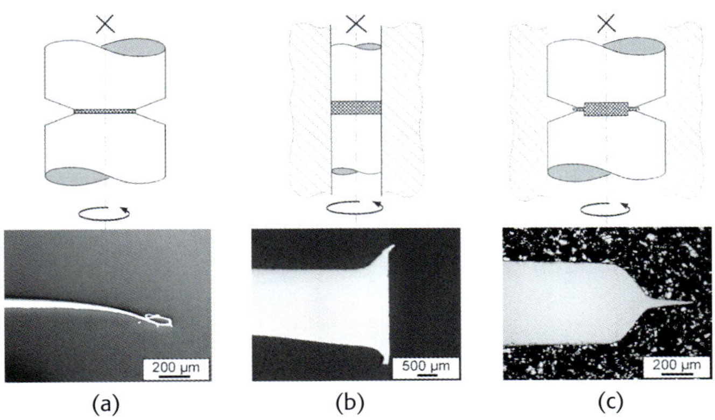

Figure 4.6 Schematic illustration of HPT setups found in literature and examples of disc cross-sections obtained using different setups: (a) idealised unconstrained HPT; (b) fully constrained HPT; (c) quasi-constrained HPT.

Hardness measurements taken at low strains clearly prove this linear dependence (Figure 4.2(a)). In order to allow a quantitative comparison of the applied strain with other deformation techniques or deformation routes, a criterion for equivalent strain has to be chosen. A common criterion for the estimation of an equivalent strain is the von Mises equivalent strain, based on the equivalence of imposed deformation work (Stüwe, 2003)

$$\varepsilon_{vM} = \frac{\gamma}{\sqrt{3}} \tag{4.2}$$

Although the applicability of this criterion for polycrystalline materials is questionable, it enables a direct comparison of materials with regard to the applied strain and the resulting differences in properties. Frequently only the number of rotations is quoted. This, however, is often not conclusive for direct comparisons regarding results obtained by different groups of researchers since the applied strain depends strongly on the specimen's geometry as Equation (4.1) shows, and the specimens are not standardised.

To start an experiment, regardless of the HPT setup used, a sample in the shape of a disc is placed between two anvils. The specimen is loaded with a certain pressure before one of the anvils is rotated against the other. This idealised HPT version is macroscopically unconstrained, which makes it technically difficult to put into practice. The problem is that under the pressures that are typically used, far beyond the plastic yield point of the material, the specimen can flow freely in the RD, especially when the material is soft, and the friction coefficient between the anvil material and specimen is low. This can result in the thickness of the specimen being reduced. The bottom of Figure 4.6(a) shows the outer region of a cross-section through the middle of the specimen deformed by such a setup. As shown, the sample loses its thickness with increasing radius, becomes fairly thin at the edge and its shape is poorly defined. The consequence of this behaviour is that the applied pressure decreases significantly as the disc area increases. Even more important, it should be noted that the thinning of the specimens can cause severe contact between the rotating anvils. This leads to additional friction which causes a complex and inhomogeneous pressure distribution depending mainly on the radius of the disc and the flow stress. Furthermore, in such a case the torque cannot be used as a measure of the flow stress because a large unknown contribution would come from the additional friction induced by the unintentional tool contact. Lastly, the additional contact could lead to a total failure of the anvils. The problem of excessive thinning of the specimen can be avoided by using a fully constrained HPT setup.

4.3.2 Fully constrained HPT

A schematic of a fully constrained HPT setup is shown in Figure 4.6(b). In principle this setup is very similar to the last one, but there is also an outer cylinder around the anvils. This arrangement induces hydrostatic pressure and the geometry is not altered during the experiment. However, friction between the outer cylinder and the rotating specimen is present in this setup, which again complicates the stress state and the entire shear deformation process. As the schematic suggests, this ideal case would not allow any dimensional tolerances, which must be included for tool assembly and operational reasons. As a consequence of the technical realisation of this setup, the anvil will have a slightly smaller diameter than the outer cylinder so that the assembly of the tool and the rotation are both enabled. Because of this technical necessity the processed material can flow into the gap between the anvil and the outer cylinder, which is also shown on the cross-section of a sample in Figure 4.6(b). The excess material in the gap can drastically increase the required torque for torsional deformation. Thus, the measured torque cannot be used as a measure for the flow stress due to the undefined friction effects. Another disadvantage is that the sample and the anvils have the same diameter. Thus, the maximum compressive stress onto the sample is restricted by the compressive strength of the anvils in this design.

4.3.3 Quasi-constrained HPT

At present, the so-called quasi-constrained HPT setup (Figure 4.6(c)) is used in the majority of cases, as it tackles some of the issues discussed in Sections 4.3.1 and 4.3.2. In this setup both anvils possess a central cavity or depression. The cavity has two uses. First, it permits the central positioning of the sample, which is of great significance for an accurate torsional deformation process. If the specimen is off-centre an undefined, irregularly shaped specimen could develop during rotation. Secondly, the cavity enables the formation of a hydrostatic compressive stress state during compression. For that to occur, the sum of both cavities must be marginally smaller than the thickness of the specimen. When the specimen is loaded, a small amount of material, limited by friction, will flow into the gap between the anvils creating the so-called burr zone. Thus, a backpressure in the radial direction is exerted and a hydrostatic compressive stress component evolves. The excess material also prevents direct contact between the anvils, which could cause wear or even their failure, and so it acts as a solid lubricant. A cross-section of a sample processed with this version of HPT is shown in Figure 4.6(c). The sample has a clearly defined shape compared to the other two examples and allows reproducible results to be obtained. Some loss of thickness during straining must be taken into account, especially if the material is soft. The typical amount of compression deformation is of the order of 5–10% and can be neglected compared to the typical amount of shear strain produced during torsion. The technical realisation of a quasi-constrained HPT and the technical details of such a tool will be discussed in Section 4.4.

4.4 Design criteria for building an HPT device

4.4.1 The test rig

In principle, HPT can be performed by adapting any conventional testing rig used for tensile or compression tests. As a basic requirement, depending on the

Figure 4.7 Typical HPT setup with a maximum loading capacity of 400 kN

planned specimen size, the rig must be able to endure and exert the required axial forces onto the specimen to induce the required pressure. Whereas the axial stiffness is usually adequate, as rigs are generally designed for a certain maximum force capacity, the torsional stiffness can become an issue. In particular, two-column load frames can lack adequate torsional stiffness, which is important to guarantee the axial symmetry of the setup. A lack of stiffness can result in an undesirable lateral shift of the specimen during deformation and so an abnormal disc shape (Figure 4.9(a)). Therefore, ideally ultra-stiff four-column load frames should be chosen for adaption or acquisition, as shown in Figure 4.7, which displays a hydraulic test rig with a maximum load capacity of 400 kN.

It should be noted that the anvils in Figures 4.7 and 4.8 are placed centrally in cylindrical holding fixtures. The anvils can easily be removed from them and new anvils can be substituted whenever they are needed. In contrast to the schematic in Figure 4.6(c), in the technical realisation, the cylindrical holding fixtures rather than the anvils, are guided by an outer hollow cylinder. The intention behind this is to improve the axial symmetry during straining. The lack of an outer cylinder can lead to radial displacements and thus to misaligned rotation axes, shown on the right-hand side of Figure 4.8.

The misalignment is transferred onto the sample with the result shown in Figure 4.9. In Figure 4.9(a), the upper and lower halves of the specimen show a significant shift, which indicates that the upper and lower anvils were misaligned in terms of the rotational axes. The same misalignment may also occur using the aforementioned two-column test frame. In contrast, in Figure 4.9(b), the upper and lower halves of the specimen are axially aligned because the specimen was processed with an outer guiding cylinder. The axial symmetry is important for two reasons. First, when exact strain or radius measurements are needed, a well-aligned torsion axis is required. Otherwise, as Figure 4.9(a) shows, it is very difficult to define the centre of the specimen and so to determine an exact radius, which is especially important for small discs where it is nearly impossible to determine it. The radial displacement can also induce additional

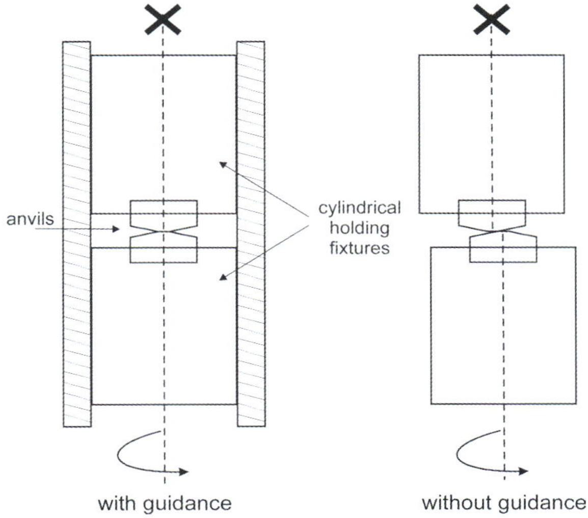

Figure 4.8 Schematic of setup indicating possible misalignment of torsion axis when no outer guidance is in use (misalignment exaggerated for viewing reasons)

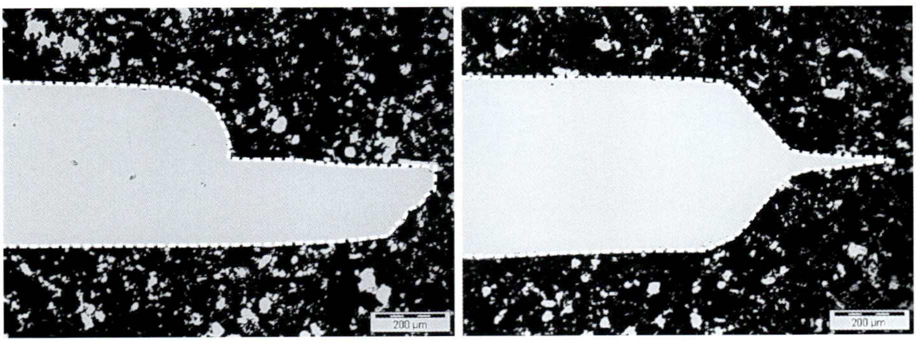

(a) (b)

Figure 4.9 Comparison of HPT disc cross-sections showing effect of a misaligned torsion axis

deformation, which may lead to the disappearance of the undeformed centre. Secondly, large radial displacements can lead to a pronounced contact between the anvils leading to a decrease in the pressure exerted on the sample and so to a slippage between the sample and anvils or even failure of the anvils.

4.4.2 Applied pressure

HPT experiments always include the following principal steps: axial loading of a specimen between two anvils, imposing the shear strain and unloading the tool. Thus, the applied pressure plays a central role in this technique and has two main functions. First, it suppresses various modes of crack formation allowing higher deformation strains. Thus, it can lead to smaller grain sizes compared to other SPD techniques. Secondly, the pressure itself may play an important intrinsic role, influencing the grain size (Zehetbauer *et al.*, 2003). However, at the start of an HPT experiment the pressure is needed to deform the specimen slightly in order to fill the cavities of the HPT anvils. The disc-shaped specimen must deform plastically so that the excess material can flow into the gap between the two closing anvils (Figure 4.10(a)). The minimum pressure needed for that procedure will be at least equivalent to the yield strength of the material being tested. In reality, this is only a lower bound and

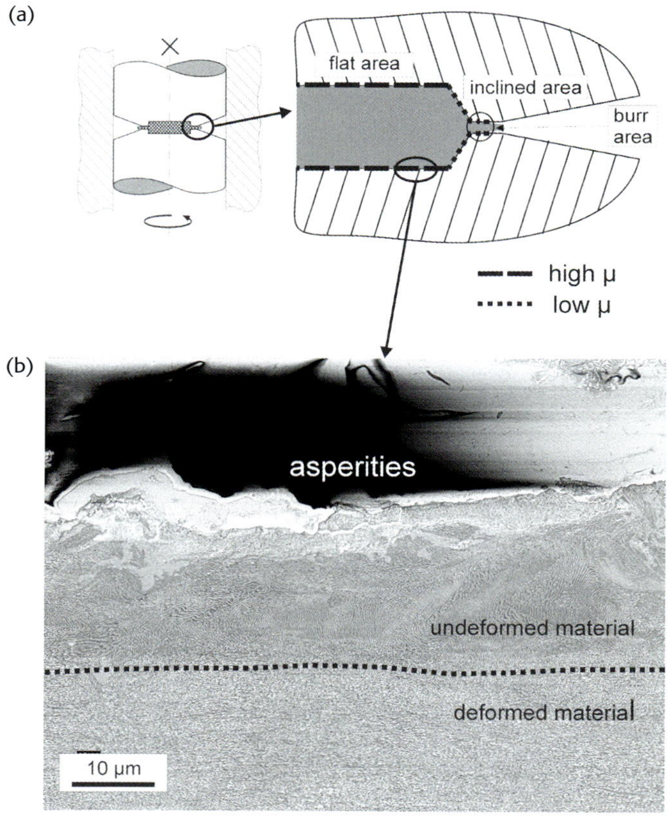

Figure 4.10 Loading situation after pre-deformation of HPT discs. (a) Anvil and specimen configuration in loaded state showing areas in cavity of tool where a high friction coefficient μ and a low one is desired. Excess material is squeezed out between anvils and forms a burr. (b) Micrograph displaying asperities on a processed sample. A thin layer beneath asperities remains undeformed (coarse-grained) since no slip between sample and anvil occurred during straining. Underneath, deformed material is visible with a much finer microstructure than undeformed one

the actual value will depend on the hardening behaviour of the material and the total geometrical changes during the compression deformation. At first glance, the filling step seems to be trivial but it can determine the feasibility of the entire HPT experiment. As already mentioned in Section 4.3.3, with the filling process a compressive deformation of several percent needs to be imposed with a low hydrostatic pressure. For brittle materials having low compressive ductility, the required compressive deformation must be kept low so that cracks cannot occur during the loading stage. Otherwise the samples become shear-deformed with pre-existing cracks.

In addition to the macroscopic filling of the cavities, the sample and anvil surfaces should have perfect contact at the microscopic scale. The cavities and the samples should normally be sandblasted. Doing this ensures that the surfaces are degreased and, more importantly, a pronounced surface roughness can be generated. This procedure allows the friction to be enhanced between the sample and anvils. Although it is normally better to avoid friction (e.g. in the walls of an ECAP tool), here high friction forces between the flat faces of the sample and the cavities are desired. In the area where excess material flows out, the burr area and also on the inclined walls of the cavities, the friction coefficient should be low so that the material is prone to slip (Figure 4.10(a)). High friction between sample and anvils, combined with the applied pressure, guarantees that the specimen will rotate without slippage and so induce the shear strain. In order to see a beneficial effect from surface roughening, the metal should flow into the asperities of the micro-roughened anvils and fill them completely (see micrograph in Figure 4.10(b)). A minimum pressure is also needed for this step. As a rule of thumb, the minimum pressure required for the plastic deformation of the metal surface and the subsequent filling of the anvil asperities will be three times the yield stress σ_y of the sample material in the undeformed state. This rule is simply inferred from the well-known Tabor rule (Bowden and Tabor, 1954). An excellent fit between disc and anvils is generated before the actual experiment starts. However, unless monolithic diamond anvils are used, there is a technical upper bound for the applicable pressure, which is about three times the yield stress of the anvils. Since the cavity in the anvil can act as a root of a notch, this value will be lower. For a modern high-speed tool steel this upper bound will be around 8–9 GPa. A finite element simulation in Figure 4.11 shows the stress distribution in a sample and an HPT anvil with a diameter of 40 mm and a cavity of 8 mm (Ecker, 2004) loaded with 350 kN. For symmetry reasons only one-quarter of the configuration had to be modelled. As Figure 4.11(a) shows, the maximum von Mises stress is already at a level of 5 GPa induced by the geometry of the cavity, although the nominal stress in the cross-section of the anvil is much lower. This value is already far beyond the compressive strength of the material and it should fail. In reality, the anvil does not fail due to the beneficial effect of the hydrostatic pressure, as shown in Figure 4.11(b), which is not only positive for the deformed material but also for the anvil itself.

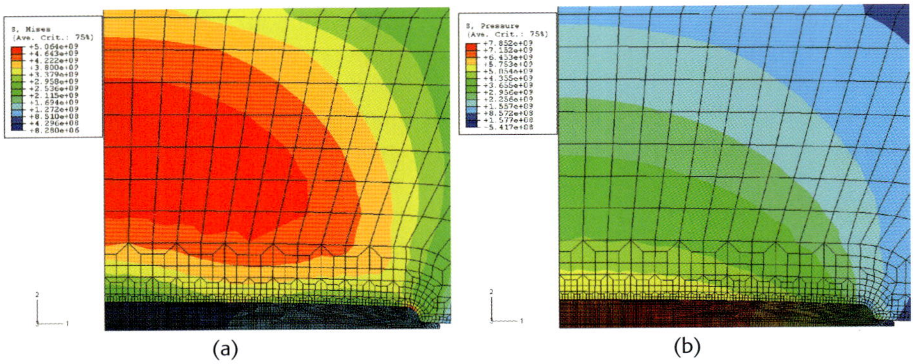

Figure 4.11 Finite element simulation of a loaded tool showing: (a) von Mises stress; (b) hydrostatic stress distribution. For reasons of symmetry, only one-quarter of configuration of anvil was modelled

The value of the pressure for the following torsional straining must be set between the described lower and upper bounds. Experience has shown that, especially for medium- and high-strength materials, the highest possible pressure is desirable. This is illustrated in Figure 4.12(a), where the *in situ* measured torque for an HPT-processed steel is plotted for different nominal

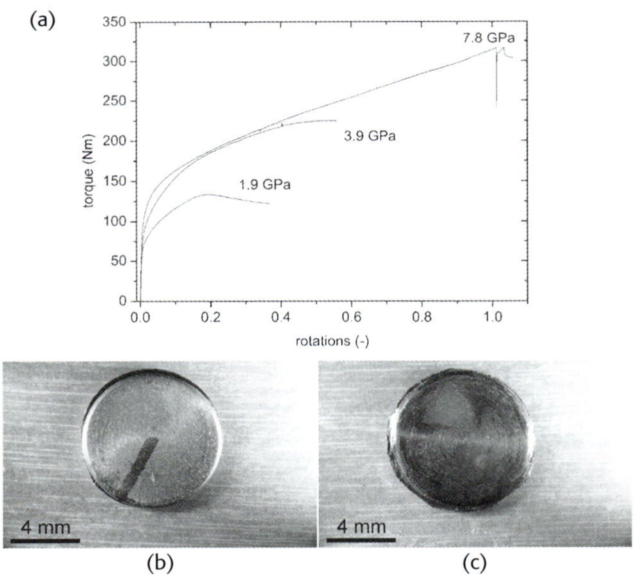

Figure 4.12 (a) Examples of torque records showing positive effect of increased pressure on maximum deformability of a steel sample. (b) Example of a specimen with a conserved marking after a HPT experiment. (c) A specimen that experienced slippage during experiment resulting in a loss of marking

pressures. Details regarding important technical aspects of the *in situ* torque measurement will be given in Section 4.4.3. Here the measured torque is used as an indicator of the maximum possible degree of deformation which can be achieved in a steel specimen. An ongoing increase in the torque as the strain increases is typical for this material. Saturation, or even a decrease in the torque, is a clear sign that the material is no longer deforming. The remaining torque stems from the relative sliding between the rotating anvil and the specimen. This can easily be proved by marks (e.g. lines), which are put onto the specimen before deformation. As long as the material is deforming without slippage, the position of the line cannot change and will be clearly visible on both sides of the specimen after the experiment (Figure 4.12(b)). Whenever a pressure-dependent maximum degree of deformation is reached, the specimen begins to slide relative to the rotating anvil and the strain will not increase in the specimen. In this case the marking on the specimen vanishes, meaning that the perfect sticking contact between the micro-roughened anvil and the sample has been lost, as can be seen in Figure 4.12(c). The specimen has a shiny, polished surface due to the relative sliding against the anvil. The incidence of sliding is of great significance and should therefore be checked, especially when new or unknown materials are being investigated. Otherwise it can represent a great source of error for any kind of post-HPT examination since no direct relation between applied strain (e.g. number of rotations) and investigated property can be drawn.

As the example in Figure 4.12(a) shows, a marked improvement of the maximum strain with increasing pressure can be reached by increasing the pressure from 1.9 GPa to 3.9 GPa. Nevertheless, even at the highest applied pressure, 7.8 GPa, the strain is limited. The graph shows that there was no levelling out of the torque, rather there was an instant drop. Again, the examination of such a specimen shows a shiny surface, indicating relative sliding between specimen and anvil but slight plastic deformations of the anvils could also be found. Compare this with Section 4.4.5 on tool failures. The observed plastic deformation of the anvils typically sets in when the deformation-induced hardening of the HPT-processed material reaches the hardness of the anvils. In this case the anvils had a hardness of 60–62 HRC (Rockwell hardness), which was also reached in the sample at the maximum radius area after approximately one rotation. Due to the multi-axial stress state in the HPT tool, samples can reach slightly higher hardness values than that of the tool. Despite that, as a rule of thumb, it can be concluded that an upper limit for HPT-induced grain refinement and hence strength enhancement is given by the resulting strength of the processed material, which should not exceed the yield strength of the anvil's material. This means that grain refinement in metals or alloys is technically limited by the strength of the anvils rather than being limited by physical factors which determine the saturation grain size. In this case the saturation grain size could be even smaller, however, due to the strength restriction of the anvils, it might not be accessible.

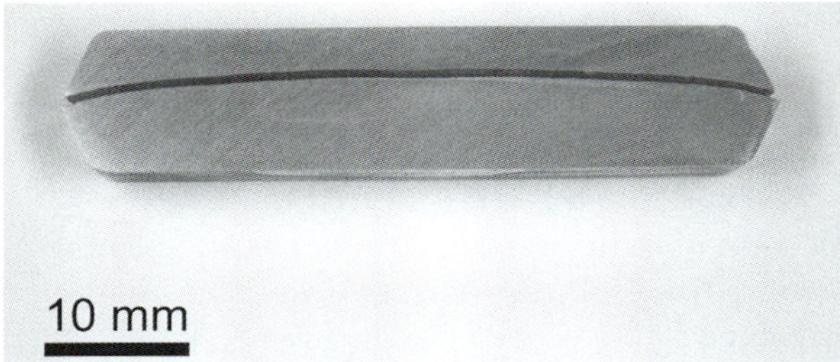

Figure 4.13 Crack formation in a HPT specimen due to shear localisation using insufficient axial pressure

The pressure increase not only prevents the specimen from sliding but can also suppress crack formation. Common sense would suggest that cracks should not occur during shear deformation of a sample under pressure. Unfortunately, that is not always the case, as the cross-section of an Al-alloy HPT disc in Figure 4.13 illustrates. It was found that during deformation strong shear localisation occurred in the middle of the disc. Due to lower compressive stresses at the outer edge of the disc, a crack initiated and propagated under Mode III loading (i.e. the crack grew in the shape of a spiral, which became longer as the number of rotations increased). After a certain number of rotations the specimen was separated into two halves, although a compressive stress component in the range of several GPa was still present. This material-dependent tendency to localised deformation and crack formation can be suppressed by increasing the pressure.

4.4.3 Applied torque

As well as an appropriate amount of pressure, an adequate motor with sufficient performance and maximum driving torque is required. From a technical perspective, this seems to be more difficult to realise than the appropriate pressure when specimen sizes are increased.

For instance, when the diameter of the sample is doubled, the load capacity of the load frame has to be quadruplicated as the area increases quadratically in order to obtain the same pressure. However, a higher torque is simultaneously required for the same material. The torque scales with the cube of the diameter. Hence doubling of the size of the specimen results in the required torque being eight times larger.

In classical torsion tests, the torque can be used to evaluate the shear stress τ for a sample with a certain radius r

$$\tau(r) = \frac{3M}{2\pi r^3} \tag{4.3}$$

However, in HPT the required torque is also strongly dependent on the geometry of the anvil. An anvil consists not only of the cavity into which the specimen

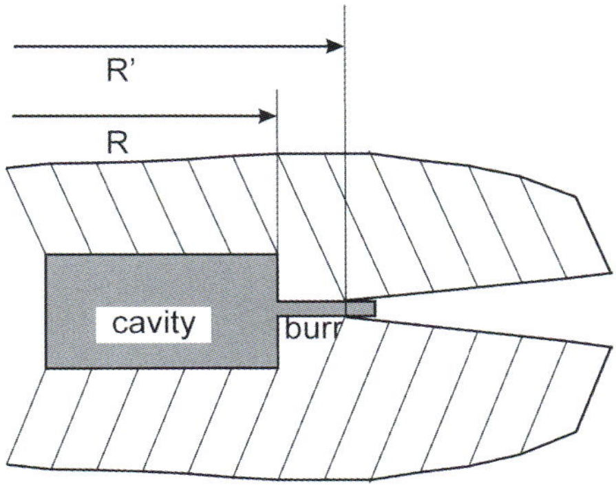

Figure 4.14 Schematic of geometry of anvil cavity and burr area with corresponding radii. R is radius of HPT disc and R' is radius including burr

is put, but it also needs to have an adequate burr zone, which is more or less a flat ring area around the cavity in which the specimen is placed. In the burr zone, the excess material can squeeze out between the rotating anvils (see Figure 4.14).

Therefore, the external measured torque M has two contributions. The torque M_{sample}, required to deform the sample and the torque M_{burr}, stemming from the burr area

$$M_{total} = M_{sample} + M_{burr} \tag{4.4}$$

Excluding the hardening behaviour of the deforming material and only focusing on the saturation regime, where $t(r)$ becomes independent from the applied strain, M_{sample} can be expressed as

$$M_{sample} = 2\pi \int_0^R \tau(r, \dot{\varepsilon}, \varepsilon, T) r^2 dr \tag{4.5}$$

Where $\tau(r)$ would depend mainly on the deformation temperature T, the radius r and in many materials also on the strain rate $\dot{\varepsilon}$. The additional torque contribution stemming from the friction in the burr zone can be written as

$$M_{burr} = 2\pi \int_R^{R'} p\mu r^2 dr \tag{4.6}$$

This assumes a relative sliding between the burr and the anvil (Vorhauer and Pippan, 2008). As can be seen, M_{burr} depends strongly on the pressure distribution $p(r)$ in the burr zone and also on the friction coefficient μ. The determination of $p(r)$ and μ is very demanding, either analytically or through modelling. In contrast, using the measured external torque for a shear flow stress evaluation by Equation (4.3) would strongly overestimate the real value. A straightforward method to measure the geometry-independent torque required solely to

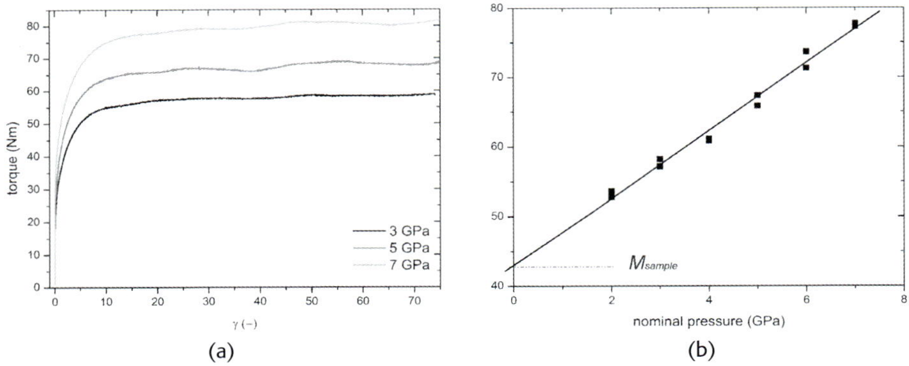

Figure 4.15 (a) Plot of torque of pure Cu discs with a diameter of 8 mm against shear angle γ as a function of different nominal pressures. Waviness of the torque is a consequence of slight imperfections in setup. (b) Linear regression of saturation torque for different applied pressures. At a virtual nominal pressure of 0 GPA torque needed to deform sample M_{sample} can be taken

deform the sample is to measure the saturation torque at different applied nominal pressures, as shown in Figure 4.15 on Cu for different pressures (Hafok, 2007). These data points can be extrapolated to a pressure equal to zero (Figure 4.15(b)) so that the corrected torque for the saturation regime M_{sample} can be found. As Figure 4.15(b) shows, the applied pressure exerts a very pronounced influence on the measured torque, which might be a reason for later misinterpretation of results. The procedure presumes that the anvil geometry will remain the same in subsequent experiments. This means that the same pair of anvils should be used in order to exclude any geometrical changes, which is difficult to achieve whenever there is massive wear in the burr zone.

As presented, torque corrections are complicated and, when different alloys are compared, very labour-intensive. Additionally, if saturation does not occur in the observed deformation range, the proposed procedure is not applicable. As a consequence, the torque curves are frequently not corrected and so should only be used as a qualitative measure to compare different alloys in terms of their hardening behaviour, or to study the influence of temperature or strain rate. Such experiments can still be very informative but it should be noted that for these experiments the chosen pressure, and so M_{burr}, must remain the same.

4.4.4 Anvil design and material selection

A typical design of an anvil for small HPT samples is shown in Figure 4.16. As is usual for quasi-constrained HPT, the anvils have a small cavity in the middle in which the specimen is placed. Additionally, there is a slight corner arc at the bottom of the cavity. It is helpful to reduce the stress in this area, which can become very large as corners act as stress risers (see also Figure 4.11(a) which shows a high von Mises stress at the bottom corner of the cavity).

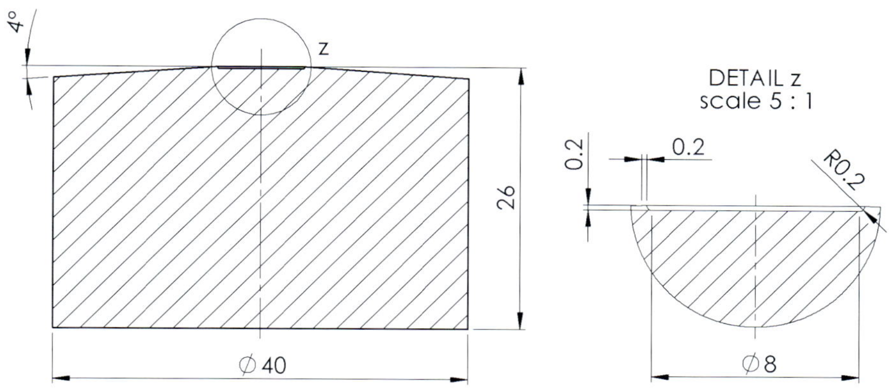

Figure 4.16 Dimensions of an HPT anvil

Some aspects of the selection of the materials for anvils have already been discussed in Section 4.4.2. The anvil material needs to sustain extremely high compressive stresses during the experiment and it also requires an appropriate shear strength for the torsion experiment. The shear stress at the outer diameter of the sample must be lower than the shear strength of the anvil material. Therefore, as discussed for the required torque, the diameter of the anvils has to be increased when the size of the sample size increases. Additionally, properties such as ductility and fracture toughness are essential for choosing the anvils. So far, the best results for producing HPT anvils have been obtained with the following classes of material.

4.4.4.1 High-speed and hot-work tool steels

High-speed tool steels produced via the powder metallurgy processing route are an excellent choice. They are also used in industry for similar applications, for instance for dies or punches. The advantage of steels produced by powder metallurgy over cast ones is a more homogeneous microstructure. An excellent strength in the range 60–65 HRC with considerable toughness and ductility can be obtained. The heat treatment to achieve the desired hardness consists of hardening and tempering and can be performed in simple laboratory furnaces. Since the material still remains machinable, the final contour of the anvils can be introduced after heat treatment. Even more importantly, the anvils can be re-machined after use when the contour has worn off. Another advantage of this class of materials is their thermal stability up to 400–500°C. For experiments performed at elevated temperatures, it is necessary to ensure that the material does not soften excessively, which could lead to macroscopic plastic deformation of the anvils. Hot-work tool steels are also suitable for experiments at elevated temperatures. Generally, these have a lower maximum obtainable hardness, but higher toughness and better temperature stability.

4.4.4.2 Cold-work tool steels

Cold-work tool steels are attractive due to their higher toughness and a similar strength capacity compared to high-speed tool steels. Nevertheless, they can only be used for HPT applications at room temperature as their strength reduces considerably at elevated temperatures.

4.4.4.3 Hard metals

WC–Co hard metals allow the use of higher hydrostatic pressures and higher deformation temperatures than can be used with steels. A huge drawback is their restricted machinability so that turning operations cannot be performed on anvils. The cavity must be shaped using other methods, for example, by arc erosion. Due to the low intrinsic fracture toughness the damage tolerance is very low and the anvils lack a long life. Additionally the price of the material is considerably higher than that of steels.

4.4.5 Tool failures

4.4.5.1 Slipping of a sample

Slipping of a sample leads to wear and shearing of the sandblasted surface in the anvil's cavity. In Figure 4. 17(a) a shiny, worn off surface in the anvil's cavity is visible, which can be compared to Figure 4.17(b) which shows a new sandblasted cavity. In the case of slipping one cannot be sure about the applied strain. The number of externally applied rotations does not necessarily correspond to the number of rotations experienced by the specimen due to a relative movement

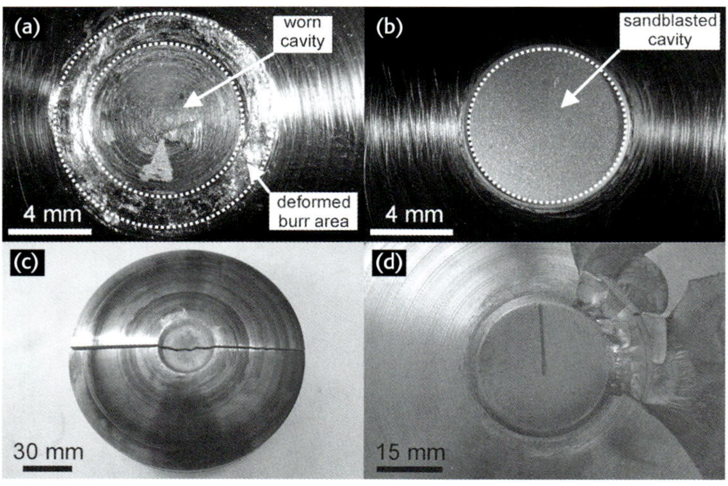

Figure 4.17 Examples of tool damage. Comparison between an improperly used anvil: (a) showing characteristics of disc slipping with a shiny surface and plastic deformation in burr area; (b) an unused sandblasted one; (c) radial crack cleaving an anvil; (d) spalling of edge pieces

of the specimen and anvils. Slipping may occur when the pressure is too low but also during plastic deformation of the anvils as will now be explained. Proper sandblasting of the anvils or any other micro-roughening procedure is normally sufficient to allow the anvils to be re-used, as long as no plastic deformation has occurred.

4.4.5.2 Plastic deformation of the anvil

Materials showing ongoing hardening during HPT deformation can lead to a stress comparable to, or higher than, the strength of the material used for the anvils. In such cases the burr area of the anvil, which is a rather thin ring, begins to deform plastically (see Figure 4.17(a)). A strong misalignment of the anvil axes may also lead to excessive contact between the anvils and thus to plastic deformation in the burr area. The contact between anvils in this area can also cause a substantial decrease in pressure and lead to the sample slipping. Plastically deformed anvils should be re-machined before further use.

4.4.5.3 Radial cracks

The most severe type of failure is a total disintegration of the tool. This can occur by the propagation of radial cracks (Figure 4.17(c)). An example of such failure has been observed during experiments at ambient temperatures with anvils that had previously been used for experiments at elevated temperatures. In this case an anvil can fail, even during the preliminary loading of the specimen. A simple explanation is based on the introduction of internal stresses. Even high-speed tool steels lose a certain amount of strength at elevated temperatures compared to their strength at room temperature. The volume beneath the cavity can become plastically compressed during experiments at elevated temperatures. The plastically compressed volume in the axial direction experiences a radial expansion and so residual tangential stresses are introduced. In a consecutive experiment at room temperature, the loading stress is superimposed onto the residual stresses, which may reach the fracture stress leading to cracking in the radial direction.

4.4.5.4 Spalling of single pieces

An example of this is shown in Figure 4.17(d), where individual pieces of material broke away from the edge of the anvil's cavity. This type of failure is frequently attributed to small pre-existing cracks at the edge of the cavity. These cracks are assumed to be induced during the heat treatment or the machining of the cavity.

4.4.6 Attachments for HPT experiments

4.4.6.1 Torque cell

As already discussed, measuring the torque during HPT experiments enables a direct comparison of the processing of different materials and can convey helpful information about the deformation behaviour for the very high strain regime.

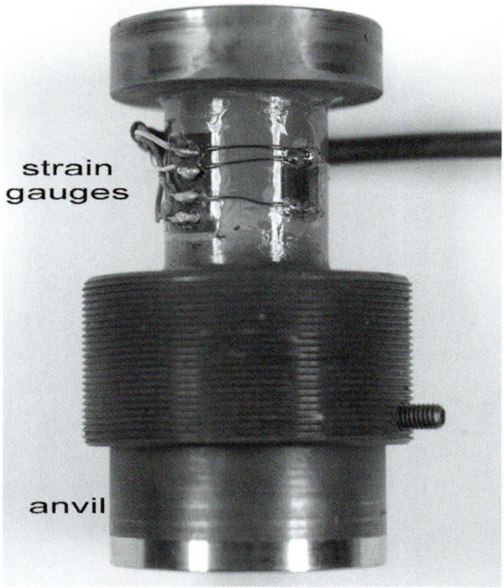

Figure 4.18 Example of a torque cell

Additionally, it can be a useful technical indicator of unwanted slippage of the specimen during the experiment. Although it is also possible to measure the driving torque of the engine, it is always preferred to measure the torque very close to the specimen, for instance on the anvil. In this way the contribution of the friction, for example arising from the gear transmission, can be excluded. A practical method for such measurements is to design a torque cell, in which an anvil can be inserted and so easily replaced (Figure 4.18). The torque cell is in the form of a shaft between the anvil and the non-rotating holding fixture, to which strain gauges are attached. Due to the mechanical equilibrium, the total torque necessary to deform the specimen equals the torque measured on this shaft with strain gauges. In order to amplify the signal mechanically, the shaft can have a lower diameter than the anvil as long as the strength is adequate.

4.4.6.2 Heating unit

An induction heating system is preferable for experiments at elevated temperatures (Figure 4.19(a)). As Figure 4.19 shows, an induction coil is placed around the upper anvil before the tool is closed. With this system, the deformation temperature can be increased up to approximately 700°C. Induction heating systems would normally have sufficient power to reach higher temperatures. However, the upper temperature limit is determined by the material used for the anvils. As can be seen in Figure 4.19(a), a pyrometer is used to control the temperature. Alternatively, a thermocouple could be welded onto the non-rotating anvil in order to adjust and control the desired deformation temperature.

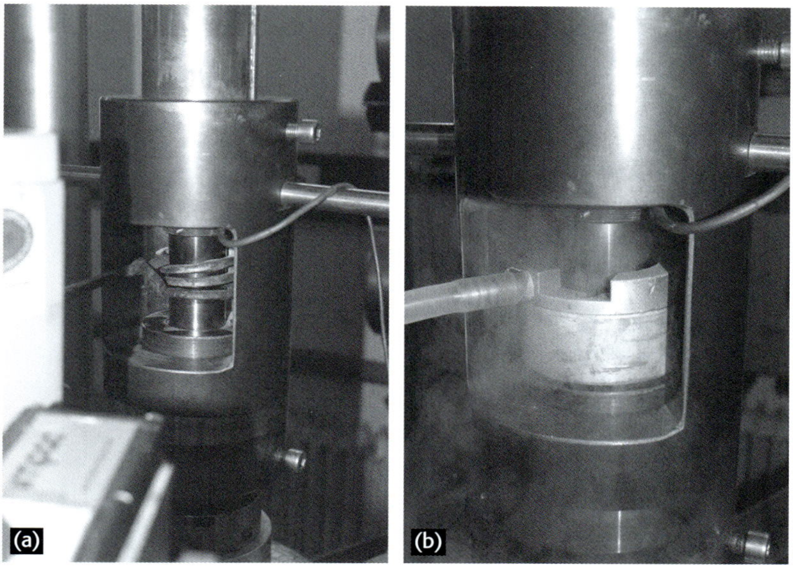

Figure 4.19 Experimental setups for experiments at: (a) elevated temperatures; (b) for low-temperature experiments

4.4.6.3 Cooling unit

A solution for performing experiments at temperatures below room temperature is presented in Figure 4.19(b). A small cup is attached to the lower anvil that can be filled with a cooling liquid such as liquid nitrogen or other cooling liquid mixtures if desired. The liquid cools the anvil and the specimen. The nominal temperature is measured by a thermocouple attached to the non-rotating anvil.

At both temperature ranges it is possible to measure the *in situ* torque during deformation. Cooling of the torque cell, for instance with pressurised air, is needed for the high-temperature regime in order to protect the strain gauges.

4.4.7 Upscaling of HPT

Technical design criteria which need to be considered when building an HPT rig for larger HPT specimens, have already been discussed in Section 4.4.3. Such considerations led our group to construct an HPT rig with a maximum nominal loading capacity of 4000 kN (~400 ton). The maximum torque provided by an electric motor, integrated with a strong transmission unit, was 13 kNm (see Figure 4.20(a)). Through this upscaling, it was possible to increase the sample diameter from 8–14 mm up to diameters in the range 30–50 mm (Figure 4. 20(b)), while keeping the pressure in the range of several GPa. Due to the much higher torques that were required, the diameter of the anvils had to be increased from 40 mm to 150 mm. Another aspect of upscaling that had to be kept in mind for enlarged specimens was the chosen rotational speed because excessive heating

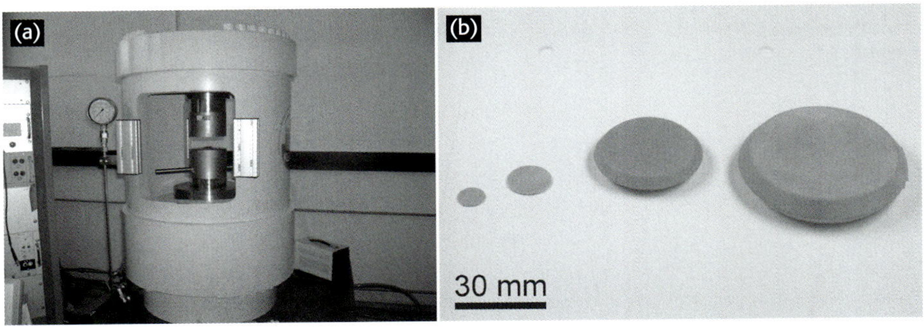

Figure 4.20 Outcome of upscaling HPT in Leoben, Austria. (a) Hydraulic HPT rig with a maximum loading capacity of 4000 kN: (b) evolution of specimen size from 8 mm to 50 mm within approximately one decade

of the specimen during the experiment, which is more severe for large rotational speeds and hard materials, can influence the results (Edalati *et al.*, 2011). It can originate from two sources: first, self-heating of the specimen produced by the intense plastic deformation; and second, heat produced by friction in the burr zone. Both contributions can be minimised if the speed is reduced and the heat that is produced can dissipate into the cooler environment of the anvils. A typical speed used for smaller specimens (8–14 mm) is ~0.2 rotations per minute. This is reduced to 0.07 rotations per minute for the larger specimens (30–50 mm). So the increase in temperature measured very close to the cavity of the anvil is usually lower than 30°C.

When test specimens are extracted from HPT discs for mechanical tests, or perhaps in future applications for industrial needs, a homogeneous microstructure and the associated hardness are demanded. Therefore, homogeneity is also a significant issue for upscaling. As was shown in Section 4.2, the radial homogeneity, along the radius of an HPT disc, is possible for a large variety of materials, provided an adequate number of rotations is applied. This leads to the smallest achievable grain size for even very small radii. However, increasing the diameter of the sample usually means increasing its thickness as well. As a result, the axial homogeneity of the HPT discs becomes important and must be considered. An excessive increase in the thickness of the sample can lead to quite inhomogenous microstructures (Sakai *et al.*, 2005) due to shear deformation concentrated in the middle plane of the specimens.

The problem of the axial homogeneity of HPT discs was first raised in a systematic study performed on Fe samples (Hohenwarter *et al.*, 2009). In this work, specimens with different aspect ratios t/d were deformed to comparable strains but far beyond saturation. Cross-sections through the centre of these specimens were then used to measure the hardness distribution along the radial direction in the middle plane, and also along the AD at certain normalised radii (see insets in Figure 4.21).

Focusing first on the two iron specimens in Figure 4.21(a), almost the same hardness was reached after deformation of the large portion of the HPT disc except its very centre. Comparing the Fe specimens, the radial homogeneity is not significantly influenced by the size of the specimen, as long as similar deformation strains are applied. Additional results for pure Cu show much lower saturation hardness. Comparing the Cu specimen with the Fe specimen of the same aspect ratio ($t/d = 1/6$) and the same diameter of 30 mm, no detectable decrease of hardness in the middle of the Cu specimen is visible. This is simply the result of applying a higher number of rotations. The situation is quite different for the axial homogeneity of the same two Fe specimens in Figure 4.21(b, c), where the hardness across the thickness for certain normalised radii was measured (0.2R is close to the centre, 0.5R is in the middle and 0.9R almost on the edge). It can clearly be seen that the degree of homogeneity strongly

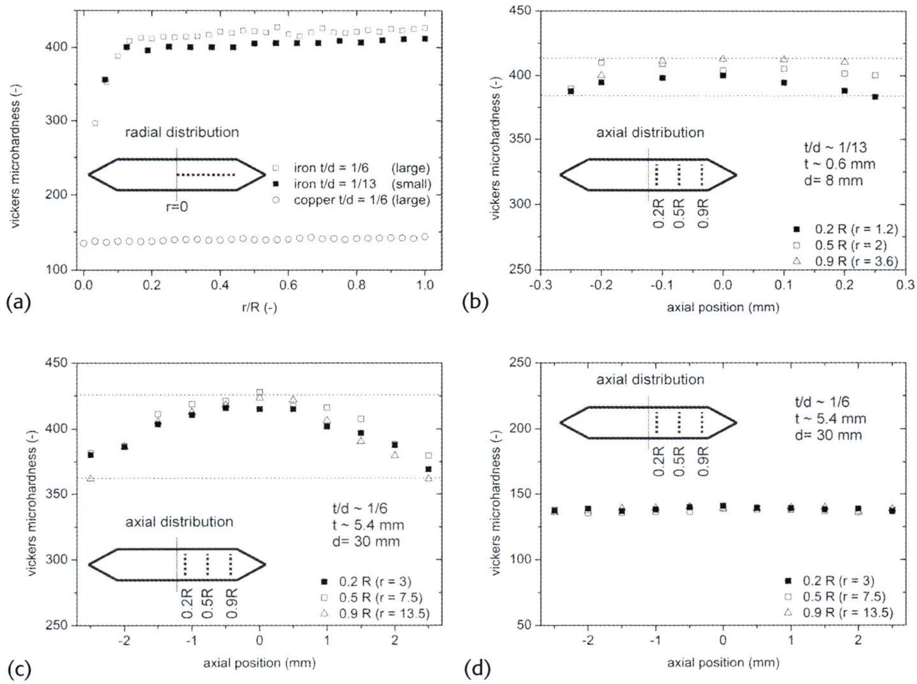

Figure 4.21 Hardness investigations on HPT-processed Fe and Cu discs along radial and axial directions taken at certain radii. (a) Hardness distribution measured in middle plane of specimens along normalised radius. For Fe, measurements for two different aspect ratios, thickness to diameter t/d are shown. (b) Axial hardness distribution of small specimen, $t/d \sim 1/13$ with considerable homogeneity. (c) Axial hardness distribution of large specimen, $t/d \sim 1/6$ with less pronounced homogeneity. (d) Axial hardness distribution of a large Cu specimen with same aspect ratio ($t/d \sim 1/6$) but also exceptional homogeneity in axial direction

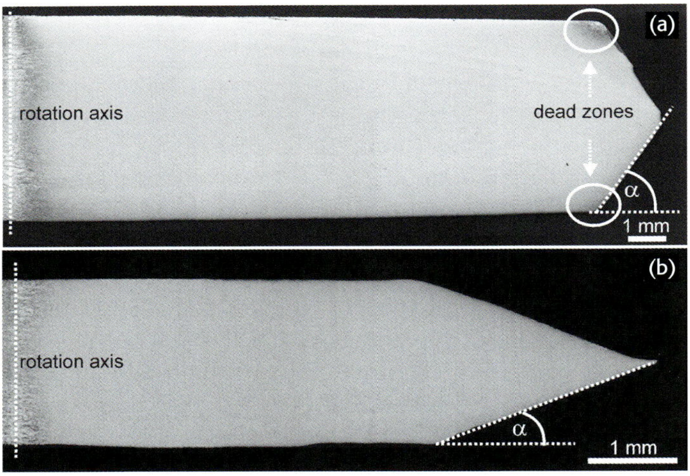

Figure 4.22 Comparison of cross-sections of HPT discs with different cavity angles α. (a) Specimen with a diameter of 30 mm and a thickness of ~5.4 mm (t/d~1/6) showing low deformed regions in area of rotation axis on left (dotted line) and on right at top and bottom edges, known as dead zones. (b) Specimen with a diameter of ~9 mm and a thickness of ~ 1.9 mm (regarding area of constant thickness) and a t/d of 1/5. Again a region of low deformation is seen near rotation axis, however, there are no dead zones

depends on the aspect ratio. For a ratio of ~1/13 (Figure 4.21(b)) the difference in maximum and minimum hardness is distinctly lower than for the larger specimen with an aspect ratio of 1/6. This shows that one should stay below a certain ratio of thickness t to diameter d when samples are enlarged in order to keep a reasonable degree of axial homogeneity. This inhomogeneity is attributed to the occurrence of dead zones, which are formed at the maximum radius at the edges of the specimen (see Figure 4.22(a)) and a more detailed view in Figure 4.23). These regions experience very little deformation with almost unaffected initial microstructure, which is similar to the area close to the rotational axis (shown on the left-hand side of Figure 4.22(a)). These regions can often be found when an anvil design typical for quasi-constrained HPT is used. The flat areas of the cavity, as mentioned in Section 4.4.2, should have a high friction coefficient in order to ensure that the sample cannot slip in this region. The friction coefficient should remain low on the inclined edges of the cavity and in the burr area so that the material can slip. However, the transition between these extremes is continuous and not abrupt. The material at the edges is constrained in its deformation, which consequently leads to low deformed regions. Such dead zones constrain the deformation not only at the edges of the disc but throughout the specimen, along the bottom and top faces of the disc towards its centre. This results in a decrease of hardness from the middle plane of the disc towards the faces, as shown for Fe in Figure 4.21(b), for small,

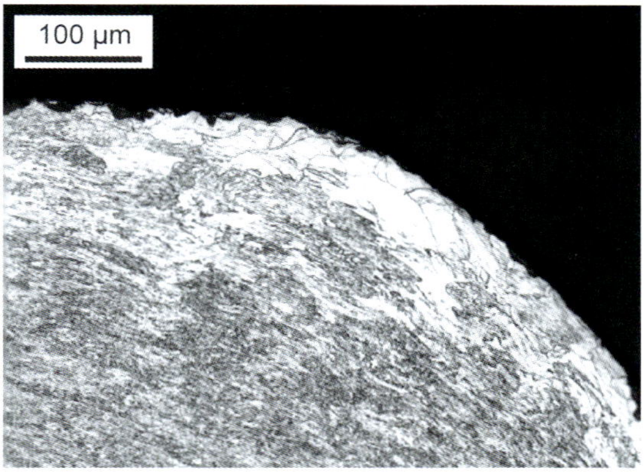

Figure 4.23 Representation of a region of low deformation, known as a dead zone, at one edge of specimen. An undeformed coarse-grained area is visible

intermediate and high radii. The size of the dead zones at the edges controls the degree of inhomogeneity in the AD.

The same axial measurements were also performed on a high purity Cu having the same specimen and anvil design (Figure 4.21(d)), which surprisingly exhibits an almost perfectly homogenous hardness distribution even in the AD. In this case it was found that the pre-deformation of the recrystallised specimen already increases the hardness from about 60 HV to 110 HV. The subsequent shear straining only increases the hardness from 110 HV to ~140 HV. Although the pre-deformation of Fe leads to comparable hardness, the increase in the hardness due to shear deformation is far more pronounced. This shows that the hardening behaviour of the material plays an important role in the axial homogeneity in the saturation regime of deformation. Pronounced strain-rate sensitivity is also considered to have a positive effect on the axial homogeneity. In the case of positive strain-rate sensitivity the volume elements at the outer radii experience a higher flow stress which could prohibit the localisation of deformation.

In the case of large specimens with low axial homogeneity, for example for Fe, the homogeneity can still be markedly improved by changing the cavity in the anvil. The angle α of the inclined edge (shown on the cross-sections in Figure 4.22) plays an important role. It is defined as the angle between the bottom of the cavity and the inclined edge of the cavity and is introduced for processing reasons. When dealing with large samples it is not necessary to have a defined corner arc at the edges of the anvil's cavity, as shown in Figure 4.16. The inclined edges have a length of several millimetres and are better defined with α. The α value should be far below 90° to circumvent problems which may occur while removing samples after the experiment if the angle is too steep. On the other hand, it cannot be 0°, which would correspond to unconstrained HPT. From experience, as shown in Figure 4.22(a), a value of about 50° is used. As already mentioned in this section, the main problem leading to disc inhomogeneity seems to be a constrained material deformation at the edges. It was found that this situation can be considerably improved by lowering α, as shown in Figure 4.22(b). For a small angle α the friction in the inclined area is still sufficient for the volume elements in this area to undergo a similar shear deformation as elements in the area of constant thickness. As a result, a dead zone cannot occur in the area where the thickness is gradually reduced. The hardness distribution in such a specimen, having a similar aspect ratio as the large Fe specimen with $t/d \sim 1/5$ shown in this section, becomes very homogenous (Hohenwarter *et al.*, 2009) so that no dead zones can be found, as shown in Figure 4.22(b). A slight drawback of such a sample is that a large part of the disc does not have a constant thickness.

4.5 Outlook

As Section 4.4 has shown, there seems to be a realistic possibility of upscaling the HPT process, which could also be of industrial interest. This is especially true when functional materials with tailored properties are required (Zehetbauer *et al.*, 2010). In contrast to structural materials, material quantities can be smaller and so could be suitable for applications in fields such as: electronics, micro-mechanics or medicine. Further upscaling of discs in the radial direction is probably impractical due to the large axial forces that are needed. It is the present authors' opinion that an even more severe problem associated with increasing the specimen diameter is that the required torque will become increasingly more difficult to provide with the conventional motors and transmissions normally available on the market. Also, with increasing diameter, a cooling of the tool would be necessary in order to allow sufficient processing efficiency. Otherwise, excessive heating could become counter-productive to grain refinement. A simple method of overcoming these problems could be based on using ring specimens (Harai *et al.*, 2008). However, such specimens may experience inadequate axial homogeneity.

Despite attempts at upscaling, HPT will, in this form, remain a batch process. This strongly restricts the economy of the process. Motivated by this, continuous processes based on traditional HPT become increasingly interesting. The most promising attempts for mass productions that have been made are known as continuous HPT (Edalati and Horita, 2010) and torsion extrusion (Ma *et al.*, 2004; Mizunuma, 2006). So far these processes have only been applied to rather soft materials and again, the lack of homogeneity could be a problem. Another, possible technique is known as incremental HPT (Hohenwarter, 2015). It is very similar to torsion extrusion, however, the deformation zone will change in a stepwise manner rather than continuously. Within one step the material becomes HPT deformed and then the processed zone is shifted in the radial direction to deform the adjacent volume of the material. This method can process materials which are hard to deform and, through changing the deformation zone, large quantities of material can be deformed. An example of successfully deformed bulk Cu sample is shown in Figure 4.24.

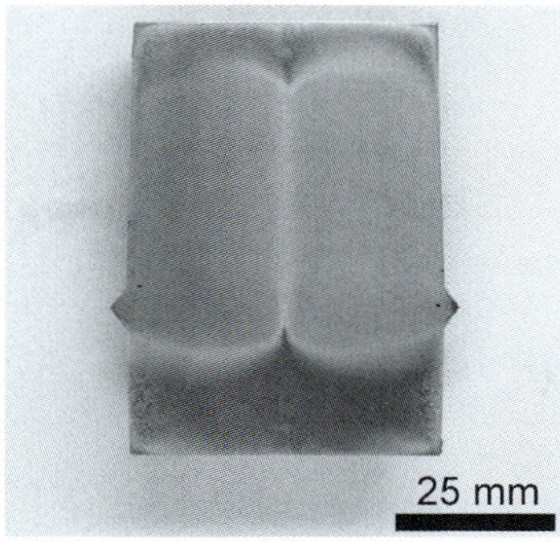

Figure 4.24 Etched cross-section of a Cu specimen deformed by HPT. Difference in brightness as a result of etching, reveals deformed zone

References

Bowden, F. B. and Tabor. D. (1954) *Friction and Lubrication of Solids*, Clarendon Press, Oxford, UK.

Bridgman, P. W. (1935) 'Effects of high shearing stress combined with high hydrostatic pressure', *Physical Review*, 48: 825–847.

Bridgman, P. W. (1943) 'On torsion combined with compression', *Journal of Applied Physics*, 14: 273–283.

Ecker, W. (2004) 'Simulation of high pressure torsion experiment to calibrate data for a railway material', Master's thesis, University of Leoben, Austria.

Edalati, K. and Horita, Z. (2010) 'Continuous high-pressure torsion', *Journal of Materials Science*, 45: 4578–4582.

Edalati, K., Miresmaeili, R., Horita, Z., Kanayama, H. and Pippan, P. (2011) 'Significance of temperature increase in processing by high-pressure torsion', *Materials Science and Engineering A*, 528: 7301–7305.

Hafok, M. (2007) 'Examination of the deformation mechanisms of severely deformed materials', PhD thesis, University of Leoben, Austria.

Harai, Y., Ito, Y. and Horita, Z. (2008) 'High-pressure torsion using ring specimens', *Scripta Materialia*, 58: 469–472.

Hohenwarter, A. (2015) 'Incremental high pressure torsion as a novel severe plastic deformation process: Processing features and application to copper', *Materials Science and Enineering A*, 626: 80–85.

Hohenwarter, A., Bachmaier, A., Gludovatz, B., Scheriau, S. and Pippan, R. (2009) 'Technical parameters affecting grain refinement by high pressure torsion', *International Journal of Materials Research*, 100: 1653–1661.

Hohenwarter, A. and Pippan, R. (2010) 'Anisotropic fracture behavior of ultrafine-grained iron', *Materials Science and Engineering A*, 527: 2649–2656.

Kawasaki, M. and Langdon, T. G. (2008) 'The significance of strain reversals during processing by high-pressure torsion', *Materials Science and Enineering A*, 498: 341–348.

Langdon, T. G. (2009) 'Research on bulk nanostructured materials in Ufa, Twenty years of scientific achievements', *Materials Science and Engineering A*, 503: 141–144.

Ma, X., Barnett, M. R. and Kim, Y. H. (2004) 'Forward extrusion through steadily rotating conical dies: Part 1: Experiments', *International Journal of Mechanical Sciences*, 46: 449–464.

Mizunuma, S. (2006) 'Large straining behavior and microstructure refinement of several metals by torsion extrusion process', *Materials Science Forum*, 503/4: 185–190.

Pippan, R., Scheriau, S., Hohenwarter, A. and Hafok, M. (2008) 'Advantages and limitations of HTP: A review', *Materials Science Forum*, 584/6: 16–21.

Pippan, R., Scheriau, S., Taylor, A., Hafok, M., Hohenwarter, A. and Bachmaier, A. (2010) 'Saturation of fragmentation during severe plastic deformation', *Annual Review of Materials Research*, 40: 319–343.

Sakai, G., Nakamura, K., Horita, Z. and Langdon, T. G. (2005) 'Developing high-pressure torsion for use with bulk samples', *Materials Science and Engineering A*, 406: 268–273.

Stüwe, H. P. (2003) 'Equivalent strains in severe plastic deformation', *Advanced Engineering Materials*, 5: 291–295.

Valiev, R. Z., Islamgaliev, R. K. and Alexandrov, I. V. (2000) 'Bulk nanostructured materials from severe plastic deformation', *Progress in Materials Science*, 45: 103–189.

Valiev, R. Z., Xia, K. and Langdon, T. G. (2009) 'Processing by severe plastic deformation: an ancient skill adapted for the modern world', *International Journal of Materials Research*, 100: 1623–1631.

Vorhauer, A. and Pippan, R. (2008) 'On the onset of a steady state in body-centered deformation at low homologous temperatures', *Metallurgical and Materials Transactions A*, 39: 417–429.

Wetscher, F. and Pippan, R. (2006) 'Cyclic high-pressure torsion of nickel and Armco iron', *Philosophical Magazine*, 86: 5867–5883.

Zehetbauer, M., Grössinger, R., Krenn, H., Krystian, M., Pippan, R., Rogl, P., Waitz, T. and Würschum, R. (2010) 'Bulk nanostructured functional materials by severe plastic deformation', *Advanced Engineering Materials*, 12: 692–700.

Zehetbauer, M. J., Stüwe, H. P., Vorhauer, A., Schafler, E. and Kohout, J. (2003) 'The role of hydrostatic pressure in severe plastic deformation', *Advanced Engineering Materials*, 5: 330–337.

Zhilyaev, A. P. and Langdon, T. G. (2008) 'Using high-pressure torsion for metal-processing: fundamentals and applications', *Progress in Materials Science*, 53: 893–979.

Zhu, Y. T. and Langdon, T. G. (2004) 'The fundamentals of nanostructured materials processed by severe plastic deformation', *JOM Journal of the Minerals, Metals and Materials Society*, 56: 58–63.

Cyclic extrusion–compression (CEC)

CHAPTER 5

Jan Richert

5.1 Initial CEC device for unlimited deformation

The cyclic extrusion–compression (CEC) method, developed and patented in Poland in 1979 (Richert *et al.*, 1979; Richert and Richert, 1986), is widely used in investigations of the microstructure and evolution of mechanical properties under severe plastic deformation (SPD) conditions. A characteristic feature of SPD is the possibility of preserving the primary billet geometry at all stages of deformation. Another very important characteristic of CEC is the possibility of exerting high hydrostatic pressures, which enable a very large plastic deformation to be obtained without loss of cohesion in the processed material (Lapovok, 2006; Agnew *et al.*, 2005; Semiatin and DeLo, 2000; Bowen *et al.*, 2000: Rosochowski, 2005).

The first device used for unlimited deformation of metals and alloys by the CEC method is shown in Figure 5.1. The equipment was adapted from a

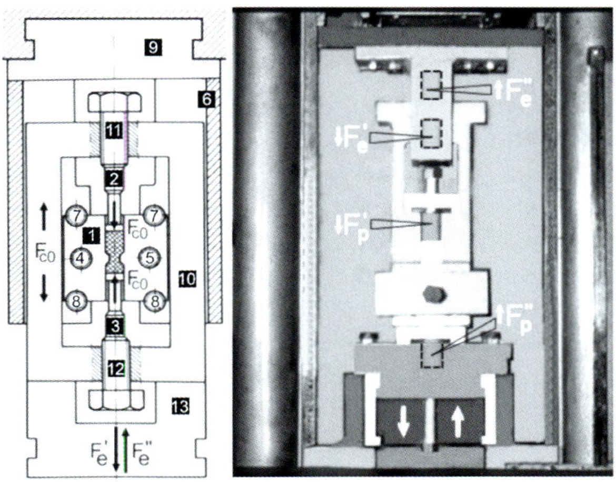

(a) (b)

Figure 5.1 First CEC device for unlimited deformation of metals and alloys: (a) original device construction; (b) modified device working on universal testing machine (dotted line indicates position of force sensors), F_{c0} = pre-compression force, F_e', F_e'' = CEC process forces, F_p', F_p'' = punch forces universal testing machine, with a load capacity of 100 kN. The principle of CEC results from a combination of extrusion and compression processes. The basic part of device, in the laboratory version, is a split die (1) consisting of upper and lower chamber of equal diameters d_0. A narrow channel of diameter d_m connects the two chambers. The two halves of the split die (1) are joined by the screws (4) and (5) and are fixed to the frame (6) by bolts (7) and (8). The frame (6) is connected via the head (9) to the stationary beam of the testing machine. The inner frame (10) with the thrust screws (11) and (12) is connected via the head (13) to the crosshead of the machine. Deformation proceeds by a cyclic flow of the metal from one chamber to the other. For example in a particular cycle, the sample is extruded from the upper to the lower chamber by the upper punch (2). Here, under the action of the lower punch (3), compression occurs simultaneously with the extrusion, so that the sample is restored to its initial size. Before the deformation process starts, the sample is subjected to pre-compression by the thrust screws (11) and (12) with a force F_{c0}.

In this way, the material is deformed cyclically in the presence of a hydrostatic pressure $p_h = -\sigma_m$ ($\sigma_m < 0$ is a mean stress), permitting arbitrarily high deformations without the development of cracks. The magnitude of the cumulated true strain is approximately

$$\varepsilon = 2\, n_c \ln (S_0/S_m) = 4\, n_c \ln (d_0/d_m) \tag{5.1}$$

Where n_c is the number of CEC cycles, S_0 is the cross-sectional area of the chamber orifice, S_m is the cross-sectional area of the die orifice, d_0 is the diameter of the chamber orifice, and d_m is the diameter of the die orifice.

5.2 Microstructural evolution during CEC

Due to hydrostatic pressure, extremely large deformations were achieved, up to $\varepsilon = 90$ of the true strain in aluminium (Korbel *et al.*, 1981). Numerous shear bands and micro-shear bands were found in the microstructure of the aluminium alloys deformed by the CEC method (Korbel and Richert, 1985; Richert and Korbel, 1995; Richert *et al.*, 1998a,b; Richert *et al.*, 2001). Intensive shearing, which occurs in these processes, creates a characteristic chessboard-like microstructure, which can be observed even through a light microscope, particularly in the case of a coarse-grained billet of pure aluminium (Figures 5.2(a,b)). The chessboard-like microstructure is typical for the saturation state in all grades of aluminium, whatever the purity (Richert and Richert, 1986; Richert, 1990). A considerable amount of energy is stored in the shear bands. Therefore, recrystallisation is observed in high purity aluminium, even at room temperature. New grains are usually formed on the boundaries of old grains, especially in places crossed by shear bands and inside the grains in the areas occupied by shear bands (Figure 5.2(b)) (Richert, 1990). Investigations have shown that the development of shear bands is very important in the formation of the nanograin microstructure (Kurzydłowski *et al.*, 2004). Studies of strain localisation

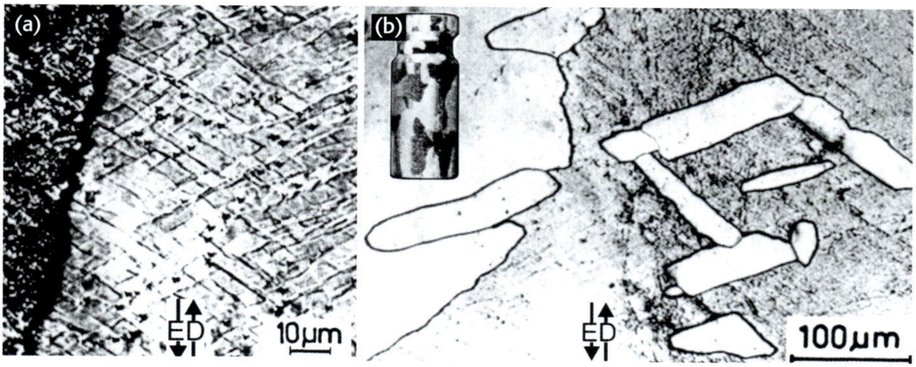

Figure 5.2 Microstructure of Al99.99 after CEC processing at room temperature: (a) chessboard-like microstructure after true strain of $\varepsilon = 3.6$; (b) colony of new grains after true strain of $\varepsilon = 4.5$. ED is extrusion direction

and the development of the shear band microstructure in the deformed materials help us to understand the nature of grain refinement during SPD.

CEC experiments show that the microstructure evolves from a cell-block structure at lower strains (Figure 5.3(a)) into an equiaxed cell structure at large strains (Figure 5.3(b)). This structural transformation is assisted by localised glide and shear. In the first few cycles, a very fast increase in the number of shear bands transforms the grain structure into very elongated sub-grains. Fragmentation of the sub-grains occurs during the second stage of the evolution of the microstructure. It results from the crossing of shear bands. In effect, the sub-grains take up shapes leading to a chessboard-like microstructure.

The localisation of deformation in shear bands leads to a plateau on the strain-hardening curve (Figure 5.4). The most important feature of these bands is their deep penetration through the structure. The homogeneous microstructure is formed by mutual crossing of long-range shear bands and microbands.

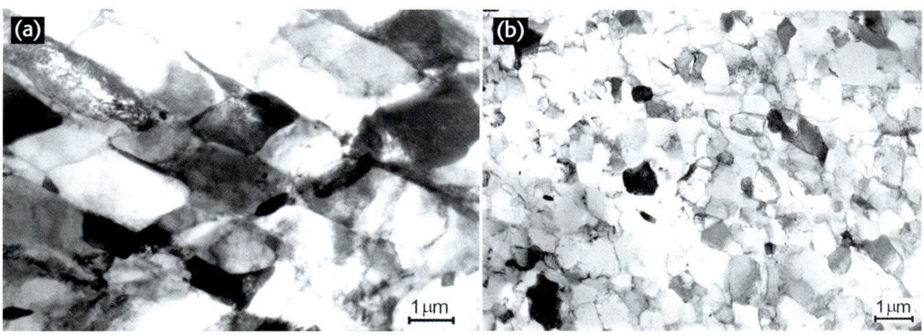

Figure 5.3 Microstructure of Al99.5 after cold CEC processes with large true strains: (a) $\varepsilon = 9$ (10 cycles of CEC); (b) $\varepsilon = 90$ (100 cycles of CEC)

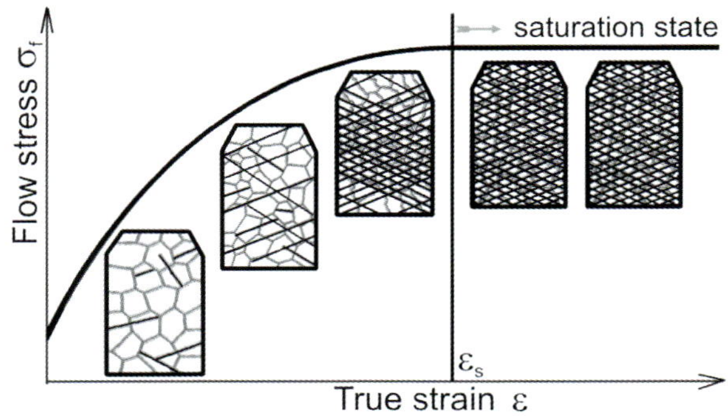

Figure 5.4 Strain-hardening curve for very large deformation and accompanying microstructure evolution

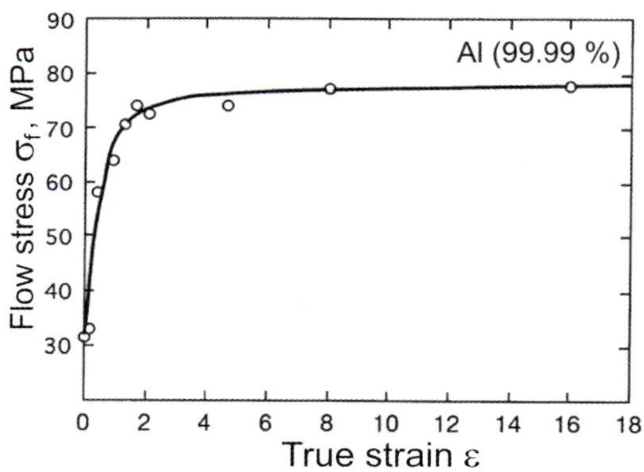

Figure 5.5 Flow stress for Al99.99% as a function of accumulated true strain

A typical flow stress curve for pure aluminium is shown in Figure 5.5 as a function of the true strain.

5.3 Force parameters of CEC processes

In order to obtain appropriately high hydrostatic pressure, the billet located in a chamber between two punches is subjected to a pre-compression force F_{c0} before the process is started. In the course of turning the pre-compression screws, the device's frame undergoes tensile elastic deformation. The elastic energy stored in the frame provides the required pre-compression force F_{c0} (Figures. 5.1 and 5.6). When the process is started, the pre-compression force F_{c0} becomes the counterforce F_c. However, under the influence of an additional elastic extension of the frame by the process forces F_e, a decrease in the counterforce takes place immediately after the beginning of the process (i.e. $F_c < F_{c0}$) (Figure 5.6). For this reason it is necessary to apply an appropriately larger

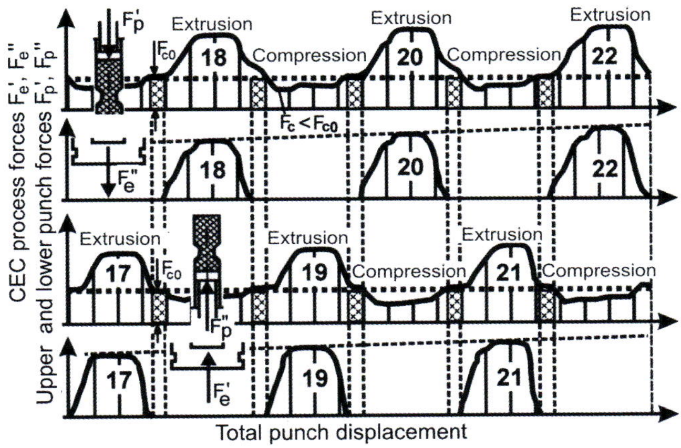

Figure 5.6 Fragmentary diagram of forces (F_e', F_e'', F_p', F_p'') measured by four special sensors during 17–22 CEC cycles; F_{c0} is pre-compression force

pre-compression force F_{c0}. The counterforce F_c should guarantee complete filling of the tool space behind the die in all the deformation cycles that are carried out, and at the same time ensure sufficiently high hydrostatic pressure within the deformation zone. When the punches are moving, the friction force in front of the die decreases while it increases behind the die. However, the friction force will only increase if the tool space behind the die is completely filled. If the applied counterforce F_c does not guarantee that the space is sufficiently filled, the process force F_e is reduced, especially at the final stages of the deformation cycle (Figure 5.6) (Richert, 2001).

Recording of the process force F_e enables continuous control of the quality of the deformation that develops (Figure. 5.7). The process forces F_e measured

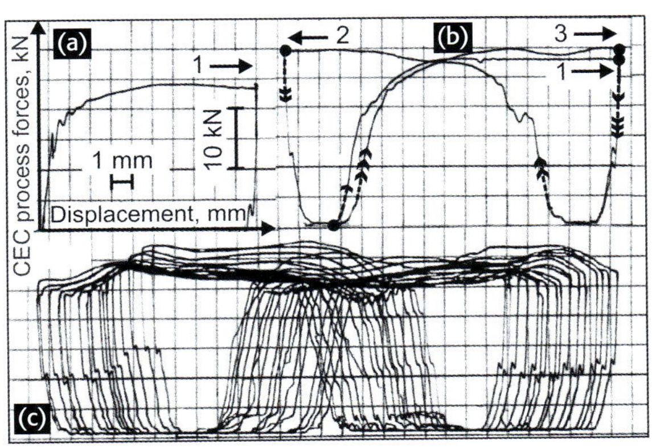

Figure 5.7 Force–displacement diagrams plotted by a universal testing machine during CEC processing of Al99.5: (a) after one cycle; (b) after three cycles; (c) after 34 cycles

in a given cycle should not decrease significantly. The occurrence of a distinct decrease in the process force F_e is evidence that the tool space behind the die was not completely filled. Such adverse deformation conditions must not occur, even during the first cycle, where they might result from applying an insufficiently high pre-compression force F_{c0} on the material. If the decrease in the F_e force occurs in the subsequent cycles of the process, it is usually due to the escape of lubricant, which is too fluid, from the closed tool space. When the material is pre-compressed by means of screws (Figure 5.1), even the smallest outflow of the lubricant leads to a considerable reduction in the hydrostatic pressure p_h. Under such conditions the material rapidly loses its cohesion, especially behind the die, where deformation of the material takes place through radial extrusion (Figure 5.8). In the absence of hydrostatic pressure ($\sigma_m \geq 0$), the plastic deformation of the metal takes place in the presence of tensile circumferential stresses ($\sigma_\theta > 0$).

In order to reduce the friction it is necessary to use lubricants which have no tendency to escape from the closed tool space, even under very high hydrostatic pressures. For this reason, the most appropriate lubricants are those which have a paste-like consistency. Their density should correspond to the type of deformed metal and the applied hydrostatic pressure. In practice, the lubricants can only be selected by means of preliminary tests of the CEC process.

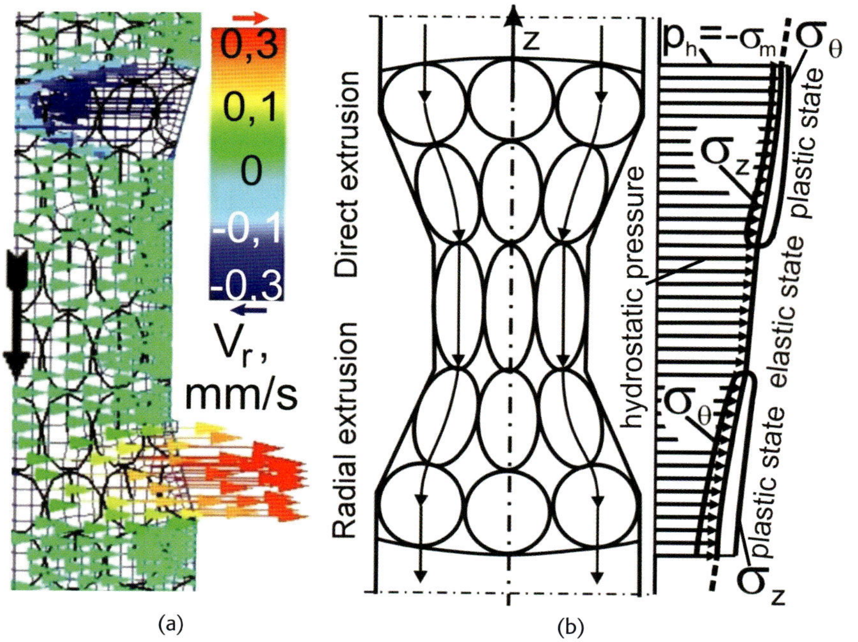

Figure 5.8 Changes in radial velocity and strain–stress conditions along CEC deformation zone: (a) finite element distribution of radial velocity; (b) distribution of axial stress σ_z, circumferential stress σ_θ and hydrostatic pressure p_h

5.4 Effect of tool geometry on formation of shear bands

It should be noted that despite the commonly used name for the process (cyclic extrusion–compression (CEC)), the plastic deformation realised behind the die is actually the result of compression rather than radial extrusion. However, for radial extrusion to occur behind the die, it is necessary to use an appropriately large counterforce (i.e. a compressive force directed opposite to the outflow of metal from the die).

The shape of the working surface of the die orifice, in particular its inclination in the entrance and the exit part of the die, has the greatest influence on the strain–stress state of the deformed material. As a result of the dual symmetry of the die, the conditions for metal flow for direct extrusion and then for radial extrusion, should be similar. Substantial differences may refer only to the sign of the strain components (i.e. to their positive or negative values). For example, if at direct extrusion some strain component is negative, then on the other side of the die, where radial extrusion occurs, the same strain component should have a positive value.

It is known from many experimental studies that direct extrusion also guarantees an advantageous state of stress at an arbitrary shape of die when a flat face die is applied. On the other hand, in the case of radial extrusion only a few shapes of the die guarantee proper conditions for plastic deformation (Richert, 1991, 1988).

The relationship between the principal strains can be used to determine three profile shapes (L1, L2, L3) of the working surfaces associated with three characteristic strain–stress states (Figure 5.9)

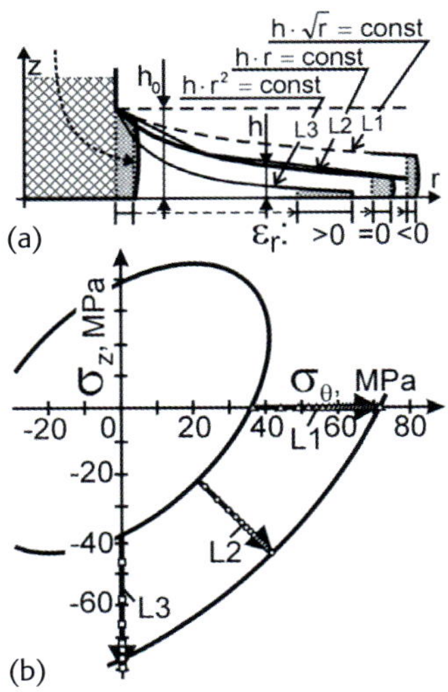

Figure 5.9 Influence of die shape on stress conditions of radial extrusion of Al 99.99: (a) die profiles (L1, L2, L3) forcing metal flow in under three different stress states; (b) stress paths of circumferential tension (L1), circumferential–axial shear (L2) and axial compression (L3) (Richert, 1991)

L1: for the circumferential tension ($\varepsilon_z = -0.5\varepsilon_\theta$, $\varepsilon_r = -0.5\varepsilon_\theta$, $\sigma_\theta > 0$, $\sigma_z = \sigma_r = 0$)

$$\varepsilon_z = -0.5\varepsilon_\theta \tag{5.2}$$

$$\text{In}\,\frac{h}{h_0} = -0.5\ \text{In}\,\frac{r}{r_0} \tag{5.3}$$

Thus

$$h = h_0\sqrt{r}_0\frac{1}{\sqrt{r}} \tag{5.4a}$$

or

$$h\sqrt{r} = \text{const} \tag{5.4b}$$

L2: for the circumferential–axial shear ($\varepsilon_z = -\varepsilon_\theta$, $\varepsilon_r = 0$, $\sigma_\theta = -\sigma_z$, $\sigma_r = 0$)

$$\varepsilon_z = -\varepsilon_\theta \tag{5.5}$$

$$\text{In}\,\frac{h}{h_0} = -\frac{r}{r_0} \tag{5.6}$$

Thus

$$h = h_0 r_0\frac{1}{r} \tag{56.7a}$$

or

$$hr = \text{const} \tag{5.7b}$$

L3: for the axial compression ($\varepsilon_z = -2\varepsilon_\theta$, $\varepsilon_r = -2\varepsilon_\theta$, $\sigma_z < 0$, $\sigma_\theta = \sigma_r = 0$)

$$\varepsilon_z = -2\varepsilon_\theta \tag{5.8}$$

$$\text{In}\,\frac{h}{h_0} = -2\ \text{In}\,\frac{r}{r_0} \tag{5.9}$$

Thus

$$h = h_0 r_0^2\frac{1}{r^2} \tag{5.10a}$$

or

$$hr^2 = \text{const} \tag{5.10b}$$

The graphic representation of the strain–stress states by Pełczynski's three-arm stars, corresponding to the experimental conditions of radial extrusion, is shown in Figure 5.10 (Richert, 1988). Figure 5.10 also shows the well-known Mohr circles (dotted lines) which, regardless of the previous approach, explain the changes in the states of strain and stress.

It is necessary to emphasise that the geometric representation of the strain–stress states by Pełczynski's three-arm stars are based on trigonometric

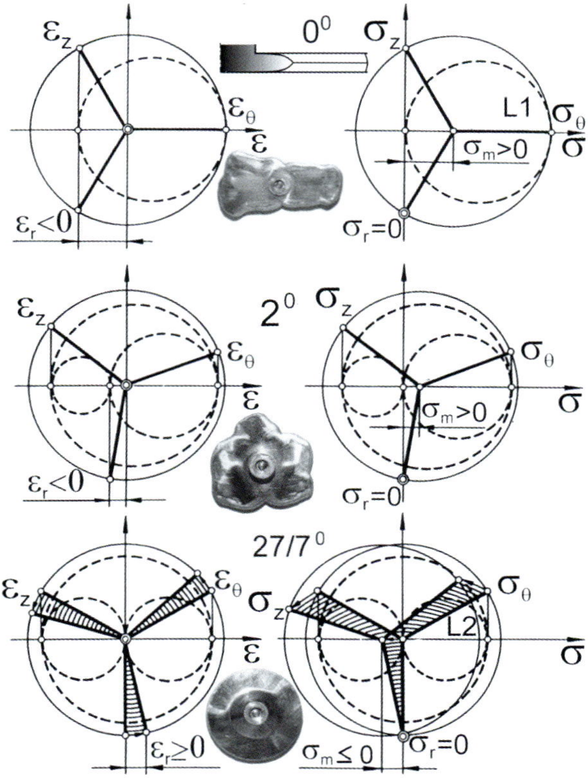

Figure 5.10 Pełczynski's three-arm stars (solid lines) and Mohr's circles (dotted lines) used for graphic representation of strain–stress states, corresponding to experimental conditions of radial extrusion of Al99.99 (Richert, 1988)

solution of cubic equations (Viète's method). For the cubic equation $ax^3 + bx^2 + cx + d = 0$ with three real roots, the roots form an equilateral triangle with vertices A, B and C in a circle (Figure 5.11).

For rigid perfectly plastic incompressible materials (constant-volume relationship $\varepsilon_1 + \varepsilon_2 + \varepsilon_3 = 0$, mean strain of $\varepsilon_m = 0$), the roots of the strain cubic equations $a\varepsilon^3 + b\varepsilon^2 + c\varepsilon + d = 0$ are the principal strains, which can be represented in the following trigonometric form

$$\varepsilon_\theta(\varepsilon_1) = \varepsilon_i \cos \varphi \tag{5.11}$$

$$\varepsilon_r(\varepsilon_2) = \varepsilon_i \cos (\varphi + 240^\circ) \tag{5.12}$$

$$\varepsilon_z(\varepsilon_3) = \varepsilon_i \cos (\varphi + 120^\circ) \tag{5.13}$$

Where ε_i is the effective strain (strain intensity) of the deformed material.

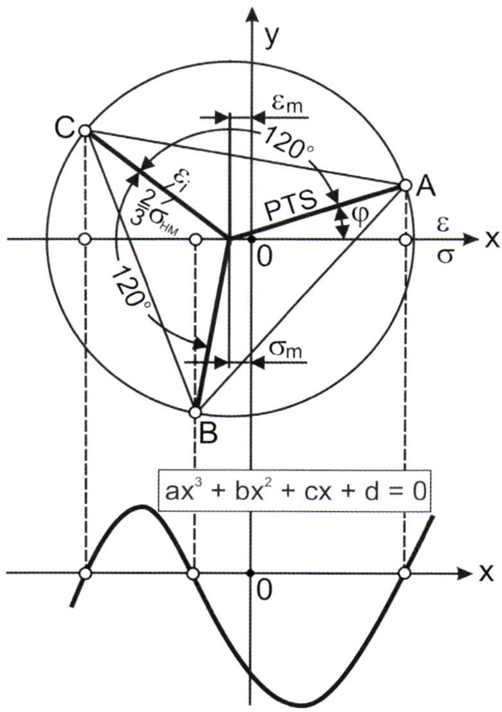

Figure 5.11 Viète's trigonometric solution of cubic equations used by Pełczynski for graphic representation of strain–stress states by means of three-arm stars

The roots of the stress cubic equations $a\sigma^3 + b\sigma^2 + c\sigma + d = 0$ are the principal stresses, which can be represented in the following trigonometric form

$$\sigma_\theta(\sigma_1) = \sigma_m + \frac{2}{3}\sigma_{HM}\cos\varphi = \sigma_m + \frac{2}{3}\sigma_f\cos\varphi \tag{5.14}$$

$$\sigma_1(\sigma_2) = \sigma_m + \frac{2}{3}\sigma_{HM}\cos(\varphi + 240°) = \sigma_m + \frac{2}{3}\sigma_f\cos(\varphi + 240°) \tag{5.15}$$

$$\sigma_z(\sigma_3) = \sigma_m + \frac{2}{3}\sigma_{HM}\cos(\varphi + 120°) = \sigma_m + \frac{2}{3}\sigma_f\cos(\varphi + 120°) \tag{5.16}$$

Where σ_m is the mean stress, σ_{HM} is the the Huber–Mises effective stress (stress intensity) which is equal to the flow stress σ_f at the current value of the effective strain (strain intensity) ε_i.

An example using Pełczynski's stars to estimate the usefulness of tools for radial extrusion is presented in Figure 5.12 (Richert, 1991). It should be noted that a very dangerous stress state (circumferential tensile stress $\sigma_\theta > 0$, $\sigma_m > 0$, gradual reduction in the radial thickness of elements) appears on the outer

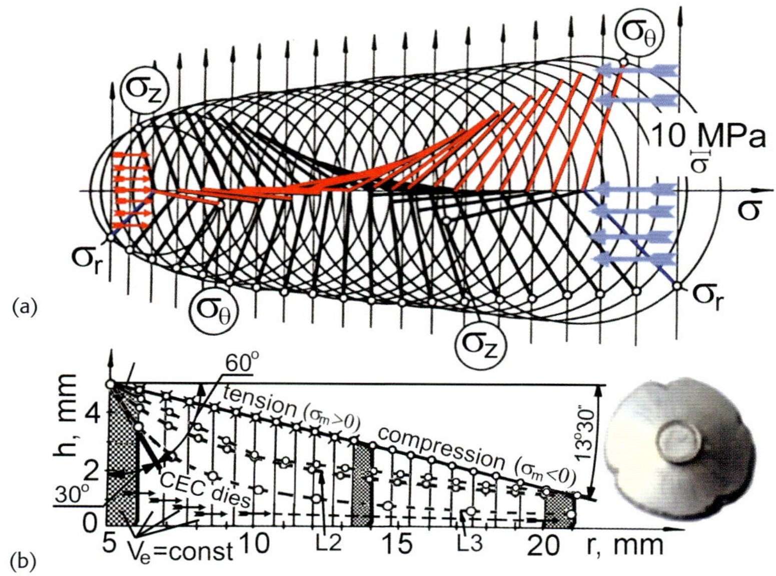

Figure 5.12 Changes of stress states during radial extrusion of Al99.99 calculated numerically by using constant-volume element method developed by present author (Richert, 1991): (a) graphic representation of results; (b) radial flow of constant-volume elements in die clearance

surface ($\sigma_r = 0$) of the deformed metal in the first part of the die clearance. Fully ductile materials with a reduction area (*RA*) of 100% subjected to circumferential tension undergo radial plastic separation (for example in Al99.99). Plastic fracture occurs in conventional materials with $RA < 100\%$ (for example in Al99.5). Such effects were observed during many different experimental tests (Richert, 1991, 1988). The stress state improves in the second part of the die due to the presence of axial compressive stresses ($\sigma_z < 0$, $\sigma_m < 0$, gradual increase in the radial thickness of elements). However, a die with that shape of working surface is useless because the defects formed earlier in the deformed metal cannot be removed. On the basis of theoretical considerations and experimental tests it was found that the application of a die with double inclination of its work surface at 27° and 7°, adapted to the curve of hyperbola L2 ($hr = \text{const}$), ensures the correct conditions for the radial extrusion processes (Figures 5.9(a) and 5.10(c)). In the case of CEC dies with a very small narrowing, the die inclination angles should be considerably larger than 27°, and adapted to the initial point of hyperbola L2 or L3 (45°–60°).

The deviatoric polar angle φ is known as the Lode angle after the German engineer W. Lode who first used it in 1926. The Lode angle measures the angle between the deviatoric projection of the σ_1 axis and the radius vector of the current stress point. The cylindrical coordinates of hydrostatic length,

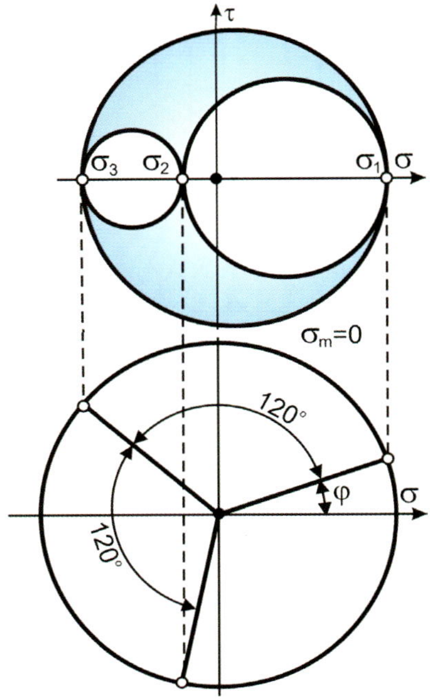

Figure 5.13 Relationship between Mohr circles and Lode angle (in deviatoric plane)

deviatoric length and Lode angle define the plastic yield surfaces in the stress space. The relationship between the Mohr circles and Lode angle is shown in Figure 5.13.

An increase in the diameter of the material that is extruded radially, when a die with a flat surface is used, induces unavoidable cracking or plastic separation of the deformed material, as a large positive circumferential strain is induced by tensile circumferential stress. On the basis of theoretical analysis and experimental tests, it has been proved that only an appropriate inclination of the die surface will guarantee that the deformed material retains its cohesion. It has been found that the correct deformation conditions can be obtained by using dies, whose surfaces are inclined at an angle equal to, or greater than, the inclination of the tangent to the hyperbola L2 (Figures 5.9 and 5.10). The negative values of the average stress (i.e. the hydrostatic pressure) $\sigma_m < 0$, appear at such an inclination (Figure 5.10(c)). Maintaining correct deformation conditions was also analysed by means of experimental studies with the application of a die with double inclination of its surface at 27° and 7°, adjusted to the curve of the hyperbola L2 (Figures 5.9(a) and 5.10(c)). It has been proved that in such cases numerous shear bands appear on the outer surface, inclined at the angle of $\omega = 45°$ to the circumferential stresses (Figure 5.14), in which plastic separation of the outer layer of the extruded sample does not occur, even at very high true plastic strains, amounting to $\varepsilon = 2.9$.

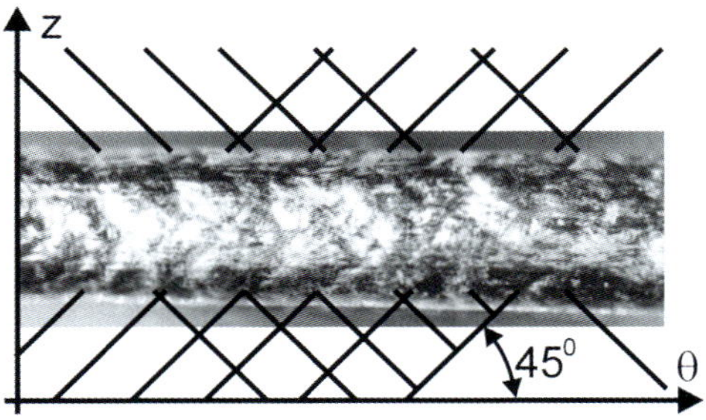

Figure 5.14 Shear bands visible on outer surface of samples extruded radially (Richert, 1988)

Such an inclination of the shear bands ($\omega = 45°$) is in perfect agreement with theoretical predictions (Figure 5.15(b)). It can be assumed that the CEC processes are characterised by similar deformation conditions for the metal, especially behind the die, where metal flow is induced by radial extrusion. The differences may only be in the hydrostatic pressure p_h, which is much higher for CEC processes (Figure 5.15).

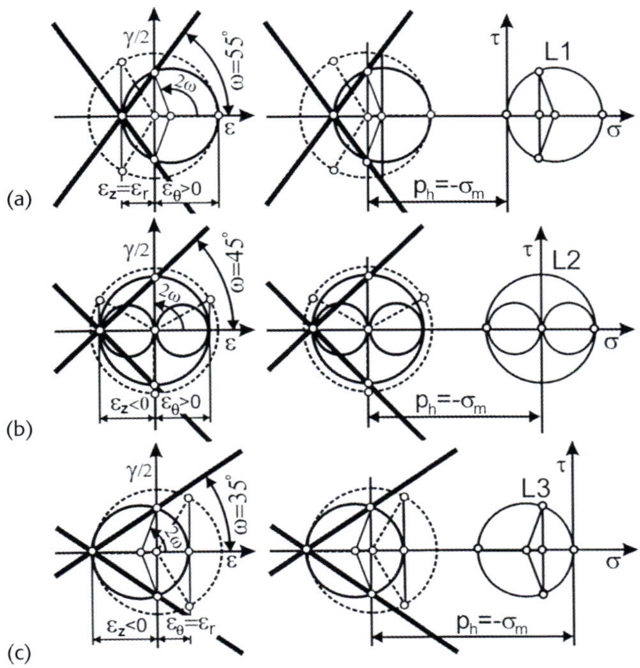

Figure 5.15 Mohr circle representation (solid lines) of shear band inclination angle: (a) $\omega = 55°$; (b) $\omega = 45°$; (c) $\omega = 35°$, during radial extrusion with three different die profiles L1, L2, L3 (shown in Figure 5.9(a))

5.5 Special hydraulic press for CEC processes

A unique hydraulic press with a capacity of 600 kN, was designed and constructed for CEC processing (Figure 5.16) (Richert, 2006, 2010a,b). The press was equipped with a microprocessor control system for the CEC processing with an automatic (intelligent) backpressure control. The microprocessor cooperates with a personal computer, which displays and stores all the process parameters.

In the experimental studies, conducted on the new CEC machine, tools composed of two punches, two chambers and exchangeable dies (Figure 5.17(a)) were used. A dense lubricant prepared from many components, mainly graphite and molybdenum disulfide (MoS_2) was used to lubricate the working surfaces of the tools. To prevent the loss of tightness between the contact surfaces of two chambers and the die, a considerable hold-down force, equal to $F_t = 140$ kN was exerted (Figure 5.17(a)). All CEC processes were performed at a very low speed ($v = 0.1$ mm/s), to ensure that the deformation temperature remained close to room temperature.

The microprocessor control system of the press enables the CEC process to be operated using the counterforce F_c, selected both manually (by introducing the appropriate value of this parameter to the microprocessor memory), as well as automatically (by selecting an appropriate microprocessor program). A standard data acquisition system, supported by a personal computer, was used to capture and record the process data including the process force F_e resulting

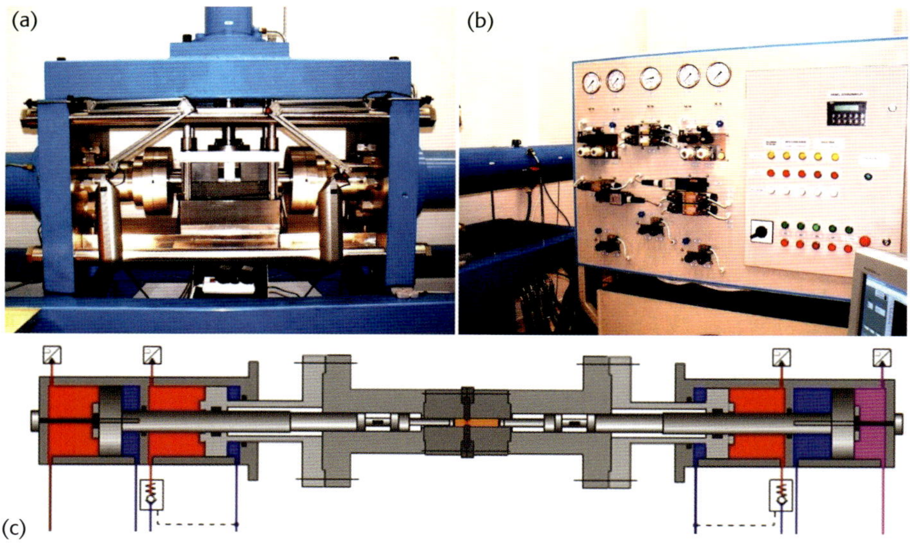

Figure 5.16 Five-cylinder 600 kN CEC press with automatic control of backpressure (Richert, 2006): (a) front view of press; (b) separate control panel connected with computer; (c) hydraulic system of main cylinders

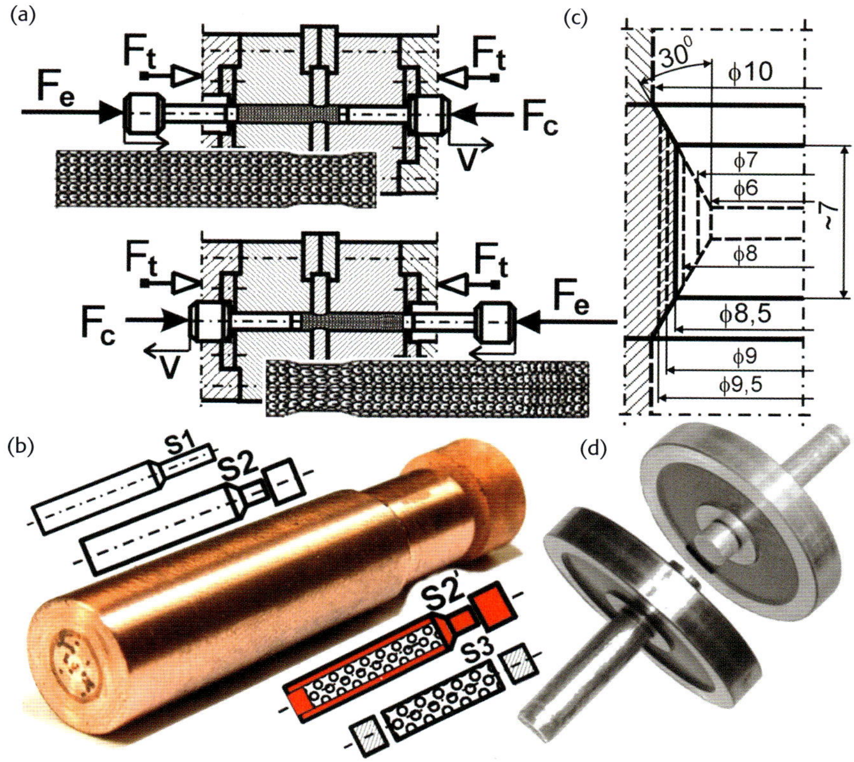

Figure 5.17 Working conditions of CEC process: (a) changes in tool configuration during two successive cycles: F_e = process forces, F_c = counterforces, F_t = hold-down tool forces, v = punch speed; (b) examples of initial samples; (c) die orifice dimensions; (d) steel dies with samples

from the counter force F_c as well as other process conditions. Additionally, the relationships between forces F_e and F_c were plotted on the computer screen as a function of time.

5.6 Grain refinement in Al6082 alloy

Samples of diameter d_0 = 10 mm and length l_0 = 50 mm, prepared from rods produced by industrial hot extrusion of the Al6082 alloy (AlMgSi) having the chemical composition of Al–1.0Mg–1.18Si–0.66Mn–0.04Cu–0.04Zn–0.03Cr–0.03Ti (in wt.%), were used in the experiment. Before starting the process, the samples were solution heat-treated at 525°C for 30min and then quenched in water.

Samples consisting of one part (S1), as well as samples consisting of two parts (S2) were used (Figure 5.17(b)). The divided configuration of the initial samples (S2, S3) guarantees faster and easier adjustment of the volume of the material to the working space of the tools at the beginning of the first cycle of deformation. For this reason after the preliminary tests all subsequent tests were performed using the divided samples (S2). Samples in the shape (S1) require preliminary compression before starting the first cycle. Samples of the three-part divided configuration (S3), with two soft-metal ends, can be applied when brittle materials, in particular the powder materials, are processed.

Dies with various diameters of the orifice within the range 6.0–9.5 mm were prepared for the CEC processes performed on the new hydraulic press (Figure 5.17(c)). A 30° inclination of the entrance and exit working surfaces was applied to all dies. A die with an orifice diameter $d_m = 8.5$ mm and whose die bearing was about 7 mm in length, was selected (Figure 5.17(c)) from among the many dies prepared for the experimental texts. According to Equation (5.1), the true strain obtained in one cycle ($n_c = 1$) is $\varepsilon_1 = 0.65$. The principal aim of the present study is to determine the stress conditions for shear band formation, since these conditions have a major influence on the possibility of obtaining high plastic deformations and the nanocrystalline structure of the processed material. From the many realised tests connected with this problem, only three processes have been chosen, characterised by completely different conditions regarding hydrostatic pressure in the deformation zone. These processes allow a clear and concise presentation of the problems being analysed.

5.6.1 Evolution of shear bands under varying backpressure

Favourable conditions for deformation by the CEC method are obtained when the smallest possible counterforce F_c is applied. However, that guarantees a hydrostatic pressure within the deformation zone that enables the cohesion of the processed material in all cycles of the process to be preserved. Due to the application of the smallest possible counterforce F_c, it is possible to obtain the required high plastic deformation with a minimal force F_e, thus preventing rapid damage to the tools. It is difficult to select the appropriate counterforce without the aid of a suitable computer program.

A correct choice of hydrostatic pressure in the first cycle is very important (Figure 5.18, Table 5.1), since already at the start of the process when the material is not yet work hardened, the tool space must be completely filled with the deformed metal. At a low backpressure the chamber situated behind the die is not completely filled, and the conditions for radial extrusion in the presence of tensile circumferential stresses are very disadvantageous. Hydrostatic pressure does not develop during such deformation since the values of the mean stress behind the die are greater than zero (i.e. $\sigma_m > 0$). When the metal is being displaced behind the die it does not adhere completely to the walls of the chamber. In such a case the process force F_e is mainly determined by the deformation conditions in front of the die, and not behind it. In front of the die, with a reduction in the length of the extruded metal, the friction force at the chamber wall becomes smaller. However, at a low counterforce, this friction force is not balanced by an appropriate increase in the friction force behind the die. In that

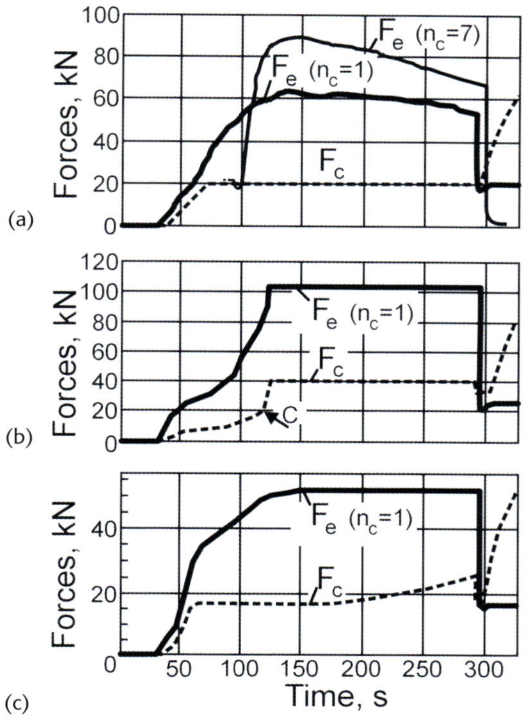

Figure 5.18 Forces during CEC processing for different counterforces: (a) too small; (b) excessively large; (c) optimal

Table 5.1 Experimental conditions for CEC processes

Test number	Number of cycles / True strain / n_c / ε_c	Counterforce / Backpressure / F_c / p_c	Extrusion force / Extrusion pressure / F_e / p_e	Method of selecting counterforce (evaluation of backpressure values)	Evaluation of preservation of cohesion; quality of obtained alloy
1	7 4.6	20 kN (const) 255 MPa	62–52 kN 790–660 MPa	Manually (backpressure too low)	Incorrect, cracks (Fig. 5.19(a))
2	9 5.9	40 kN (const) 510 MPa	105 kN 1340 MPa	Manually (backpressure too high)	Correct, no cracks (Fig. 5.19(b))
3	15 9.8	18–26 kN 229–331 MPa	52.5 kN 668 MPa	Automatically (optimal backpressure)	Correct, no cracks (Fig. 5.19(c))

case the process force F_e becomes smaller, which is evidence that the CEC process is being carried out under highly inappropriate stress conditions, resulting from a counterforce F_c which is too small.

On the basis of the graphs recorded by the computer, it can be seen that the application of the counterforce F_c = 20 kN leads to a considerable decrease in the process forces F_e during the first cycle (Figure 5.18(a), Table 5.1). It should be noted that with increasing work hardening of Al6082 alloy this decrease becomes larger, and in the last cycle (n_c = 7) it was very distinct. Numerous shear cracks could be observed on the surfaces of the processed Al6082 alloy (Figure 5.19(a)) when the deformed sample was removed from the chambers. These cracks are inclined at an angle of $\omega = 45°$ to the direction of extrusion. The counterforce of 20 kN is too small and the hydrostatic pressure within the deformation zone is also too low to prevent cracking. Cracks appear in the areas with the largest accumulation of plastic deformations (i.e. within the shear bands). For this reason, cracks have been formed at the inclination angles of $\omega = 45°$ to the extrusion direction.

Thus, it has been concluded that the counterforce must be increased in order to retain the cohesion of the Al6082 alloy. Since the value of a sufficient counterforce was not known, it was decided to carry out a deformation test applying a counterforce that was twice as large (i.e. F_c = 40 kN) (Figure 5.18(b),

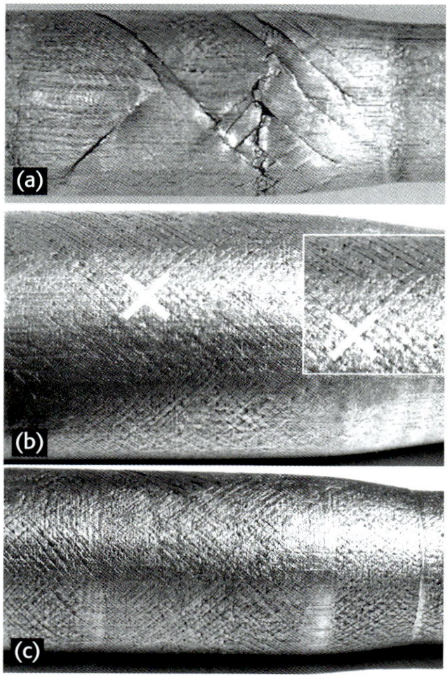

Figure 5.19 Surface traces of shear bands observed in Al6082 alloy CEC processed using different backpressures: (a) too low; (b) excessively high; (c) optimum

Table 5.1). It has been found that when applying this counterforce, the process force remains at a constant level, however, it reaches an excessively high value ($F_c = 105\,\text{kN}$). Such great force occurring already in the first cycle creates a danger of damage to the tools, especially to the punches. For this reason, after temporarily stopping the process, the next cycles were carried out at a lower counterforce ($F_c = 25\,\text{kN}$). In the case of the new hydraulic press, it is quick and easy to change the value of the counterforce. The required configuration of the press control parameters is entered into the microprocessor by means of a display panel equipped with appropriate push buttons and an LCD monitor (Figure 5.16(b)). The processed sample was taken out of the hydraulic press after nine deformation cycles had been completed. After removing the lubricant layer it was found that, due to the increased backpressure, the cohesion of the Al6082 alloy had been preserved. However, a much greater and positive surprise was the unexpected look of the outer surface of the deformed sample, where a very distinct relief could be observed even without any etching. Its characteristic shape suggests that it was created by the formation of a great number of intersecting shear bands (Figure 5.19(b)). All the shear bands are inclined at an angle of $\omega = 45°$ to the extrusion direction and intersect at 90°.

The clear visibility of the localised shear strain is due to reversible cyclic deformation, leading to banded surface indentations and protrusions, which are similar to the intrusions and extrusions observed in an experimental study of the fatigue strength of metals (Polak *et al.*, 2003; Wu *et al.*, 2004). It should be noted that there is an essential difference between the formation mechanisms of the above-mentioned surface relief for non-monotonic loading conditions. In the first case, under the influence of a considerable, cyclic, reversible plastic deformation, a deep, clearly visible, surface relief is formed (Figure 5.19), corresponding to the directions in which shear bands were formed. In the second case, under the influence of small, cyclic elastic-plastic straining, a micro-relief is formed, corresponding to the location of persistent slip bands (Polak *et al.*, 2003; Wu *et al.*, 2004).

The above method of selecting the counterforce is incidental and cannot be accepted in practice. Thus, a complex program was loaded into the memory of the microprocessor of the hydraulic press, which carries out complex deductions which allows the counterforce to be automatically changed and for its optimal value to be applied during the process. First, the regularity that was noticed earlier has been used to elaborate the procedure to determine the required value of a counterforce. As observed earlier, the processed material does not lose its cohesion when the process force in the particular deformation cycles is kept at nearly the same level (i.e. at $F_e = \text{const}$).

The results of running the first program, which guarantees automatic selection of the counterforce of the punch, are shown in Figure 5.18(c). A relatively low value of the initial counterforce $F_c = 18\,\text{kN}$ was deliberately applied. It is smaller than the value adopted in the first process $F_c = 20\,\text{kN}$ (Figure 5.18(a)). From the results obtained in the second process, it can be concluded that the volume of the processed material is perfectly adjusted to the tool space at a counterforce of about 20 kN, as indicated by the arrow C (Figure 5.18(b)). When this value is exceeded the forces (F_c, F_e) increase rapidly. In the case of a microprocessor control system it has been found that after preliminary deformation of the initial sample, in which the volume of the deformed material adjusts to the tool space, which is limited on both sides by the front surfaces of two working punches, the process force F_e attains its maximal value $F_e = 52.5$ kN (Table 5.1). It is not very high as it is attained at a relatively low counterforce ($F_c = 18$ kN). It should be noted that the process force is lower than the maximal force $F_e = 62$ kN, which occurs in the first CEC process carried out at the counterforce $F_c = 20$ kN (Table 5.1). It was found that, despite the application of such a dangerously low value for the counterforce ($F_c = 18$ kN), the obtained sample was of excellent quality (Figure 5.19(c)) even after 15 deformation cycles.

The positive result of the process was obtained due to the application of the microprocessor control system. The smallest force decrease in relation to the maximal force of the process, occurring at the beginning of each cycle, is immediately revealed by the appropriate program, and quickly corrected by a proper increase in the counterforce. The greatest increase in the counterforce always occurs at the final stage of each deformation cycle. Thus, a suitably high hydrostatic pressure can be continuously maintained in the deformation zone which is situated far from the direct action of the counterforce. It has been found that the counterforce increased from 18 kN to 26 kN in the first cycle (Figure 5.18(c), Table 5.1). Generally, the graphs of the forces F_c and F_e in all cycles were similar to the first cycle, for example, the same initial counterforce ($F_c = 18$ kN) was maintained. Essential differences relate only to the maximal force F_e, occurring in different deformation cycles. In each successive cycle a gradual increase in the level of this force occurs, namely from 52.5 kN ($n_c = 1$) to 88 kN ($n_c = 15$). It has been found that the values of the counterforce F_c, which were selected by a microprocessor, did not exceed 30 kN in any of the cycles.

5.6.2 Evolution of microstructure of Al6082 alloy

The microstructure of the Al6082 alloy was investigated by a FluoView FV300 laser confocal microscope and a transmission electron microscope (JEM2010 ARP). The mean size of the observed nano- and ultra-fine grains was determined by the mean chord parameter. The grain boundary misorientation angles and microtexture were determined with a special software package (KILIN), on the base of the Kikuchi line diffractions (Richert *et al.*, 2009).

It was found that with the increase of strain, the grain size of the AlMgSi alloy deformed by the CEC method diminished (Figure 5.20). After straining the material to $\varepsilon = 16.3$ (25 CEC cycles), the mean grain size was 75 nm, which indicated the generation of a nanostructure in the alloy. The observations

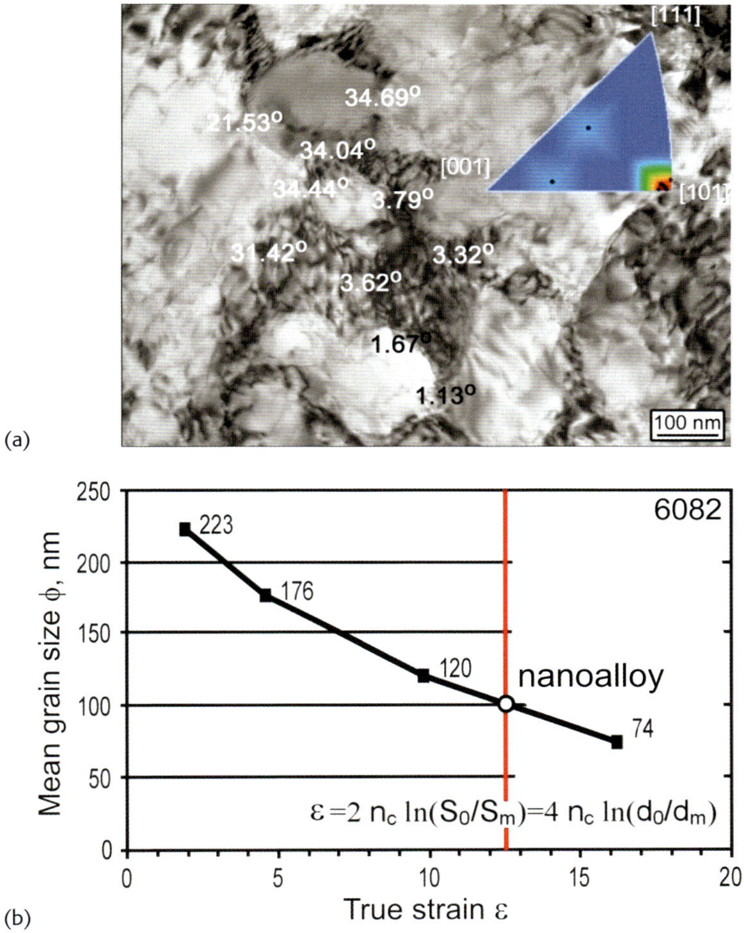

Figure 5.20 Microstructure evolution of Al6082 alloy during CEC: (a) grain boundary misorientation angles after true strain of $\varepsilon = 16.3$; (b) diagram of relationship between mean grain size ϕ and true strain ε

showed that the nanostructure appeared in about 75% volume of the sample (Richert *et al.*, 2008, 2009).

5.6.3 Finite element simulation of CEC processes

DEFORM 2D v.8.1 software was used for the simulation of plastic flow in the CEC process (Richert *et al.*, 2009). The most important parameter is the hydrostatic pressure, which prevents an increase in the dangerous tensile stresses. In order for the extruded material to completely fill the tool space behind the die, an increase in the hydrostatic pressure is needed. This can be achieved by increasing the counterforce F_c. Based on the data from a numerical simulation

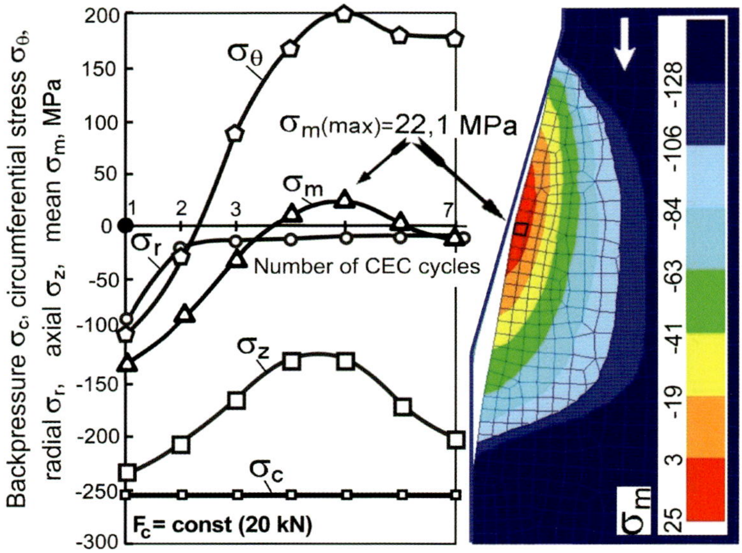

(a) (b)

Figure 5.21 Incorrect stress conditions for realisation of CEC processes at counterforce F_c = 20 kN (backpressure σ_c = −254 MPa): (a) diagram of stress changes during 1–7 cycles; (b) finite element diagram illustrating non-filling of die chamber after five cycles

of the CEC processes a special diagram has been prepared, which facilitates a suitable choice of the counterforce F_c (Figure 5.21). In the first case analysed, a stable counterforce $F_c = 20$ kN was established in each CEC cycle. It leads to a stable value of $\sigma_c = -254$ MPa and thus ignores the increasing hardening of the deformed Al6082 alloy (Figure 5.21(a)).

The conditions of the plastic flow are worst just behind the die clearance due to the gradual hardening of the Al6082 alloy. The tool space is not fully filled by the extruded material. A high circumferential stress ($\sigma_\theta = 168$ MPa) appeared in the fourth cycle, while the mean stresses achieved a positive value ($\sigma_m = 9.4$ MPa). In the fifth cycle the mean stresses obtained their highest positive value ($\sigma_m = 22.1$ MPa) (Figure 5.21(b)).

The results of the modelling show that in the fifth cycle, an empty volume appears just behind the die clearance, leading to an unstable plastic flow. In such situations the working surfaces of the tools do not restrict the plastic flow and cracking of the Al6082 alloy becomes inevitable. This simulation indicates that a higher counterforce should be applied in the following CEC cycles.

In the next variant of simulation the value of the counterforce was selected proportionally to the increasing plastic flow stress of the Al6082 alloy. The range of the counterforce applied was F_c = 18–41.4 kN. The results of the simulation are shown in Figure 5.22.

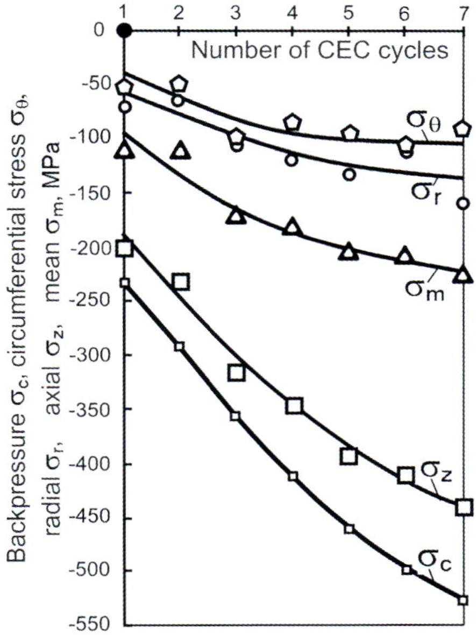

Figure 5.22 Diagram of proper stress conditions for realisation of CEC processes at counterforce $F_c = 18–41.4$ kN

It was found that in the all cycles that were investigated, there were very positive conditions for plastic flow. This proves that the finite element simulation enables a proper selection of the counterforce in the CEC processes (Figure 5.23).

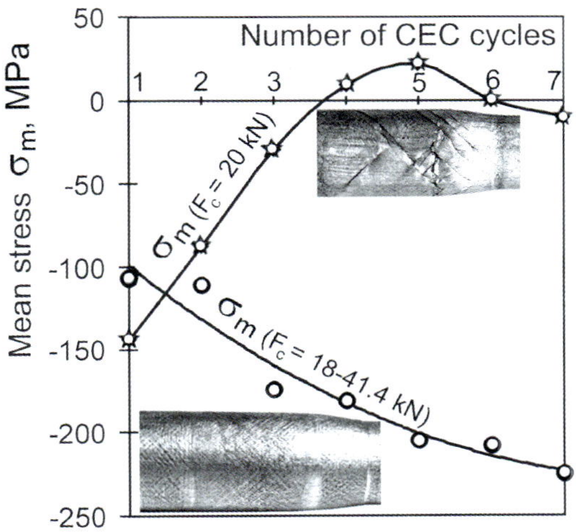

Figure 5.23 Correct and incorrect choice of counterforces for CEC of Al6082 alloy

5.6.4 Stress path analysis in CEC processes

An identical yield criterion for the initiation of plastic flow was proposed independently by Huber (1904) and von Mises (1913). It is based on the observation that a hydrostatic pressure cannot cause plastic yielding of the material. Only the energy of distortion influences the transition from the elastic to the plastic state. The yield surface of the Huber–Mises yield criterion (hypothesis) in a stress space is shown in Figure 5.24.

The Huber–Mises cylinder creates the elliptical shape of the yield surface on the three planes $\sigma_1-\sigma_2$, $\sigma_2-\sigma_3$ and $\sigma_1-\sigma_3$. The most relevant ellipse is the one which characterises two extreme stresses having the algebraically largest σ_1 and the smallest σ_3 principal stresses. This ellipse enables allows us to calculate the maximum shear stress

$$\tau_{\max} = \frac{(\sigma_1-\sigma_3)}{2} = \frac{(\sigma_\theta-\sigma_z)}{2} \tag{5.17}$$

For example, in the case of Al6082 alloys with an initial flow stress $\sigma_f = 90\,\text{MPa}$, the maximum shear stress in the first cycle of CEC processes for a simple shear deformation ($\varepsilon_\theta = -\varepsilon_z$, $\varphi = 30°$) can be calculated using Equations (5.14) and (5.16)

$$\tau_{\max} = \frac{(\sigma_\theta-\sigma_z)}{2} = \frac{\left(\sigma_m + \frac{2}{3}\sigma_f\cos\varphi\right) - \left[\sigma_m + \frac{2}{3}\sigma_f\cos(\varphi+120°)\right]}{2}$$

$$\tau_{\max} = \frac{2}{3}\sigma_f\cos 30° = \frac{2}{3}\sigma_f\frac{\sqrt{3}}{2} = \frac{\sqrt{3}}{3}\sigma_f$$

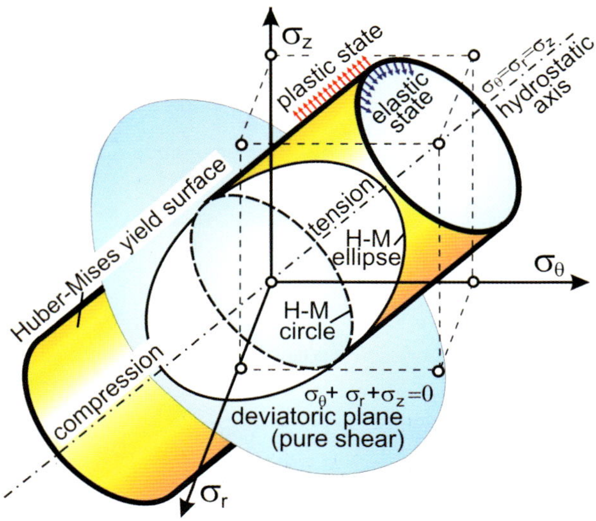

Figure 5.24 Yield surface of Huber–von Mises criterion in stress space

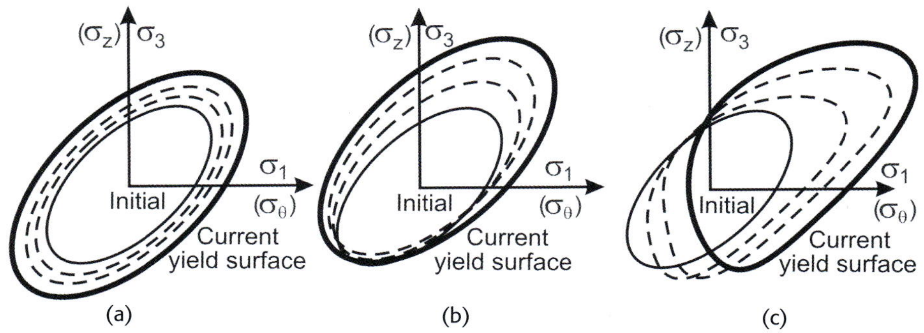

Figure 5.25 Schematics of three deformation-hardening types: (a) isotropic; (b) isotropic and kinematic; (c) anisotropic

Thus

$$\tau_{max} = \frac{(\sigma_\theta - \sigma_z)}{2} = 0.577\sigma_f = 52\,\text{MPa} \tag{5.18}$$

If loading occurs during plastic deformation, the state of stress moves outwards from the current yield surface to a new yield surface. The three types of deformation hardening which may occur are: isotropic, kinematic and anisotropic hardening. In isotropic work (deformation) hardening, the yield surface remains centred about its initial centreline and expands in size as the plastic strains develop. For materials with isotropic plastic behaviour this is termed isotropic hardening (Figure 5.25(a)). Kinematic hardening assumes that the yield surface translates in stress space with progressive yielding. Figure 5.25(b) illustrates a combination of isotropic and kinematic hardening. In the case of anisotropic hardening the yield surface undergoes expansion, translation and distortion due to the evolution of the anisotropy caused by the development of the texture in plastically deformed polycrystalline materials (Figure 5.25(c)).

The main differences between the stress conditions of early CEC processes and those performed now are shown in Figure 5.26(a,b). The arrows OA, overlapping from seven deformation cycles, show a temporary drop in the pre-compression of the deformed material caused by the elastic tension of the frame of the device at the start of each cycle (Figure 5.6). An elastic increase in the compressive force (arrows AB), which is necessary to start plastic deformation of the material by direct extrusion, leads to an increase in the hydrostatic pressure in the deformation zone. The highest value for the hydrostatic pressure is obtained at the start of the direct extrusion (Figure 5.8(b)). While the material particles are displaced along the flow lines they are work hardened with a simultaneous decrease in hydrostatic pressure (BC arrows). Due to work hardening, in accordance with the flow criterion, the stress path goes through various yield surfaces. When the material approaches the die bearing, the plastic deformation (CD arrows) is temporarily stopped. While the particles are displaced in the bearing section a further drop in the hydrostatic pressure (DE arrows) occurs. The second stage of deformation, in which the initial shape of the particles is renovated by means of radial extrusion (EF-FG-GO arrows), begins behind the die bearing. During this deformation path there is a further decrease in the hydrostatic pressure p_h (Figures 5.8(b), 5.26(a)).

In the case of the special hydraulic press for the CEC processes, the technical possibilities to ensure appropriately high hydrostatic pressure in the deformation zone are more advantageous (Figure 5.26(b)) and the dangerous distressing of the material at the start of the process does not occur (OA arrows – Figure 5.26(a)). The CEC processes can be started at low hydrostatic pressure (A_1 – first cycle) and only with increasing work hardening is the hydrostatic pressure (A_1–A_7) appropriately increased.

The hydrostatic pressure has no influence on the directions of the shear bands. However, the application of low hydrostatic pressures increases deformation inhomogeneity and the occurrence of cracks in the shear bands (Figure 5.19(a)). It has been found that for CEC processes performed on a

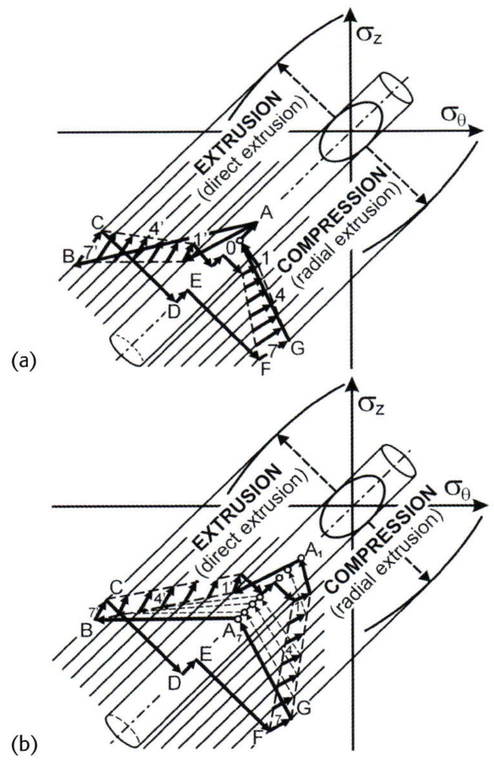

Figure 5.26 Stress path in seven cycles of CEC processes using: (a) initial device without counterforce control; (b) special hydraulic press with counterforce control

hydraulic press, with the application of a die with 30° inclination of the entrance and exit working surfaces (Figure 5.17(c)), all the shear bands are inclined at an angle of $\omega = 45°$ to the axis of the processed sample and these bands intersect at an angle of 90°. It has been noticed, however, that the actual shape of the narrowed conical area of the deformed samples (Figure 5.19) does not correspond to the shape of the die. Such an effect in the CEC process can be attributed to the fact that, given good lubrication of the material present in a tightly closed interior tool space, there is a possibility of optimal, spontaneous configuration of the deformation zone. This means that the material being processed chooses a flow path through the die orifice, which requires the least energy of plastic deformation. Stable wedges of lubricant form in front of the die bearing, which cause considerable elongation of the conical shape of the deformation zone. As a result, the inclination angle of the side surface of the zone is much smaller than that of the die (Figure 5.19). Hence it could be assumed that the CEC processes, performed on the new hydraulic press, guarantee the deformation of metal under conditions of hydrodynamic friction (i.e. the most advantageous of the known types of friction). Under such lubrication conditions there is no direct contact between the flowing metal and the tool walls. Accordingly, even when there are many cycles ($n_c = 15$), the outer surface of the deformed samples retains its original relief, induced by localised straining in the form of shear bands (Figure 5.19(c)). It can be seen that the characteristic relief is visible not only on the surface of a cylindrical sample, but also in the deformation zone and the die bearing. This is evidence that, even in the die bearing area, hydrodynamic friction is maintained until the end of the process. It should be assumed that such favourable conditions for friction in the die bearing are maintained due to its rather great length (ca. 7 mm), which corresponds approximately to the diameter of the die orifice (Figure 5.17(c)).

5.6.5 Stress state in the deformation zone

Comparing the directions of the shear bands, determined theoretically for the particular states of strain and stress (Figure 5.15), with the directions observed experimentally (Figure 5.19), a very important final conclusion can be drawn: that in the deformation zone of the CEC processes there is a strain state corresponding to axial-circumferential and circumferential–axial simple shear, in which the radial strain assumes a value equal to zero ($\varepsilon_r = 0$), and the circumferential and the axial strains have the same absolute value ($|\varepsilon_\theta| = |\varepsilon_z|$) both in front of and behind the die (Figures 5.15(b), 5.27).

Similar deformation conditions, with simple shear deformation dominating, were found in other studies (Segal *et al.*, 1981; Segal, 2002; Jining *et al.*, 2004).

The components of the principal stress tensor and the values of hydrostatic pressure can be determined using the basic theoretical dependences of plastic flow and the data obtained during the performed experimental tests. In the CEC processes the plastic deformations are performed under conditions of

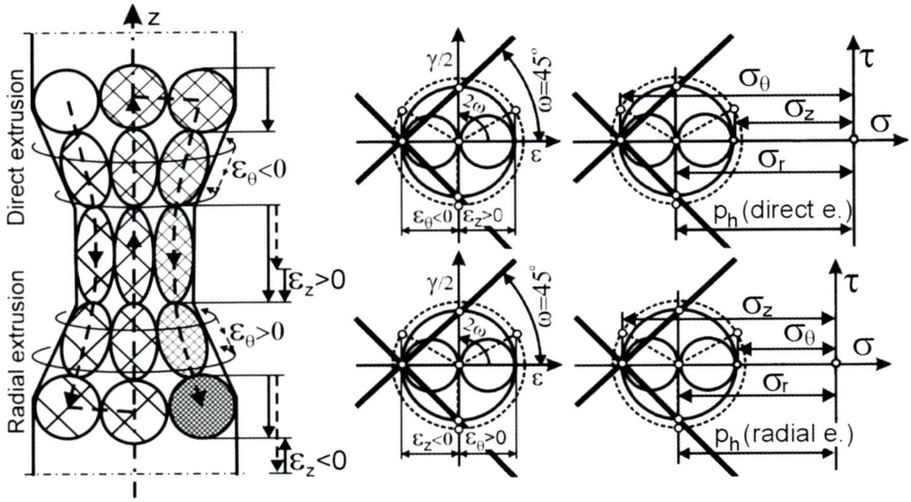

Figure 5.27 Strain–stress conditions of shear band formation in deformation zone of CEC processes: (a) schematic of shear band evolution; (b) graphic representation of shear band directions

multiaxial compression ($\sigma_\theta < 0$, $\sigma_r < 0$, $\sigma_z < 0$). It has been found experimentally that shear bands are formed at an angle of 45° to the principal strain directions. It follows from the above statement that radial stress takes the mean value of the circumferential and the axial stresses (Figure 5.27)

$$\sigma_r = \frac{1}{2}(\sigma_\theta + \sigma_z) \tag{5.19}$$

The value given in Equation (5.19) is simultaneously the mean value of the three principal stresses

$$\sigma_m = \frac{1}{3}(\sigma_\theta + \sigma_r + \sigma_z) = \frac{1}{2}(\sigma_\theta + \sigma_z) = \sigma_r \tag{5.20}$$

Thus the hydrostatic pressure occurring in CEC processes can be defined by the following equation

$$p_h = -\sigma_m = \frac{1}{2}|(\sigma_\theta + \sigma_z)| \tag{5.21}$$

The axial stress σ_z changes along the deformation zone (Figure 5.8(b)), and behind the die its value depends on the backpressure p_c resulting from the applied counterforce F_c, while in front of the die it depends on the extrusion pressure p_e resulting from the extrusion forces F_e. The values of all the components of the stress state depend mainly on the flow stress σ_f. To formulate the condition for the plastic flow of metal, it is necessary to calculate an equivalent stress σ_v, which can be compared with the uniaxial stress. In the forming technology the distortion energy hypothesis (Huber and von Mises) is used for this purpose. Thus, the plastic deformation of metals occurs when the equivalent stress σ_{HM} reaches the value of the flow stress σ_f

$$\sigma_{HM} = \sqrt{\frac{1}{2}\left[(\sigma_\theta - \sigma_r)^2 + (\sigma_r - \sigma_z)^2 + (\sigma_z - \sigma_\theta)^2\right]} = \sigma_f \tag{5.22}$$

After substituting the radial stresses (Equation (5.19)) in Equation (5.22) a much simpler relationship (Equation (5.23)) is obtained

$$\sigma_{HM} = \sqrt{\frac{1}{2}\left[\left(\sigma_\theta - \frac{1}{2}\sigma_\theta - \frac{1}{2}\sigma_z\right)^2 + \left(\frac{1}{2}\sigma_\theta - \frac{1}{2}\sigma_z - \sigma_z\right)^2 + (\sigma_z - \sigma_\theta)^2\right]} = \sigma_f$$

$$\sigma_{HM} = \sqrt{\frac{1}{2}\left\{\left[\frac{1}{2}(\sigma_\theta - \sigma_z)\right]^2 + \left[\frac{1}{2}(\sigma_\theta - \sigma_z)\right]^2 + [-(\sigma_\theta - \sigma_z)]^2\right\}} = \sigma_f$$

$$\sigma_{HM} = \sqrt{\frac{3}{4}(\sigma_\theta - \sigma_z)^2} = \sigma_f$$

Thus

$$|(\sigma_\theta - \sigma_z)| = 1.15\sigma_f \tag{5.23}$$

While the metal is flowing along the deformation zone it becomes increasingly work hardened, as is shown by a gradual increase in the flow stress σ_f. For the same reason the difference between the circumferential and the axial stresses is also increasing (Figure 5.8(b)). The dependence (Equation (5.23)) allows us to find one component σ_θ from the two unknown components of the stress state (σ_θ, σ_z). After its calculation and substitution into Equation (5.23) the final form of the equation is obtained, which allows us to calculate the hydrostatic pressures which occur in the deformation zone of the CEC processes

$$p_h = |\sigma_z| \pm 0.577\sigma_f \tag{5.24}$$

Where the + sign refers to direct extrusion, and the – sign to radial extrusion.

At the beginning of the process, the flow stress of the solution heat-treated Al6082 alloy (W temper) is lowest, with a value of $\sigma_f = 90\,\text{MPa}$. The absolute value of the axial stress $|\sigma_z|$ in the deformation zone (Table 5.1, Figure 5.19(c)), situated in front of the die ($F_e = 52.5\,\text{kN}$), is close to the extrusion pressure ($|\sigma_z| = p_e = 668\,\text{MPa}$), whereas behind the die ($F_c = 18\,\text{kN}$) it is close to the backpressure ($|\sigma_z| = p_c = 229\,\text{MPa}$). Thus, according to Equation (5.24) the smallest hydrostatic pressure p_h, occurring in the deformation zone at the beginning of the first cycle, is equal to $p_h = 668 + 0.577 \times 90 = 720\,\text{MPa}$ ($\sigma_m = -720\,\text{MPa}$) in front of the die, and $p_h = 229 - 0.577 \times 90 = 177\,\text{MPa}$ ($\sigma_m = -177\,\text{MPa}$) behind the die. The components of the principal stress tensor $\mathbf{T}_\sigma$ and its constituents: hydrostatic tensor $\mathbf{A}_\sigma$ and deviatoric tensor $\mathbf{D}_\sigma$ at the initial stage of the CEC process assume the following values

(a) for deformation by direct extrusion ($\sigma_z = -668\,\text{MPa}$, $\sigma_r = -720\,\text{MPa}$, $\sigma_\theta = -772\,\text{MPa}$, $\sigma_m = -720\,\text{MPa}$)

$$\mathbf{T}_\sigma = \mathbf{A}_\sigma + \mathbf{D}_\sigma \tag{5.25}$$

$$\begin{pmatrix} \sigma_z & 0 & 0 \\ 0 & \sigma_r & 0 \\ 0 & 0 & \sigma_\theta \end{pmatrix} = \begin{pmatrix} \sigma_m & 0 & 0 \\ 0 & \sigma_m & 0 \\ 0 & 0 & \sigma_m \end{pmatrix} + \begin{pmatrix} \sigma_z - \sigma_m & 0 & 0 \\ 0 & \sigma_r - \sigma_m & 0 \\ 0 & 0 & \sigma_\theta - \sigma_m \end{pmatrix} \tag{5.26}$$

$$\begin{pmatrix} -668 & 0 & 0 \\ 0 & -720 & 0 \\ 0 & 0 & -720 \end{pmatrix} = \begin{pmatrix} -720 & 0 & 0 \\ 0 & -720 & 0 \\ 0 & 0 & -720 \end{pmatrix} + \begin{pmatrix} 52 & 0 & 0 \\ 0 & 0 & 0 \\ 0 & 0 & -52 \end{pmatrix}$$

(b) for deformation by radial extrusion ($\sigma_\theta = -125\,\text{MPa}$, $\sigma_r = -177\,\text{MPa}$, $\sigma_z = -229\,\text{MPa}$, $\sigma_m = -177\,\text{MPa}$)

$$\mathbf{T}_\sigma = \mathbf{A}_\sigma + \mathbf{D}_\sigma \tag{5.27}$$

$$\begin{pmatrix} \sigma_\theta & 0 & 0 \\ 0 & \sigma_r & 0 \\ 0 & 0 & \sigma_z \end{pmatrix} = \begin{pmatrix} \sigma_m & 0 & 0 \\ 0 & \sigma_m & 0 \\ 0 & 0 & \sigma_m \end{pmatrix} + \begin{pmatrix} \sigma_\theta - \sigma_m & 0 & 0 \\ 0 & \sigma_r - \sigma_m & 0 \\ 0 & 0 & \sigma_z - \sigma_m \end{pmatrix} \tag{5.28}$$

$$\begin{pmatrix} -125 & 0 & 0 \\ 0 & -177 & 0 \\ 0 & 0 & -229 \end{pmatrix} = \begin{pmatrix} -177 & 0 & 0 \\ 0 & -177 & 0 \\ 0 & 0 & -177 \end{pmatrix} + \begin{pmatrix} 52 & 0 & 0 \\ 0 & 0 & 0 \\ 0 & 0 & -52 \end{pmatrix}$$

It follows that radial extrusion is realised at a hydrostatic pressure which is considerably lower than that for direct extrusion. Hence the first cracks may appear behind, and not in front of, the die.

Theoretically, plastic deformation by the CEC method in frictionless conditions can be realised at a minimal backpressure, equal to $p_{c\_min} = \sigma_f$. However, this value is too small, as it is only sufficient for frictionless compression of the material flowing out of the die. From the presented experiment it can be seen that the required backpressure ($p_c = 229\,\text{MPa}$) is considerably greater than the flow stress ($\sigma_f = 90\,\text{MPa}$). Such a difference between these parameters ($p_c >> \sigma_f$) results from the fact that the applied backpressure should not only completely fill the tool space behind the die, but should also induce an appropriately high hydrostatic pressure, without allowing loss of cohesion in the material being processed. At high hydrostatic pressure, the conditions for uniform deformation as well as formation of a great number of intersecting shear bands are favourable. Thanks to these conditions, which characterise the CEC deformation method, it is possible to obtain material of the same dimensions as the initial ones, but with a completely different structure and very advantageous mechanical properties.

5.7 Plastic consolidation of metallic powder materials

Two types of processes can be used to consolidate the powder materials: either conventionally (thermal consolidation by diffusion welding) or unconventionally (plastic consolidation by adhesive bonding in the presence of a high hydrostatic pressure and a large plastic deformation). Thermal consolidation is achieved by techniques such as: cold pressing and sintering, cold isostatic pressing and sintering, hot isostatic pressing, explosive compaction, and hot extrusion. The plastic consolidation of metallic powder materials is mainly realised by SPD methods (Alexandrov *et al.*, 1998; Zhang *et al.*, 2006; Senkov *et al.*, 2005; Xia, 2010).

Silver-based refractory materials produced by powder metallurgy are used extensively as electrical contact materials due to their high conductivity, good resistance to welding, corrosion properties and hardness (Findik and Uzun, 2003). There are several successful methods including powder metallurgy and internal oxidation which can be used to prepare contact materials. The CEC method was applied to cold plastic consolidation of silver alloy powders (AgSn7.5Bi0.5 and AgNi10) which are used in the production of electric contacts (Richert *et al.*, 2010, 2011, 2012). The metallic powders were deformed in specially prepared copper containers (Figure 5.28).

The copper containers were filled successively. Each small portion of the loosely poured powder was pressed with a force of about 400N by a hard steel

(a) (b) (c)

Figure 5.28 Copper containers used for plastic consolidation of powders: (a) container construction; (b) filling containers and gradually pressing powder; (c) ejection of deformed material from die

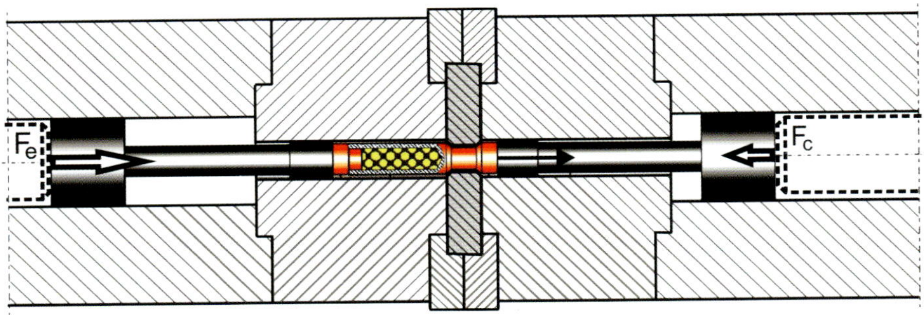

Figure 5.29 Tool configuration for powder consolidation using CEC hydraulic press: F_e = force of CEC process, F_c = counterforce

tool. The fully filled container was then tightly closed by pressing a copper disc with a diameter of 8 mm into the orifice with a force of about 500 N. The filled copper container was then inserted into the tool workspace. Additional copper discs with a diameter of 10 mm and thickness of 4 mm were located on both sides of this filled container. Thus, it was possible to easily remove the deformed material from the die orifice after a CEC process (Figure 5.28(c)) had been completed.

The CEC processing was performed on a hydraulic press with 600 kN capacity, with the working punches at a constant velocity of 0.15 mm/s. The tools used were similar to those used for monolithic materials (Figure 5.29). For simplicity, a constant counterforce of F_c = 22 kN was applied to the counterpunch.

Silver alloy powders (AgSn7.5Bi0.5 and AgNi10), with a mean particle size of 40–120 µm, were deformed in the range of true strain $\varepsilon = 0.42–25.3$, which corresponded to 1–60 CEC cycles of the process (Figures 5.30 and 5.31). In a single CEC cycle the true strain of $\varepsilon = 0.42$ was applied, which resulted from

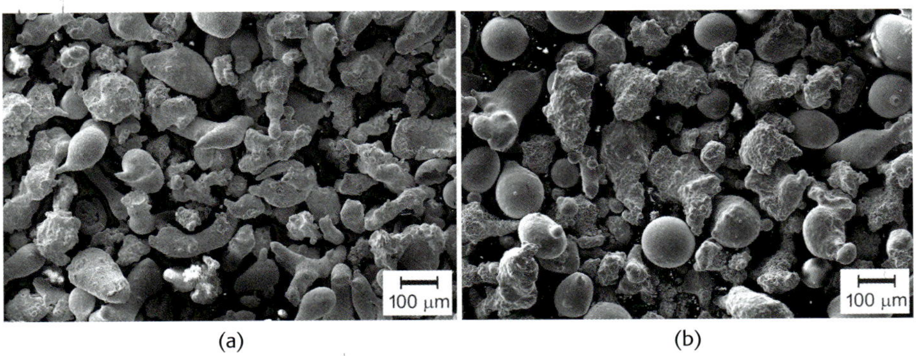

Figure 5.30 Scanning electron micrographs of silver alloy powders: (a) AgSn7.5Bi0.5; (b) AgNi10

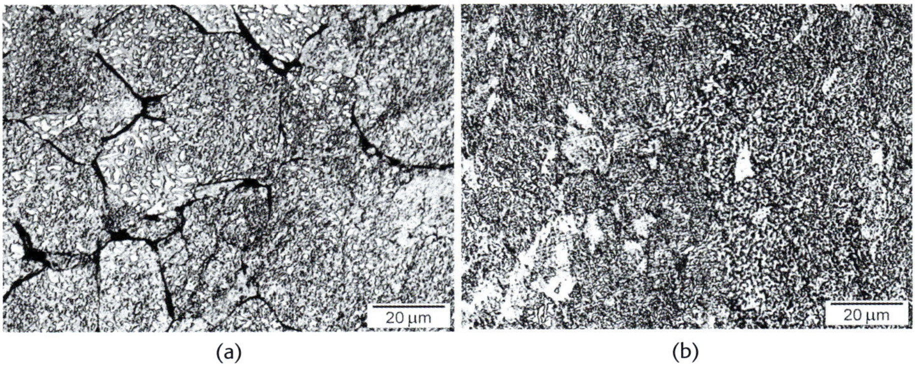

Figure 5.31 Microstructure of AgSn7.5Bi0.5 powder after true strain of: (a) $\varepsilon = 0.42$ (1 cycle); (b) $\varepsilon = 25.3$ (60 cycles)

two deformation stages coded 10–9–10 mm (the 10 mm diameter of the chamber orifice and the 9 mm diameter of the die orifice).

At the start of the CEC process the initial material (a copper container with metallic powder) does not fill the tools' entire workspace. The transverse dimensions of the initial material are somewhat smaller than those of the tool space. For this reason, in the first cycle of the CEC process, the punch displacement is markedly larger (Figure 5.32) and the punch displacement time is longest (Figure 5.33).

During the CEC process the distance between the working punches gradually decreases. Successive cycles of the CEC process cause increased densification of the powdered material. Under the influence of a high hydrostatic pressure and large plastic deformation, the pores gradually disappear. The mass

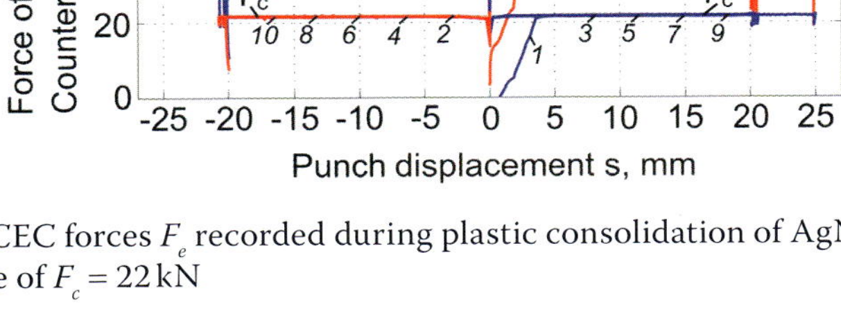

Figure 5.32 CEC forces F_e recorded during plastic consolidation of AgNi10 powder at a counterforce of $F_c = 22\,kN$

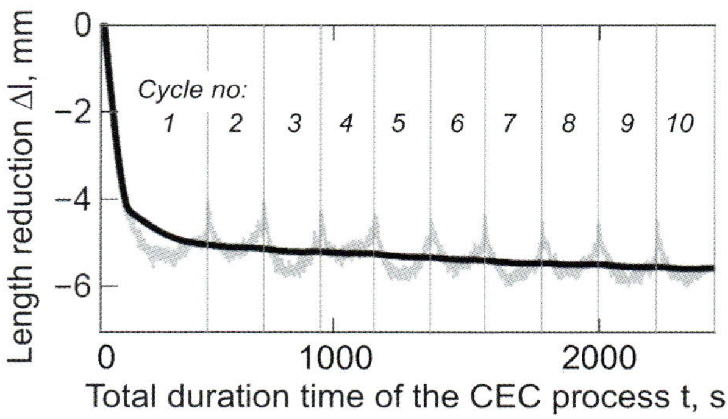

Figure 5.33 Length reduction during plastic consolidation of AgNi10 powder depending on duration of CEC process

density of the powdered material increases because the pore volume decreases. The pressed material becomes shorter, but in the second and following cycles, its length does not become considerably smaller (Figures. 5.32 and 5.33).

The preservation of a constant distance between the punches provides evidence of a complete elimination of porosity. However, an additional deformation of the early-integrated powder material is very favourable. Under such conditions the adhesion forces become even larger. Very large plastic strain leads to very good consolidation.

The transmission electron microscopy investigations of AgSn7.5Bi0.5 show nanometric sub-grains/grains inside the consolidated granules. The numerous nano-twins clearly visible inside the sub-grains were a characteristic feature of the microstructure. The statistical evaluation indicated that after the true strain was in the range $\varepsilon = 13.5–23.5$ (32–60 CEC cycles) the average size of nano-grains was reduced to 50–80nm (Figure 5.34).

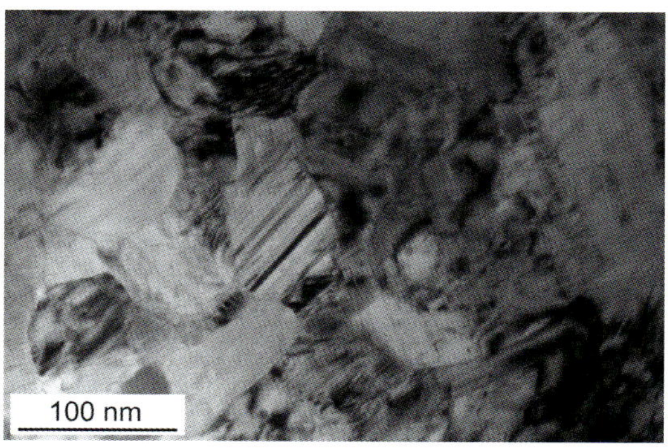

Figure 5.34 Nano-grains and nano-twins in AgSn7.5Bi0.5 granules after true strain of $\varepsilon = 13.5$ (32 CEC cycles)

The CEC method not only allows full densification, but also the plastic consolidation of metallic powders. SPD ensures a very high degree of surface restoration of powdered material particles. The clean physical contact of the metallic surfaces, in the presence of a high hydrostatic pressure, enables the creation of perfect adhesive bonds.

References

Agnew, S. R., Mehrotra, P., Lillo, T. M., Stoica, G. M. and Liaw, P. K. (2005) 'Texture evolution of five wrought magnesium alloys during route a equal channel angular extrusion: Experiments and simulations', *Acta Materialia*, 53: 3135–3146.

Alexandrov, I. V., Zhu, Y. T., Lowe, T. C., Islamgaliev, R. K. and Valiev, R. Z. (1998) 'Consolidation of nanometer sized powders using severe plastic torsional straining', *Nanostructured Materials*, 10: 45–54.

Bowen, J. R., Gholinia, A., Roberts, S. M. and Prangnell, P. B. (2000) 'Analysis of the billet deformation behaviour in equal channel angular extrusion', *Materials Science and Engineering A*, 287: 87–99.

Findik, F. and Uzun, H. (2003) 'Microstructure, hardness and electrical properties of silver-based refractory contact materials', *Materials and Design*, 24: 489–492.

Jining, Q., Han, J. H., Guoding, Z. and Lee, J. C. (2004) 'Characteristic of textures evolution induced by equal channel angular pressing in 6061 aluminum sheets', *Scripta Materialia*, 51: 185–189.

Korbel, A. and Richert, M. (1985) 'Formation of shear bands during cyclic deformation of aluminium', *Acta Metallurgica*, 33: 1971–1978.

Korbel, A., Richert, M. and Richert, J. (1981) 'The effects of very high cumulative deformation on structure and mechanical properties of aluminium', *Proceedings of 2nd Riso International Symposium on Metallurgy and Materials Science*, held 14–18 September 1981, Riso, Denmark, pp. 445–450, Riso National Laboratory, Roskilde, Denmark.

Kurzydłowski, K. J., Richert, M., Richert, J., Zasadziński, J. and Suś-Ryszkowska, M. (2004) 'Effect of non-conventional large deformations on the formation of nanostructure in materials', *Solid State Phenomena*, 101/2: 31–36.

Lapovok, R. Ye. (2006) 'The positive role of back-pressure in equal channel angular extrusion', *Materials Science Forum*, 503/4: 37–44.

Polak, J., Man, J. and Obrtlik, K. (2003) 'AFM evidence of surface relief formation and models of fatigue crack nucleation', *International Journal of Fatigue*, 25: 1027–1036.

Richert, J. (1991) 'The analysis of the effect of the shape of tools on the stability conservation of the plastic flow during radial extrusion of metals', *Metallurgy and Foundry Practice*, no. 135 (special bulletin) (in Polish).

Richert, J. (1988) 'Stability conditions of metal flow in radial extrusion', *Zeitschrift fur Metalkunde*, 79: 248251.

Richert, J. (2001) 'Optimal conditions of plastic working of materials by using cyclic extrusion compression method (CEC)', *Inżynieria Materiałowa*, 117: 156–160 (in Polish).

Richert, J. (2006) 'Method and tools and press for reciprocating extrusion', Patent PL 211118.

Richert, J. (2010a) *Innovative Methods of Metal Plastic Forming*, AGH University of Science and Technology Press, Krakow, Poland (in Polish).

Richert, J. (2010b) 'Strain–stress conditions of shear band formation during CEC processing on a new machine with control back-pressure', *Archives of Metallurgy*, 55(2): 391–408.

Richert, J. and Richert, M. (1986) 'A new method for unlimited deformation of metals and alloys'. *Aluminium*, 62: 604–607.

Richert, J., Richert, M. and Mroczkowski, M. (2008) 'New possibilities for intense plastic deformation of aluminium alloys on a special CEC press', *International Journal of Material Forming*, 1(Suppl. 1) (ESAFORM 2008): 479–482.

Richert, J., Richert, M. and Mroczkowski, M. (2009) 'The role of hydrostatic pressure in intense plastic deformation of materials by the innovative new CEC hydro-press', *Inżynieria Materiałowa*, 30(1): 12–16.

Richert, J., Richert, M., Zasadziński, J. and Korbel, A. (1979) 'Method and device for plastic processing of metals and alloys', Polish Patent 123026.

Richert, M. (1990) 'The effect of unlimited cumulation of large plastic strains on the structure-softening processes of 99.999 Al', *Materials Science and Engineering A*, 129: 1–10.

Richert, M. and Korbel, A. (1995) 'The effect of strain localization on mechanical properties of Al99,992 in the range of large deformations', *Journal of Materials Processing Technology*, 53: 331–340.

Richert, M., Hansen, N., Richert, J., Jensen, D. J., Liu, Q. and Godfrey, A. (1998) 'Formation of fine grains in aluminium deformed to large strains', *Inżynieria Materiałowa*, 19: 502–505.

Richert, M., McQueen, H. J. and Richert, J. (1998) 'Microband formation in cyclic extrusion compression of aluminium', *Canadian Metallurgical Quarterly*, 37: 449–457.

Richert, M., Richert, J., Hotloś, A., Pałka, P., Pachla, W. and Perek, M. (2011) 'Ag powders consolidated by reciprocating extrusion (CEC)', *Materials Science Forum*, 667/9: 145–150.

Richert, M., Richert, J., Książek, M., Hotloś, A., Pałka, P., Perek, M. and Maślanka, M. (2012) 'Microstructure of AgNi and AgSnBi powders consolidated by CEC', *Solid State Phenomena*, 186: 130–134.

Richert, M., Richert, J., Leszczyńska-Madej, B., Hotloś, A., Maślanka, M., Pachla, W. and Skiba, J. (2010) 'AgSnBi powder consolidated by composite mode of deformation', *Journal of Achievements in Materials and Manufacturing Engineering*, 39: 161–167.

Richert, M., Stüve, H. P., Richert, J., Pippan, R. and Motz, Ch. (2001) 'Characteristic features of microstructure of AlMg5 deformed to large plastic strains', *Materials Science and Engineering A*, 301: 237–243.

Rosochowski, A. (2005) 'Processing metals by severe plastic deformation', *Solids State Phenomena*, 101/2: 13–22.

Segal, V. M. (2002) 'Severe plastic deformation: Simple shear versus pure shear', *Materials Science and Engineer*ing A, 338: 331–344.

Segal, V. M., Reznikov, V. I., Drobyshevkiy, A. E. and Kopylov, V. I. (1981) 'Plastic working of metals by simple shear', *Russian Metallurgy*, 1: 99–105.

Semiatin, S. L. and DeLo, D. P. (2000) 'Equal channel angular extrusion of difficult-to-work alloys', *Materials and Design*, 21: 311–322.

Senkov, O. N., Senkova, S. V., Scott, J. M. and Miracle, D. B. (2005) 'Compaction of amorphous aluminium alloy powder by direct extrusion and equal channel angular extrusion', *Materials Science and Engineering A*, 393: 12–21.

Wu, S. D., Wang, Z. G., Jiang, C. B., Li, G. Y., Alexandrov, I. V. and Valiev, R. Z. (2004) 'Shear bands in cyclically deformed ultrafine grained copper processed by ECAP', *Materials Science and Engineering A*, 387/9: 560–564.

Xia, K. (2010) 'Consolidation of particles by severe plastic deformation: mechanism and applications in processing bulk ultrafine and nanostructured alloys and composites', *Advanced Engineering Materials*, 12(8): 724–729.

Zhang, L. C. Z., Xu, J. and Ma, E. (2006) 'Consolidation and properties of ball-milled $Ti_{50}Cu_{18}Ni_{22}Al_4Sn_6$ glassy alloy by equal channel angular extrusion', *Materials Science and Engineering A*, 434: 280–288.

CHAPTER

6

Twist extrusion (TE)

Yan Beygelzimer, Victor Varyukhin, Roman Kulagin and Dmytro Orlov

6.1 Introduction and historical retrospective

Twist extrusion (TE) is a technique in which a billet is extruded through a constant cross-section die having two prismatic sections separated by a helical one. This allows the accumulation of plastic strain in a billet without any alteration in its net shape. Such a property makes possible the accumulation of very large (or severe) strains when multiple-pass extrusion is used. The principal scheme of a channel in a die for TE is shown in Figure 6.1 along with a billet before and after a passage through a TE die. The profile of a TE die cross-section can be arbitrary (Figure 6.2).

The principal ability of TE is that it can process bars that have a circular cross-section profile. This can be achieved when the axis of extrusion is shifted away from the axis of symmetry in a channel. Figure 6.2(c) shows the extrusion axis (indicated by '+'), which is located away from the centre of symmetry of the channel (in the middle of cross-section). Moreover, tubular billets with a hollow centre can also be processed by TE (Figure 6.2(d)).

The idea of TE was proposed by Beygelzimer in 1999. Thereafter, the idea was implemented in the Department of High-Pressure Physics and Advanced Technologies, Donetsk Institute for Physics and Engineering, National Academy of Sciences of Ukraine led by Varyukhin. The team working on TE also included Synkov and Sapronov (Beygelzimer *et al.*, 1999), and the development of TE technique was further carried out during a series of PhD projects under Beygelzimer's supervision from Orlov to Kulagin (Orlov, 2003; Reshetov, 2006; Prokof'eva, 2008; Synkov, 2011; Kulagin, 2014).

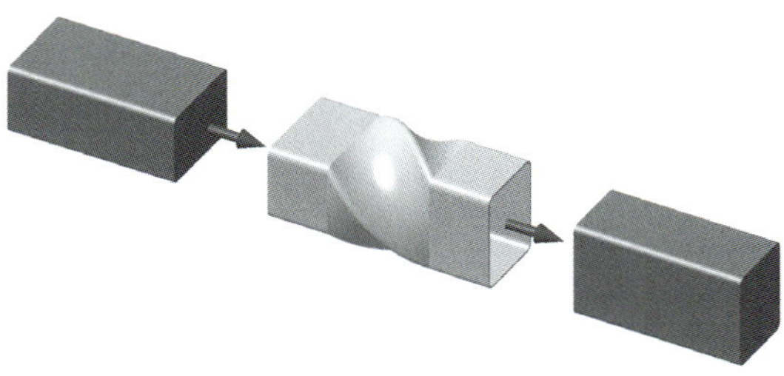

Figure 6.1 Principal scheme of a die channel for TE and a billet before and after one pass through a TE die

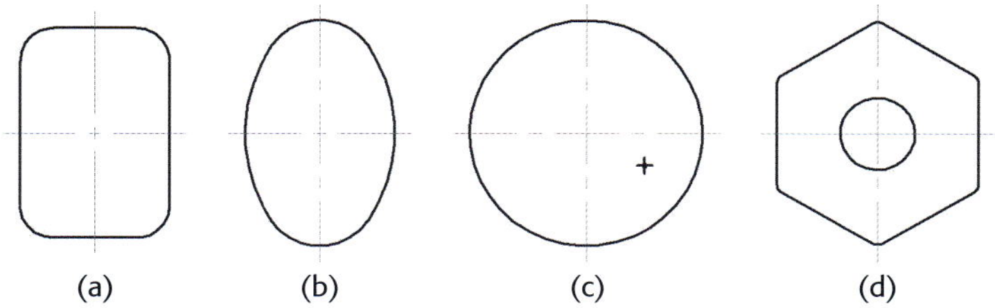

Figure 6.2 Examples of possible profiles of TE die cross sections: (a) rectangular; (b) elliptical; (c) circular–extrusion axis, indicated by '+', is located away from centre of channel symmetry, in the middle of cross-section; (d) hexagonal billets with a hollow centre

Due to a wide previous experience of working with hydro extrusion, the first tool set implementing the idea of TE was based on a hydro-extrusion machine. A pre-existing tool set for hydro extrusion was converted into a tool set for hydro-mechanical TE (Figure 6.3). A brief description of this tool set is as follows. A billet (3) with rectangular cross-section 14 mm × 16 mm and length 60 mm is extruded through a TE die (4). The die sits on a supporting liner (5), which stays on a conical die (7). This conical die is used for the direct extrusion

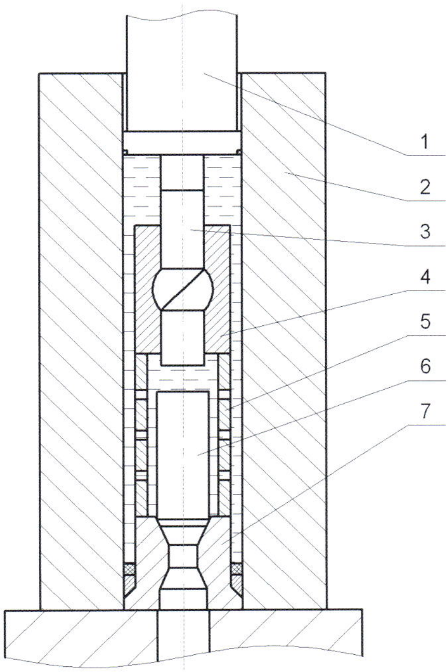

Figure 6.3 Tool set for hydro-mechanical TE: 1 - punch, 2 - container, 3 - billet, 4 - die, 5 - liner, 6 - dummy billet, 7 - conical die

of a dummy billet (6). The channel of a high-pressure container (2) is filled with a working fluid (mineral oil I-20). Punch (1) compresses the working fluid thus increasing hydrostatic pressure in the container. At a pressure of 600–700 MPa, the dummy billet begins being extruded through the conical die (7), while the punch pushes the billet (3) through the TE die (4) while moving along the container axis. The process stops when the punch touches the top edge of the TE die. A new dummy billet is then installed in the conical die, the TE die is flipped by 180°, and the process is repeated.

The main advantage of hydro-mechanical TE is that the process can be carried out under high hydrostatic pressure with no friction of a billet on channel walls. The main disadvantage of this process is its very low productivity. This tool set was therefore re-designed, and a new tool set for a mechanical TE more similar to conventional direct extrusion was adopted (Figure 6.4). It consists of a two-ring upper container (2) and a two-ring TE die (6) having the same entry cross-section. The die has a slope of 30–60° between the twist line and the extrusion axis. Its exit cross-section is twisted with regard to the entrance for 30–120°. The bottom part of the TE die has a minor reduction in the channel cross-section to compensate for residual elastic stresses and other distortion defects in the billet being processed. Receiving container (8),

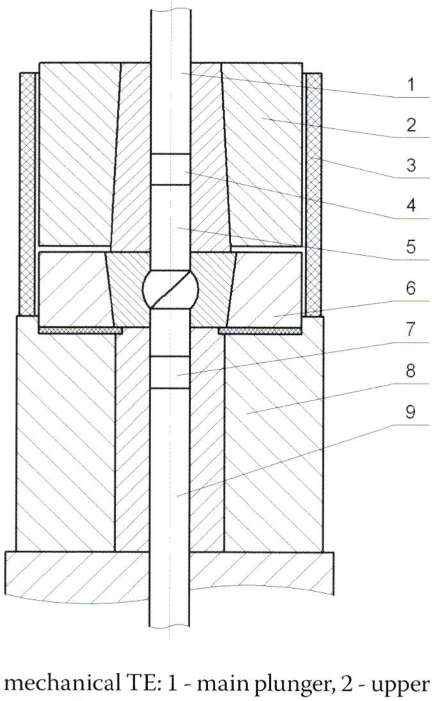

Figure 6.4 Tool set for mechanical TE: 1 - main plunger, 2 - upper container, 3 - resistance heater, 4 - dummy billet, 5 - billet, 6 - die, 7 - dummy billet, 8 - bottom container, 9 - back pressure plunger

having the same cross-section, is located underneath the die. It also prevents plastic deformation of the billet outside the die when a back pressure is used. The upper container and the die are encased in a resistance heater (3), which allows processing at temperatures of up to 450°C.

A typical extrusion cycle is as follows. Back-pressure plunger (9) is inserted through bottom container (8) up to an exit from the helical part of the TE die (6). A dummy billet is pressed into the helical part of the die. A billet to be processed (5) and another dummy billet (4) are loaded into the upper container (2). All the billets are then extruded through the die by main plunger (1). During extrusion, back pressure is maintained at a level of 0.5–1.0 σ_s (where σ_s is the flow stress of billet material) to ensure appropriate kinematics of plastic flow. In addition, this level of back pressure minimises the loss of ductility of a material being processed due to a high hydrostatic pressure in the centre of deformation (Bridgman, 1952), so that even materials that are hard-to-deform can be processed. The extra dummy billet (4) is needed to extrude the entire main billet through the TE die in one pass. After the extrusion cycle, the main billet (5) and the dummy billet (7) exit from the bottom container, while the dummy billet (4) remains in the helical part of the die. At this stage, the extrusion cycle can be repeated, if necessary.

Copper (Cu), zinc (Zn) and their alloys have been used as dummy billets. Lubricants were selected on the basis of the material to be processed and processing regimes. For extrusion at ambient temperatures, Teflon, beeswax, rosin and castor oils were used. For extrusion at elevated temperatures, low melting point glassy lubricants based on lead (Pb) and Zn oxides were utilised.

The results of experiments performed utilising the aforementioned tool sets as well as the first theoretical studies of TE have been rather well reported (Beyygelzimer *et al.*, 2002, 2003; Stolyarov *et al.*, 2005). The effectiveness of TE processing for microstructure and property tailoring in metallic materials have been prove for titanium alloys, commercially pure titanium (CP Ti) grades 2 and 4, VT-22, and VT3-1; Super alloy Inconel 718; Al–Mg–Sc alloys; recycled grades of copper–phosphorus and aluminium alloys AK9, AM5M2, AV87, etc.

At the moment, three directions for applications of TE processing can be highlighted

- manufacturing of bulk nano-structured materials through grain refinement (cf. Beygelzimer *et al.*, 2002, 2003, 2006, 2009a, 2015; Stolyarov *et al.*, 2005, 2010; Orlov 2003; Varyukhin *et al.*, 2006, 2008, 2011; Berta *et al.*, 2007; Orlov *et al.*, 2004, 2006, 2008, 2009a–c, 2012; Bahadori *et al.*, 2010; Latypov *et al.*, 2014, 2015; Reshetov *et al.*, 2011; Zendehdel *et al.*, 2012)
- improvement of ductility in recycled non-ferrous alloys (cf. Shevelev *et al.*, 2010)
- manufacturing of bulk nano-structured materials through powders consolidation (cf. Shpak *et al.*, 2006; Olshanetskii *et al.*, 2014; Pavlenko *et al.*, 2015)

Collaboration with industrial partners had been established in each of these directions. For example, technologies for turbine blades manufactured from nano-structured Ti alloys and chemically homogeneous Ti alloy wires for mono-wheels restoration are being developed with 'Motor Sich'. In collaboration with 'Zaporozhskiy titano-magieviy kombinat', a technology for Ti sponge consolidation into electrodes for electro slag refining has been developed. A technology for chemical homogeneity improvement in recycled non-ferrous alloys has been designed for JSC 'Donsplav'. Ti-6-4 and Inconel 718 were processed by TE for GE Global Research. In collaboration with Boeing, the properties of Al alloys for aviation were tailored by TE processing.

A pilot industrial tool set for TE processing has been established at Donetsk Institute for Physics and Engineering of the National Academy of Sciences of Ukraine. It allows the processing of 1200 kg of billets per annum with dimensions 30 mm × 40 mm × 140 mm. This pilot industrial line is designed to manufacture nano-structured titanium.

6.2 Mechanics of plastic flow in TE

We now consider the characteristic features of TE and compare them to equal channel angular extrusion (ECAE) and high-pressure torsion (HPT). The results of experimental and theoretical investigations of stress–strain state in TE processing are then presented. In particular, a minimal level of back pressure to ensure the kinematics of plastic flow in TE is determined, a vortex-like plastic flow within the cross-section of the TE channel is revealed, and the effect of strain hardening on plastic flow in TE is investigated.

6.2.1 Characteristic properties of deformation in TE

As has been revealed by comprehensive experimental and theoretical investigations (Beygelzimer *et al.*, 2002, 2003, 2009a,b; Orlov 2003; Orlov *et al.*, 2009, 2012; Varyukhin *et al.*, 2011), the major features of plastic flow during TE were grasped in early work by Beygelzimer *et al.* (2000), who proposed a simplistic kinematically admissible velocity field.

$$V_x = -\frac{yV_0 \tan\beta}{R}, \quad V_y = \frac{xV_0 \tan\beta}{R}, \quad V_z = -V_0 \tag{6.1}$$

Where x, y, z are coordinates, the z-axis coincides with the TE die axis of symmetry, V_0 is the velocity of a billet moving along the z-axis, R is the radius of circumference on the channel cross-section periphery, and β is the angle of the twist line slope on the channel cross-section periphery.

The field of velocities (Equation (6.1)) presumes that each virtual cross-sectional layer of the billet acts as a solid having translational displacements along the extrusion axis and a rotational one about it. No distortion of the layer in the z-direction is possible. This field of velocities is used to derive the following tensor of strain rates

$$\begin{aligned} \dot{e}_{xx} &= \dot{e}_{yy} = \dot{e}_{zz} = \dot{e}_{xy} = 0 \\ \dot{e}_{xz} &= -\frac{yV_0}{2R\cos^2\beta}\frac{d\beta}{dz} \\ \dot{e}_{yz} &= \frac{xV_0}{2R\cos^2\beta}\frac{d\beta}{dz} \end{aligned} \tag{6.2}$$

If one substitutes Equation (6.2) in the expression for the intensity of strain rates

$$\dot{e}_i = \frac{\sqrt{2}}{3}\sqrt{(\dot{e}_{xx}-\dot{e}_{yy})^2 + (\dot{e}_{xx}-\dot{e}_{zz})^2 + (\dot{e}_{yy}-\dot{e}_{zz})^2 + 6(\dot{e}^2_{xy}+\dot{e}^2_{xz}+\dot{e}^2_{yz})} \tag{6.3}$$

a relation for the calculation of strain rate intensity at any point of the deformation zone is obtained

$$\dot{e}_i = \frac{1}{\sqrt{3}}\frac{rV_0}{R\cos^2\beta}\left|\frac{d\beta}{dz}\right| \tag{6.4}$$

Integrating Equation (6.4) by time from the moment of a material's point of entrance to the TE die until its exit from there, an estimate for the von Mises equivalent strain is obtained

$$e_i = \frac{2}{\sqrt{3}}\frac{r}{R}\tan\beta \tag{6.5}$$

Equation (6.4) suggests that the majority of strain is generated in the thin zones of simple shear deformation located in the vicinities of entrance and exit from the helical part of TE die where $|d\beta/dz|$ has extreme values. Since the derivatives of β in these zones have opposite signs, the deformation in TE processing is principally reversed.

The simple form of Equation (6.5) suggests that the accumulated strain decreases linearly from a billet periphery towards the axis where it is equal to zero. However, later results, both experimental and theoretical, revealed that, although accumulated strain after TE processing is indeed heterogeneous, the strain at a billet axis is well above zero. These results have been partially reported elsewhere (Begelzimer *et al.*, 2000, 2002, 2003, 2006, 2009a,b; Orlov, 2003; Berta *et al.*, 2007; Orlov *et al.*, 2012), and are presented in detail in Section 6.2.2.

We now compare TE with HPT and ECAE, the other popular severe plastic deformation (SPD) techniques. In all of these processes, the shape of a billet does not change substantially during processing, and simple shear is the dominant deformation mode. However, in ECAE the shear direction is inclined at 45° to the extrusion axis. In TE the shear direction is perpendicular to the extrusion axis. In the latter process, strain is heterogeneous within the

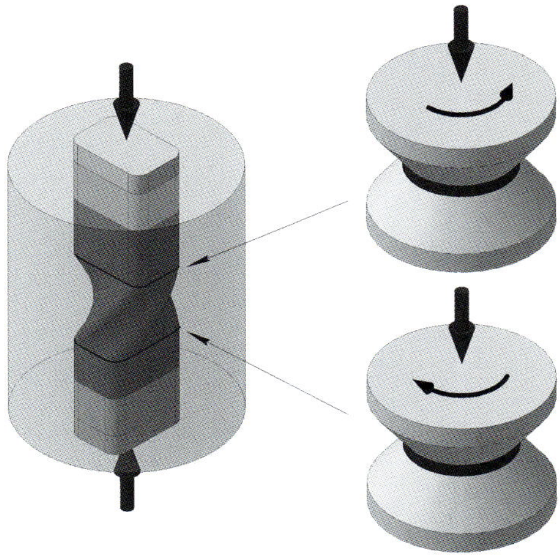

Figure 6.5 TE imagined as an extrusion through two pairs of 'transparent' HPT anvils

cross-section of the billet, which is similar to HPT. Comparing Equation (6.5) and $e_{tor} = r\varphi/\sqrt{3}h$ (where r and h are HPT sample radius and thickness, respectively, and φ is torsion angle), typically used to calculate the von Mises strain in HPT, one can see that each zone of the most intense simple shear in TE can be envisaged as a pair of HPT dies having a specific torsion angle $\varphi/h = \tan\beta/R$. Therefore, to a certain extent, TE can be imagined as an extrusion through two pairs of 'penetrable' HPT anvils (see Figure 6.5).

Dies for TE may have two different directions of twist: clockwise die (CD) and counter-clockwise die (CCD). A sequential transition from CD to CCD or vice versa allows the direction of shear to be flipped between passes, thus leading to two main routes for TE

route I: CD + CD (or CCD + CCD)

route II: CD + CCD (or CCD + CD)

It is obvious that route I imparts strain with amplitude

$$\Delta e_i^I = \frac{1}{\sqrt{3}}\frac{r}{R}\tan\beta$$

while route II

$$\Delta e_i^{II} = 2\frac{1}{\sqrt{3}}\frac{r}{R}\tan\beta$$

Last but not least, the peculiarity of TE is that billet surface increases in the helical part of the TE die. Depending on a specific die design, the area of lateral surface of a billet being processed increases by 30–50% upon entering the helical part of the TE die, and then returns to its initial surface area upon exit. After multi-pass processing, such an effect leads to a significant stirring of the inner and outer layers of the billet being processed. This can be utilised as an extra option for the modification of structure and properties if TE is carried out using surfactant fluids as lubricants (Beygelzimer *et al.*, 2009a).

6.2.2 Stress–strain state in TE

Stress–strain states during TE for different dies, processing conditions and materials have been studied theoretically (Beygelzimer *et al.*, 2002, 2003, 2011; Orlov, 2003; Latypov *et al.*, 2012, 2013, 2015; Beygelzimer and Orlov, 2002; Akbari Mousavi *et al.*, 2008; Kim *et al.*, 2015) and experimentally (Beygelzimer *et al.*, 2003, 2009b; Orlov, 2003; Orlov *et al.*, 2009b; Beygelzimer and Orlov, 2002; Reshetov, 2007). Compared to the ideas presented above, these studies significantly advanced our understanding of plastic flow and the capabilities of TE.

First, they revealed that in order to ensure the proper kinematics of plastic flow and complete filling of a channel in the TE die during processing, a back pressure of at least half the flow stress of a material being processed should be applied (Beygelzimer *et al.*, 2003, 2011; Orlov, 2003; Latypov *et al.*, 2012). The exact value of back pressure depends on the profile of TE channel cross-section, rate of material strain hardening and friction. Figure 6.6 shows photographs of CP Ti billets after TE processing with different levels of back pressure, along with their schematics, which illustrate the effect of incomplete filling of the die channel.

The second most important finding was the discovery of plastic flow in cross-sectional planes, first reported by Beygelzimer *et al.* (2002, 2003) and Orlov (2003). It is illustrated by the distortion of fibres during one TE pass as

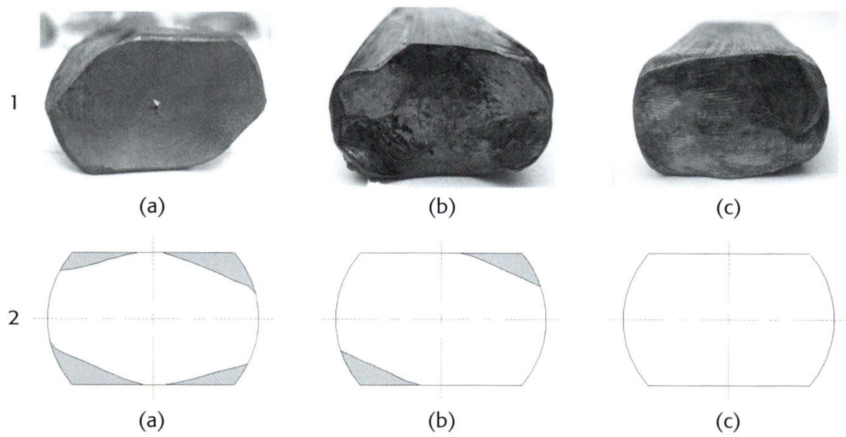

Figure 6.6 Photographs of CP Ti billets after TE processing with different levels of back pressure and corresponding schematics: (a) no back pressure; (b) back pressure ~$0.5\sigma_s$; (c) back pressure ~σ_s, where σ_s is flow stress

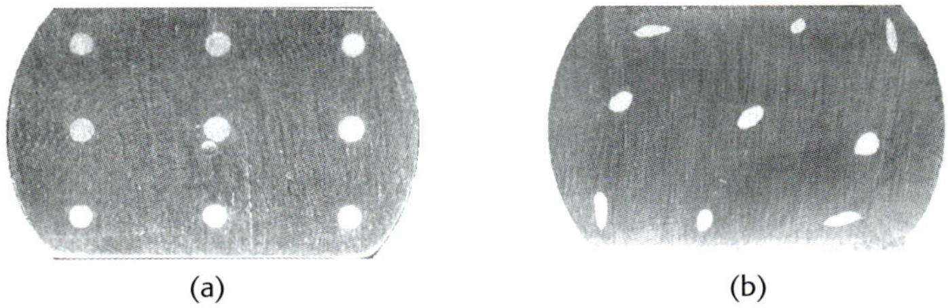

Figure 6.7 Cross-sectional view of a copper billet with embedded aluminium fibres: (a) before and (b) after a single TE pass

shown in Figure 6.7 and reported in detail elsewhere (Beygelzimer *et al.*, 2009b; Reshetov, 2007; Kulagin *et al.*, 2013).

After several passes of TE, this effect leads to a very clear vortex-like flow pattern in billet's cross-section (see Figure 6.8). Further details can be found elsewhere (Orlov *et al.*, 2009a,b, 2012; Kulagin *et al.*, 2013).

The third major improvement of the model presented in Section 6.2.1 is a quantifiable significant strain on the billet axis. Figure 6.9 shows strain distribution in the cross-section of a Cu billet after a single TE pass. It was obtained experimentally using a method described in detail by Beygelzimer *et al.* (2009b).

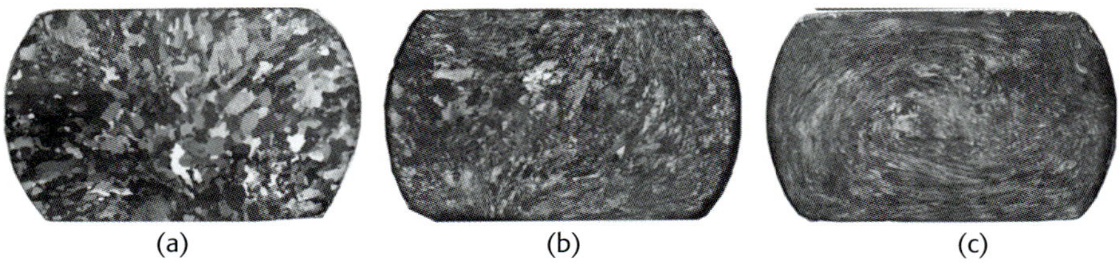

Figure 6.8 Vortex-like flow pattern visualised by optical micrographs of cross-section of pure aluminium billet: (a) before TE; (b) after one TE pass; (c) after four TE passes

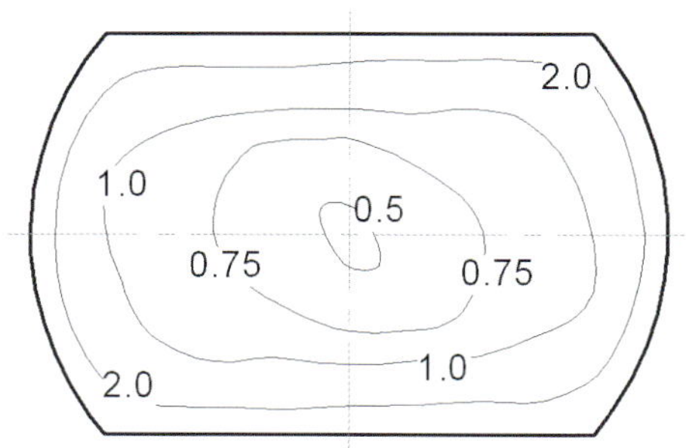

Figure 6.9 Experimental distribution of von Mises equivalent strain in cross-section of a copper billet after a single TE pass

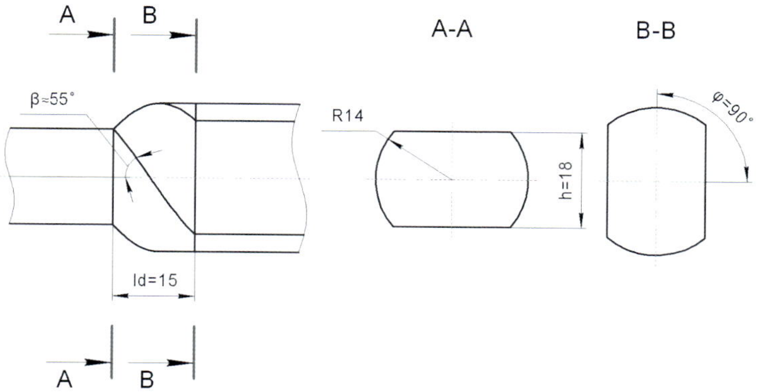

Figure 6.10 Geometry of die used in FE simulations (A-A – input cross-section, B-B – output cross-section)

A finite element code 'Deform-3D' was used to investigate the effect of strain hardening rate on strain distribution after TE processing. The stress–strain curve of a material being processed was described as $\sigma = \sigma_s(1 + A{\cdot}e)$. The coefficient A was varied in the range (0–1) to simulate different rates of strain hardening.

Die and punches were modelled with rigid elements. The Zibel law of friction $\tau = m(\sigma_s/\sqrt{3})$ (where $m = 0.25$) was used to simulate contact between the billet and a die. Back pressure was set at a level sufficient to fill the die channel. The billet material was modelled with tetrahedral elements. Adaptive meshing (automatic re-meshing) was used to accommodate large strains during the simulations. Reduced integration and hourglass control were used in the analysis. The von Mises theory of plastic flow of incompressible materials was used. The geometry of the die used in the simulations is shown in Figure 6.10.

Figure 6.11 illustrates the distribution of von Mises equivalent strain in the billet cross-section after one TE pass for no strain hardening ($A = 0$) and

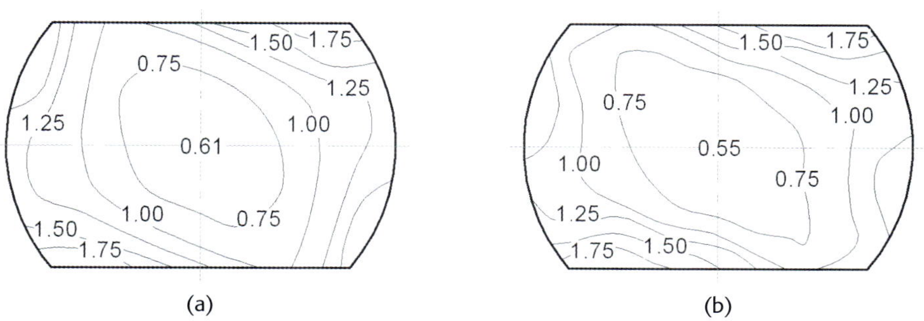

Figure 6.11 Distribution of von Mises equivalent strain in the cross-section of a billet after one TE pass for: (a) no strain hardening ($A = 0$); (b) maximal rate of strain hardening ($A = 1$)

Table 6.1 Effect of strain hardening on heterogeneity parameters in TE

A	e_{min}	k_e	e_{mean}
0.0	0.61	0.27	1.00
0.5	0.64	0.28	1.07
1.0	0.55	0.33	1.07

the maximal rate of strain hardening ($A = 1$). Analysis of this figure reveals a close similarity in the strain distributions, with small differences in strain values.

For the quantitative evaluation of the von Mises strain distribution after TE processing, the following parameters were introduced: minimum e_{min} and mean $e_{mean} = (1/S)\int_S edS$ strains as well as coefficient of strain heterogeneity $k_e = D_e/e_{mean}$, where D_e is the standard deviation of the von Mises strain. Table 6.1 summarises the values of these parameters for materials with the most representative values of strain hardening coefficient A.

These results clearly demonstrate that, with sufficient accuracy for experimental practice, stress–strain state in TE processing can be simulated using an ideally plastic model of a material being processed.

Figure 6.12 demonstrates the accumulation of the von Mises equivalent strain during one TE pass at three of the most representative points of a billet

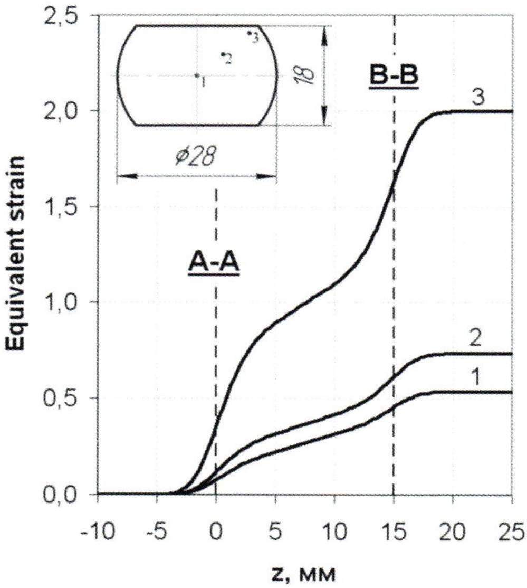

Figure 6.12 Accumulation of von Mises equivalent strain during one TE pass at three most representative points of a billet cross-section (A-A – input cross-section, B-B – output cross-section, as in Figure 6.10)

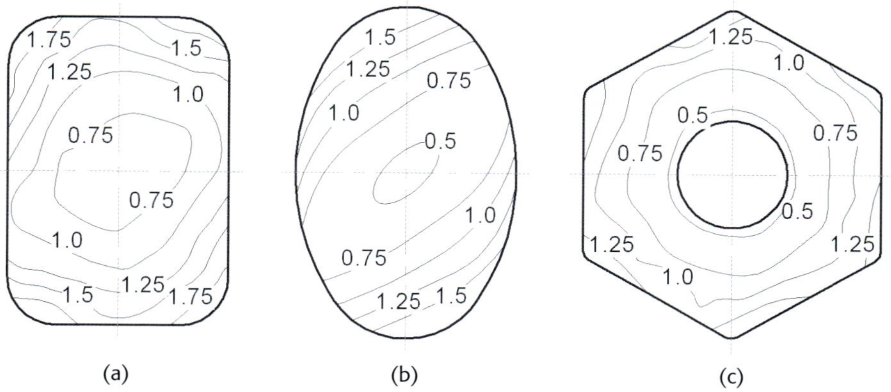

Figure 6.13 Distributions of von Mises equivalent strain in a billet cross-section having: (a) rectangular; (b) oval; (c) hexagonal with centre hole shape

cross-section. The two zones of sharp strain growth corresponding to entrance A–A and exit B–B from the helical part of TE die can easily be located on the diagram for each point. A quantitative analysis of strain accumulation reveals that the dominant mode of deformation in these zones is simple shear. In general, these results are in reasonably good agreement with the simplistic model presented in Section 6.2.1.

Final element simulations also reveal that strain distribution depends strongly on the shape of the cross-section (Figure 6.13). Figure 6.13 also shows that for any cross-section profile, minimum strain is located in the vicinity of the extrusion axis.

When the axis of symmetry of the cross-section profile coincides with the extrusion axis, multi-pass processing does not lead to an improvement in the homogeneity of strain distribution. However, when these axes are shifted away from each other, strain homogeneity can be significantly improved if route II, which involves sequential changing of CD and CCD dies, is used for TE processing. Moreover, this technique even allows samples with a circular cross-section to be processed (Beygelzimer *et al.*, 2015). The strain distribution for such a case is shown in Figure 6.14. It is evident that the homogeneity of strain distribution significantly improves after a second TE pass (Figure 6.14(c)). If the extrusion axis is shifted in each TE pass, a fairly homogeneous strain distribution can be achieved.

When using TE, it is desirable to have as homogeneous distribution of strain along the specimen's axis as possible. It has been proved that the best results are achieved when both back pressure and a dummy billet are used (i.e. when the cross-sectional profile of the TE die channel is completely filled). This result is illustrated in Figure 6.15.

Stress state in TE can be classified as non-equiaxial compression. Therefore the full stress tensor can be split into hydrostatic and deviatoric parts.

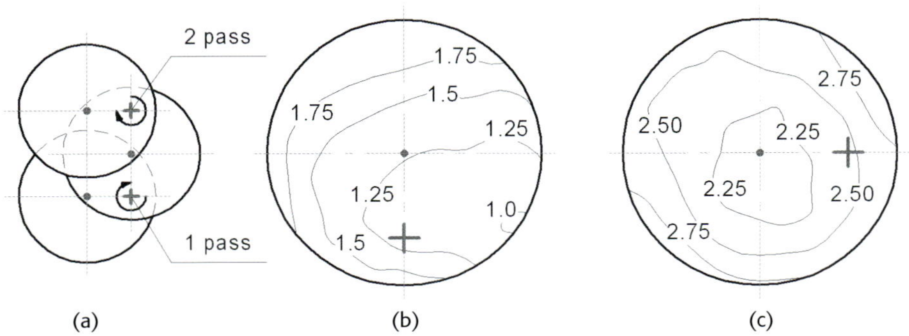

Figure 6.14 Homogeneity of strain in a circular cross-section billet processed using up to two passes of TE: (a) processing scheme characterised by extrusion axis '•' and profile axis '+'; (b) strain distribution after one pass; (c) strain distribution after two passes

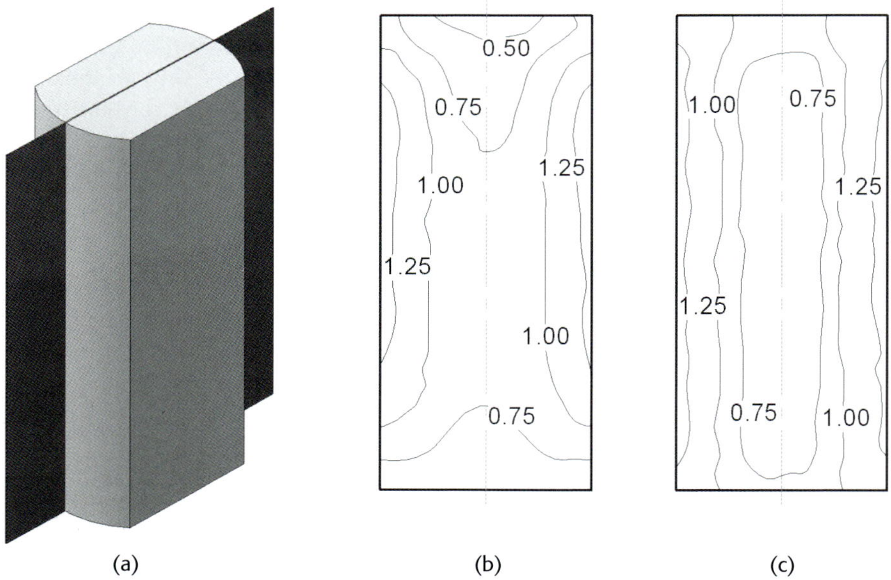

Figure 6.15 Distribution of strain in longitudinal section of a prismatic billet sectioning: (a), results without back pressure; (b), results for back pressure; (c) results for back pressure and dummy billets

The distributions of hydrostatic stress in the cross-section of a billet at entrance and exit from the helical part of TE die are shown in Figure 6.16. Large negative values of hydrostatic stress ensure the best possible ductility in a material being processed (Beygelzimer and Orlov, 2002).

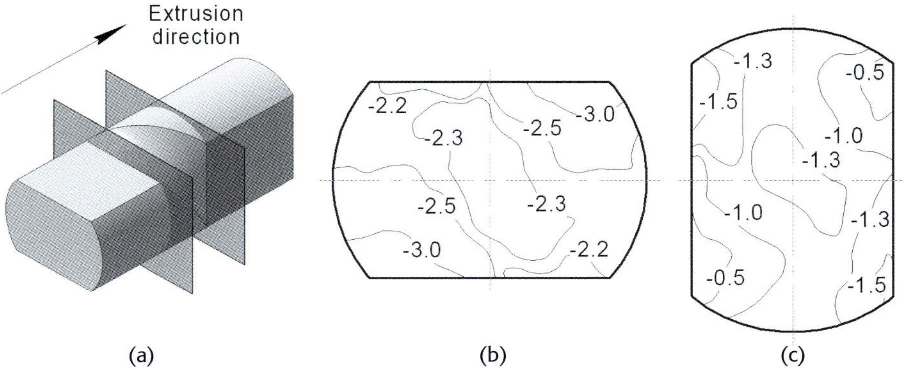

Figure 6.16 Distribution of hydrostatic pressure normalized by flow stress: (a) schematic of entrance and exist sections of TE die; (b) distributions at entrance; (c) distribution at exit

6.3 TE as a processing technique

Equations for the calculation of the main parameters of TE are now presented, and a tool set for TE located in the pilot production facility at Donetsk Institute of Physics and Engineering of the National Academy of Sciences of Ukraine is described.

6.3.1 Principal equations for estimating processing characteristics

The development of a TE technology is based on the analysis of the heterogeneity parameters that characterise the process. The following parameters describing the homogeneity of strain distribution in TE were introduced in Section 6.2.2: minimum e_{min} and mean e_{mean} strains, and coefficient of strain heterogeneity k_e. In order to obtain simple engineering equations for these parameters, a series of numerical experiments was performed using a finite element code (Deform–3D). The scheme of a TE die channel used in these experiments is shown in Figure 6.17.

The factors selected for variation were: $x_1 = hs/R$, $x_2 = ld/hs$, $x_3 = h/b$, where hs, R, ld, h, b are geometrical dimensions of the deformation zone indicated in Figure 6.17. Ranges for these factors were selected from the following considerations. Based on the experience gained during experimental work with TE, reasonable values of the twist line slope vary within $\beta = 30–60°$ while the angle of rotation between entrance and exit cross-section is $\varphi = 30–90°$. Since $\tan\beta \cdot hs = 2\pi R$, one can obtain $x_1 = hs/R = 2\pi/\tan\beta$. Substituting β with its common values, the range of variation for $x_1 = 3–11$. For x_2, one has $x_2 = ld/hs = \varphi/2\pi$, and the corresponding range of variation is $x_2 = 0.10–0.25$. The range of variation for $x_3 = 0.5–1.0$. All the factors were varied at three levels. Upper and lower levels were set by the ranges of variation, while the middle level was set as an arithmetic mean of the upper and lower levels.

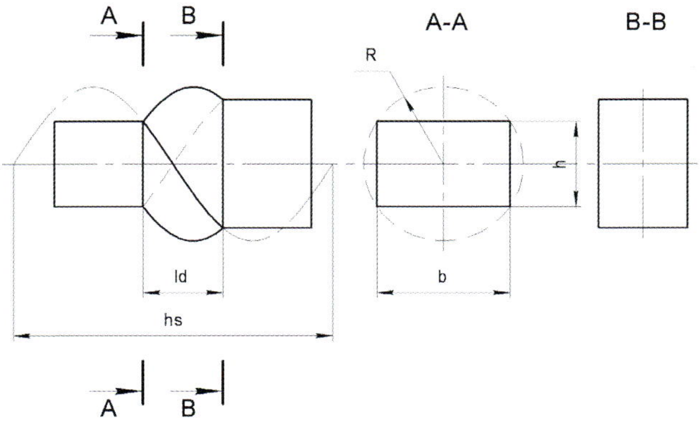

Figure 6.17 Scheme of a TE die channel used in these numerical simulations

Regression analysis of the results of numerical experiments allowed us to obtain the following equations for calculating the heterogeneity parameters for TE processing

$$e_{min} = 3.08 \cdot \left(\frac{hs}{R}\right)^{-0.65} \cdot \left(\frac{ld}{hs}\right)^{0.87} \cdot \left(\frac{h}{b}\right)^{-1.15} \cdot R^2 = 0.97 \tag{6.6}$$

$$k_e = 0.28 \cdot \left(\frac{hs}{R}\right)^{-0.06} \cdot \left(\frac{ld}{hs}\right)^{-0.39} \cdot \left(\frac{h}{b}\right)^{0.76} \cdot R^2 = 0.87 \tag{6.7}$$

$$e_{mean} = 3.46 \cdot \left(\frac{hs}{R}\right)^{-0.47} \cdot \left(\frac{ld}{hs}\right)^{0.55} \cdot \left(\frac{h}{b}\right)^{-0.56} \cdot R^2 = 0.93 \tag{6.8}$$

Where *R*2 is the coefficient of correlation.

Diagrams comparing the heterogeneity parameters of TE processing obtained in numerical experiments to the values of these parameters calculated using Equations (6.6)–(6.8) are shown in Figure 6.18. A very consistent fit of

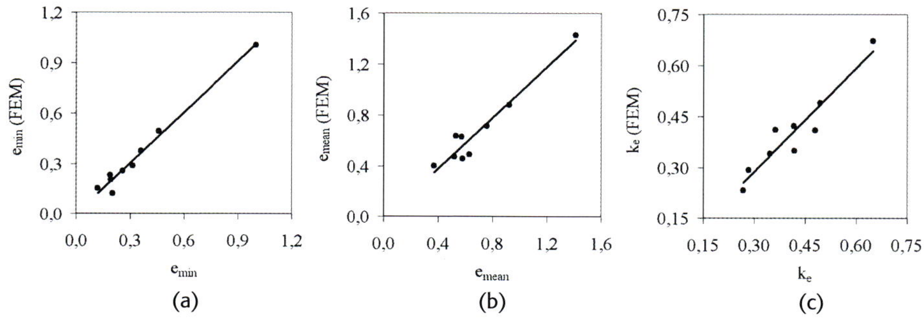

Figure 6.18 Comparison of TE heterogeneity parameters obtained in numerical simulations and calculated using Equations (6.6)–(6.8): (a) e_{min}; (b) e_{mean}; (c) k_e

the experimental data with the corresponding equations is obvious from these diagrams.

Maximum pressure during TE can be calculated using an equation from Beygelzimer *et al.* (2003)

$$p_{\max} = \sigma\left(e_{\text{mean}} + \mu\frac{2(h+b)l}{hb}\right) + p_{bp} \tag{6.9}$$

Where σ is the flow stress of a material being processed on the exit from TE die, l is the length of a billet, and p_{bp} is the back pressure.

6.3.2 Industrial prototype of TE machine

A pilot production facility was created at Donetsk Institute for Physics and Engineering of the National Academy of Sciences of Ukraine. It has a total area of 50 m^2 and includes

- equipment for thermal and chemical treatments
- furnace for preheating billets
- hydraulic press having 4 MN extrusion force capacity on which tool sets for TE and hydro extrusion are installed
- several machines for mechanical operations
- other associated equipment

Figure 6.19 shows a schematic (a) and a photograph (b) of the industrial prototype of TE machine in the pilot production facility.

The operating cycle of the industrial prototype machine is shown in Figure 6.20. Such a sequence is as follows. The receiving container is compressed against the TE die by the upward movement of the upper press ram (Figure 6.20(a)). This action is realised by moving the crosshead and tightening the double-ended bolts. The compression force is controlled by the hydraulic press. It is usually set to 0.2 MN. The position of the receiving container is set by a wedge. The back-pressure plunger is inserted up to the exit section of the helical part of the TE die, through the channel of the receiving container.

A dummy billet is inserted in the die container and pressed down into the helical part of the TE die to establish contact with the back-pressure plunger. The upper ram is lifted, and the billet to be processed (dimensions $30 \times 40 \times 140\,mm^3$) is loaded into the container. An upper dummy billet is then loaded in the container in order to extrude the billet from the helical part of the TE die in one go.

The ram then starts moving down. During this operation, the plunger is first centred by the die container, then it extrudes all the billets. Simultaneously, the extrusion of billets raises the back pressure to a preset level. This allows the kinematic conditions of plastic flow to be satisfied and significantly increases hydrostatic pressure in the deformation zone. A high level of hydrostatic pressure minimises the loss of ductility in the billet and therefore allows hard-to-deform materials to be processed.

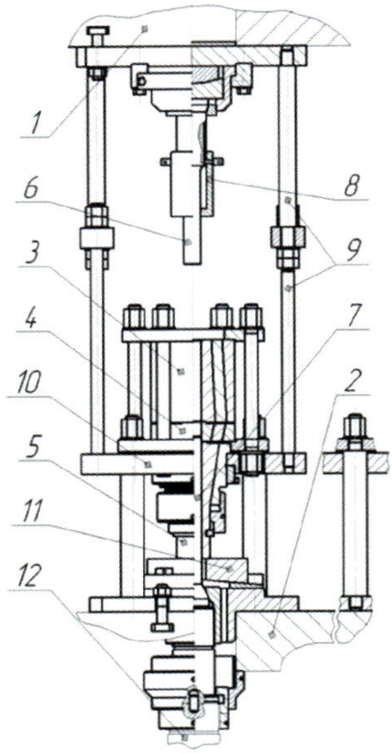

(a)

(b)

Figure 6.19 (a) Industrial prototype of TE machine: (a) assembly drawing comprising: 1 - press ram, 2 - press table, 3 - die container, 4 - die, 5 - receiving container, 6 - plunger, 7 - back-pressure plunger, 8 - stroke limiter, 9 - double-ended bolts, 10 - movable crosshead, 11 - wedge, 12 - back-pressure hydraulic cylinder; (b) photograph

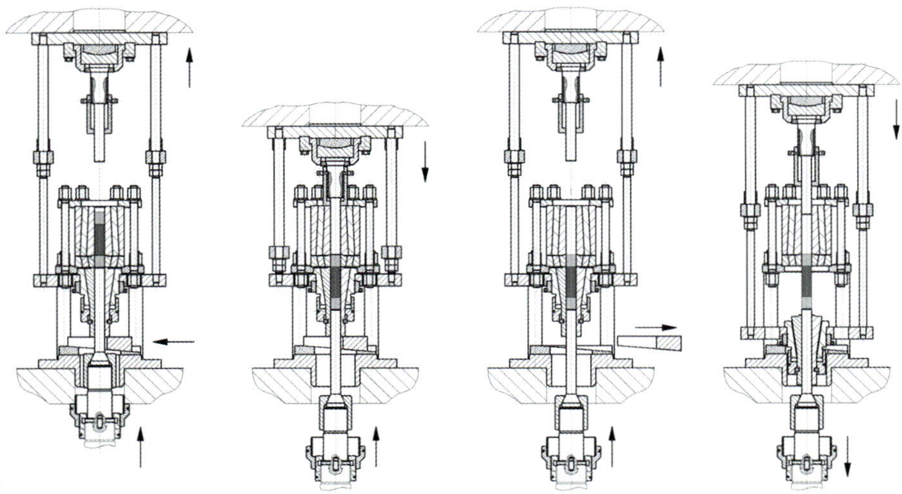

(a) (b) (c) (d)

Figure 6.20 TE machine: (a) achieving starting position; (b) twist extrusion of billet; (c) wedge removal; (d) extraction of billet

The level of back pressure is usually set to the level of flow stress of a material being deformed at the processing temperature. The back-pressure plunger moves down. It leaves space for the billets in the receiving container. At the bottom point of the ram travel, the upper dummy billet replaces the initial dummy billet in the helical part of TE die (see Figure 6.20(b)). This allows multi-pass processing.

After this operation, the receiving container is again compressed against the TE die (see Figure 6.20(c)), which unloads the wedge and allows its removal. During the working travel of the ram, the receiving container also moves down (see Figure 6.20(d)). Therefore, at the end of extrusion operation, the back-pressure plunger is moved down the channel, and the dummy billets are extracted from the receiving container. The entire tool set can be heated by an electric resistance heater when hard-to-deform materials need to be processed.

To summarise, the industrial prototype tool set for TE is installed on a hydraulic press having a capacity of 4 MN. The installation has the following parameters

- extrusion speed 3 mm/s
- maximum extrusion force 2.4 MN
- maximum back-pressure force 2.4 MN
- maximum operating temperature 400°C
- billet size: cross-section $30 \times 40\,mm^2$, length up to 140 mm

6.4 Formation of structure and properties for different applications

The effects of TE processing on the structure and properties of miscellaneous materials have been reported elsewhere (Stolyarov *et al.*, 2005, 2010; Orlov *et al.*, 2004; Begelzimer *et al.*, 2006; Berta *et al.*, 2007; Orlov *et al.*, 2008, 2009a–c, 2012; Varyukhin *et al.*, 2008, 2012; Bahadori *et al.*, 2010, 2015; Reshetov *et al.*, 2011; Zendehdel and Hassani, 2012; Shevelev *et al.*, 2010; Shpak *et al.*, 2006; Reshetov, 2007; Pashinska *et al.*, 2013; Zavdoveev *et al.*, 2015; Titov *et al.*, 2014; Latypov *et al.*, 2016; Bahadori and Akbari Mousavi, 2011; Gubicza *et al.*, 2011). Thus, only the summary of reported results, indicating the most representative data and describing potential applications of TE technique, will be presented in this section.

6.4.1 Major effects of TE

First, intensive grain refinement during TE should be emphasised. The process of grain refinement itself is not the topic of interest in the present chapter. Readers are therefore referred to specific publications dedicated to this problem (Hughes and Hansen, 1997; Kuhlmann-Wilsdorf and Hansen, 1991; Orlov *et al.*, 2012a,b; Rybin, 1986). Briefly, grain refinement is characterised by a decrease in grain size and an increase in grain boundary misorientation angles when the equivalent von Mises strain increases. In TE, grain refinement is very extensive during the first few passes. After the third or fourth pass, depending on the material being processed and processing conditions, the intensity of grain refinement substantially decreases, and mean grain size saturates at the lowest level specific to the material at a given temperature. For processing at low homologous temperatures, mean grain size at the saturation stage is usually of the order of a few hundred nanometres.

Second, the described characteristics of microstructure evolution largely affect the mechanical properties of processed materials. The rate of strain hardening decreases after the first few passes, and the flow stress saturates (Beygelzimer *et al.*, 2003, 2006; Stolyarov *et al.*, 2005; Orlov *et al.*, 2008, 2009b, 2012; Bahadori and Akbari Mousavi, 2011; Gubicza *et al.*, 2011). Such behaviour leads to the homogenisation of mechanical properties in a billet after multi-pass TE processing (Beygelzimer *et al.*, 2006; Orlov *et al.*, 2012; Reshetov *et al.*, 2011). Therefore, despite the heterogeneity of strain distribution described in Section 6.3, billets with ultra-fine grained (UFG) structure manufactured by TE have rather homogeneous mechanical properties.

The effects of strain reversal in TE on the microstructure evolution have been investigated experimentally (Orlov *et al.*, 2009c, 2012; Gubicza *et al.*, 2011). It has been shown that route II (CD + CCD, mentioned in Section 6.2.1), is more efficient for grain refinement. This result is consistent with theoretical investigations (Beygelzimer, 2005) according to which an increase in the amplitude of deformation cycle should lead to more effective grain refinement.

Third, the effect of TE processing on ductility is of interest. Beygelzimer *et al.* (2010) and Bouaziz (2010) have shown that an appropriate measure of ductility in UFG metals is the reduction of area at fracture during tensile testing. This characteristic has non-monotonic dependence on strain in TE. During the first few passes it decreases, but then increases. This effect has been shown for the TE processing of CP Ti (Beygelzimer *et al.*, 2010). A mathematical model which clarifies the physical meaning of this effect has been developed by Beygelzimer (2005).

Finally, deformation during TE processing involves significant mass transfer taking place at different length scales of the material being processed. At the macro scale, mass transfer is reflected by vortex-like plastic flow within the cross-section of the billet (cf. Section 6.2.2). At the micro scale, mass transfer is reflected by the dispersion of brittle intermetallic phases and abnormally fast diffusion (Beygelzimer *et al.*, 2003; Shevelev *et al.*, 2010; Beygelzimer, 2011).

The fast mass transfer can be explained by using the theory of two-stage simple shear (Beygelzimer, 2010a,b, 2011; Beygelzimer *et al.*, 2011). At the first stage, in the range of shear strain $0 < \gamma < \gamma_c$ where γ_c is a parameter of deformation process, the microstructure evolves in a way similar to elongation. At the second stage, when $\gamma > \gamma_c$, incidental multi-scale rotations take place in a way similar to turbulence in fluids. The physical reasons for such rotations are high pressure on the shear plane and asymmetry in the stress tensor caused by the network of high-angle grain boundaries permiting sliding.

At the second stage, large plastic deformations on the macro scale are realised through small cyclic elastic-plastic deformation of material particles inside its representative volume element. At this stage, extensive mass transfer takes place during rotations, deformation-induced dislocations move from sub-grain interiors to their boundaries, and nano-scale voids that were formed at earlier deformation stages heal. As a result, the ductility of the metal increases.

Last, but not least, TE of powder materials leads to very efficient consolidation. Compacting occurs accompanied by the formation of strong mechanical interlocking of particles. This effect has been studied experimentally (Shevelev *et al.*, 2010; Shpak *et al.*, 2006; Begelzimer *et al.*, 2008; Synkov, 2011). Calculations for TE processing of particulate materials, based on a model developed by Beygelzimer *et al.* (1994), have been reported elsewhere (Synkov 2011; Beygelzimer *et al.*, 2007).

6.4.2 Applications of TE

The effects described in Section 6.4.1 can be applied in different ways. The effect of grain refinement has been used to manufacture nano- and UFG structured materials with improved mechanical properties. The homogenisation effect has been used to improve the cast dendritic structure and to increase the ductility of recycled non-ferrous alloys. The powder consolidation effect has been used to consolidate powders and machining chips. Each of these applications is described in detail below.

6.4.2.1 Applications based on the effect of grain refinement

This effect has been used to manufacture UFG structured billets from Al–Mg-based alloys (Orlov *et al.*, 2006). The mean grain size of this material after four TE passes was reduced to 300–500 nm, (Figure 6.21).

The mechanical properties of this alloy are shown in Table 6.2. It can be seen that material processed by four passes of TE has approximately 20% improved

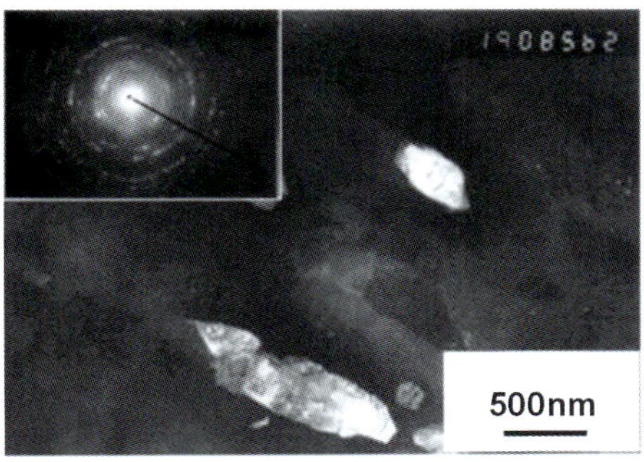

Figure 6.21 Dark field TEM micrograph of microstructure in Al-Mg-Sc-Zr alloy manufactured by TE; SAED pattern is shown in inset

Table 6.2 Mechanical properties of Al-4.45Mg-0.4Mn-0.3Sc-0.1Zr alloy (in wt. %)

Condition	YS (MPa)	UTS (MPa)	El (%)	σ_{-1} (MPa)
Initial	290	400	15	180
Four passes of TE	350	420	10	330

yield strength (YS), slightly higher ultimate tensile strength (UTS), still fairly good elongation to failure (El), and almost doubled fatigue strength (σ_{-1}). Such a material is an excellent candidate for use in the structural components of aircraft and automobiles. The utilisation of this alloy should allow weight savings, reduced fuel consumption and an increased lifetime for vehicle components.

Another material that has excellent prospects in the application of TE based on grain refinement is CP Ti for bio-medical applications. CP Ti (grade 2) has been processed by TE for up to four passes and then rolled at an ambient temperature with 70% reduction in thickness. As a result of such processing, the microstructure of this material has been refined to a sub-micron level (see micrograph in Figure 6.22). The strength increased twofold, while the elongation to failure remained almost unchanged (Table 6.3).

TE processed billets have been used to machine plates for orthopaedic applications (Figure 6.23). For such products, alloy Ti–6Al–4V modified for bio-medical applications is typically used. However, it has become possible to

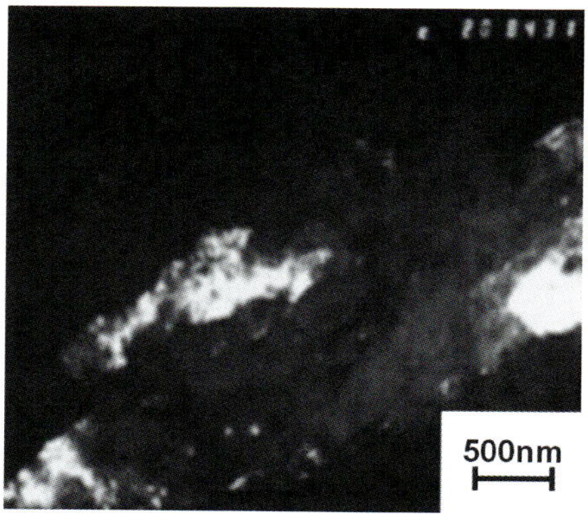

Figure 6.22 Dark field TEM micrograph of microstructure in Grade 2 CP Ti, after processing by four TE passes

Table 6.3 Mechanical properties of CP Ti, Grade 2

Condition	YS (MPa)	UTS (MPa)	El (%)
Initial	350	430	20
Four passes of TE + rolling (70%)	800	840	15

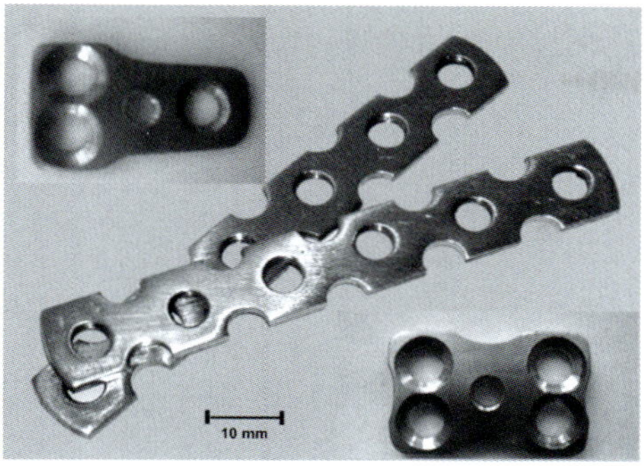

Figure 6.23 Orthopedic plates machined from UFG titanium manufactured by TE

replace it with CP Ti due to the improved mechanical properties. The benefit of using CP Ti instead of Ti–6Al–4V is better bio-compatibility due to the absence of vanadium (V), which produces toxic oxides, and aluminium (Al) which may cause Alzheimer's disease. In addition, due to the improved strength of UFG titanium, the spectrum of implant shapes can be substantially increased, including a reduction in their dimensions.

6.4.2.2 Applications based on the effect of homogenisation

Due to the vortex-like plastic flow in the cross-section of a billet and the stirring effect, TE provides intensive homogenisation of alloys being processed. These effects have been applied to process recycled non-ferrous metals and to produce precision wires for fusion welding electrodes. As can be seen in Table 6.4, the strength (YS, UTS) and ductility (E1, RA- reduction of cross-section area) of several recycled alloys were enhanced after processing by TE for homogenisation (Shevelev *et al.*, 2010).

The simultaneous improvement of both strength and ductility characteristics is explained by the dispersion of brittle intermetallics populated along grain boundaries, which is typical for these alloys (Figure 6.24).

Table 6.4 Mechanical properties of recycled non-ferrous alloys before and after processing by TE for homogenisation

Main element content	Initial state				After TE			
	UTS (MPa)	YS (MPa)	El (%)	RA (%)	UTS (MPa)	YS (MPa)	El (%)	RA (%)
Al 99.3%	96.5	84	11	17	159	137	21	70
Al 98.0%	99	81	10	15	299	263	14	30
Al 88% Si 9.5%	75.2	60	1.5	0	203	180	12	13
Al 93% Mg 2.28%	62	60	1,5	1	324	269	3.8	11
Cu 81% P 9%	–	300	1	–	–	420	11	–

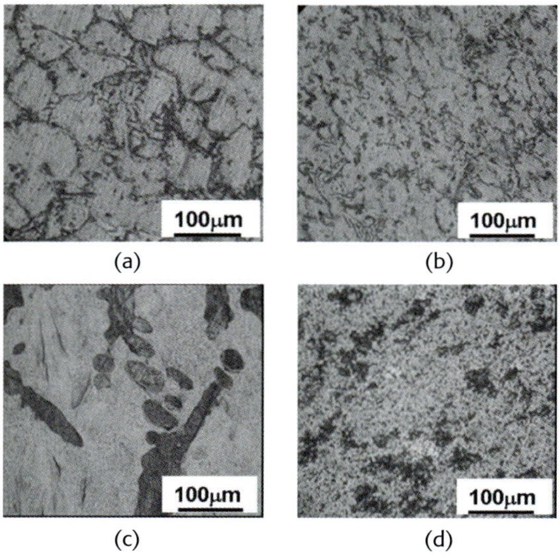

Figure 6.24 Comparison of cast and TE processed microstructures in Al alloy: (a) Al cast; (b) Al processed and Cu alloy: (c) Cu cast; (d) Cu processed

The significant improvement of ductility in these alloys after TE processing made possible further forming by plastic deformation. Hence, final products, examples of which are shown in Figure 6.25, have been manufactured by TE followed by hydro extrusion. Summarising these results, it should be noted that the processing of as-cast recycled non-ferrous metals by TE for homogenisation allows their value to be increased by making them suitable for further forming and increases the range of final products for which they can be used.

6.4.2.3 Applications based on the effect of powder consolidation

This effect has been used to manufacture long bars from magnesium (Mg), copper (Cu), and Al machining chips. An integrated tool set combining TE die followed by a conical extrusion die (Figure 6.26) was used.

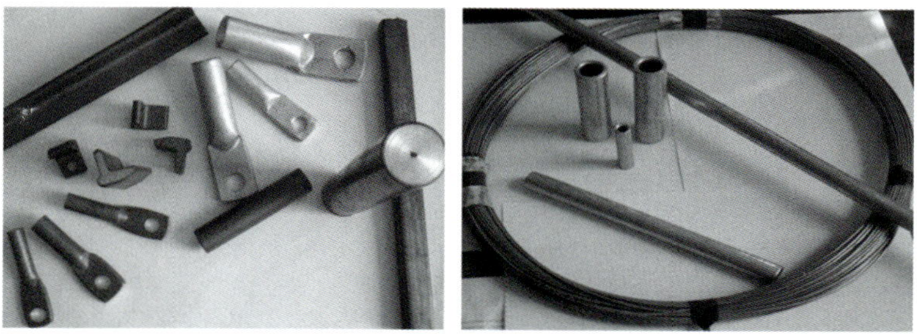

Figure 6.25 Examples of final products enabled by the TE of recycled non-ferrous metals

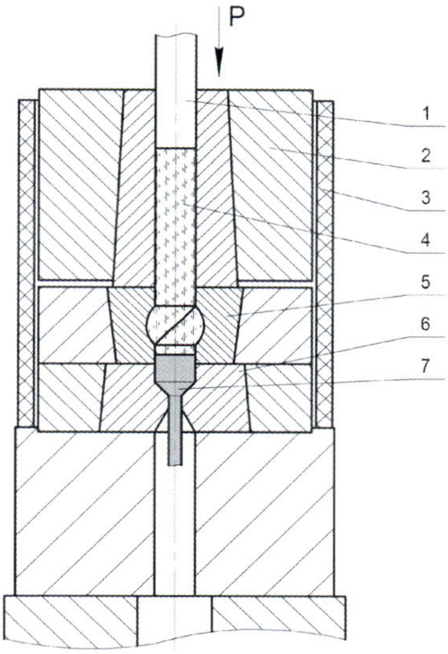

Figure 6.26 Schematic of integrated tool set combining TE die and conical extrusion die for consolidation of machining chips and powders: 1 - main plunger, 2 - container, 3 - resistance heater, 4 - metal shavings, 5 - TE die, 6 - conical die, 7 - dummy billet

Using this tool set, rods with diameter 10–15 mm and length of up to 1 m have been manufactured in a single pass from machining chips. The mechanical properties of these rods are similar to those of solid rods obtained in a traditional manner. The tool set shown in Figure 6.26 has also been used to consolidate titanium sponge into electrodes for electro slag re-melting.

6.5 Recent developments in TE and its derivatives

Two new prototypes based on the TE concept have been developed. We believe these prototypes have excellent prospects for industrial applications and therefore describe them in this section. The first is planar twist extrusion (PTE), which allows long billets with a large width-to-thickness ratio of cross-section to be processed (Beygelzimer *et al.*, 2007, 2011). The second is a laboratory horizontal tool set for TE, which allows composite and gradient materials to be manufactured.

6.5.1 Planar TE

The size of a billet for TE is limited by two factors: first, the aspect ratio (length/diameter) has to be smaller than a critical value to avoid a plunger being bent during extrusion. Second, the plunger has a limited travel distance. These are the principal limitations imposed by a processing scheme such as extrusion. This is acceptable for traditional extrusion, since a billet being extruded experiences significant elongation during processing. However, in the case of SPD processing based on the extrusion scheme (e.g. ECAE etc.), these limitations become a significant drawback since the shape of billets remains unchanged after processing.

Such a drawback can be overcome by the application of two-sided gripping of feedstock, for instance linear continuous extrusion (LINEX) (Voorhes, 1975), as well as by using containers and dies with movable walls. The latter option allows the control of magnitude and direction of friction forces on the billet–tool interface, reduces pressure applied to the plunger by partial transfer of load onto movable die walls, and allows the plunger to be designed as a more stable and stronger component. Segal (1996) proposed a design for a unique ECAE tool set in order to reduce the frictional forces. In this tool set, three lateral sides of a billet are surrounded by the plunger, and the bottom side of the die channel is free to move.

Unfortunately, this concept of tool design cannot be directly applied to TE, as the die channel has helical (i.e. non-flat) lateral surfaces. Beygelzimer *et al.* (2007) have therefore proposed TE with two parallel die walls, known as PTE. A similar scheme was subsequently proposed by Pardis and Ebrahimi (2009) under the name 'simple shear extrusion'. As can be seen (Figure 6.27), the primary difference between TE and PTE is in the shape of deforming zone in corresponding dies. In TE, all four walls of the die have the same helical shape. In PTE, only two opposite walls of the die are ruled surfaces, while the other two are flat and parallel to each other.

As can be seen in Figure 6.27(a), a PTE die channel has three distinct segments. These are straight entry and exit segments with a rectangular cross-section separated by a deforming segment in between. The cross-section of deforming segment varies along the die axis from rectangular to a parallelogram and then

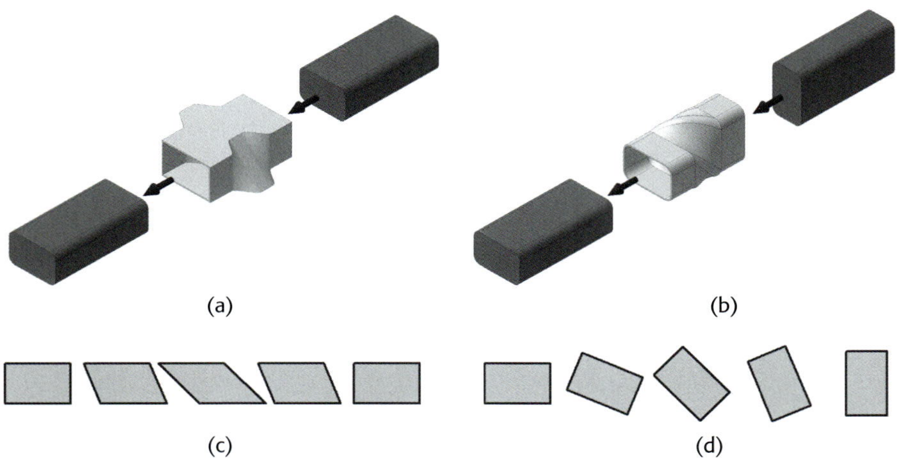

Figure 6.27 Comparison of PTE and TE processes: (a) PTE die channel; (b) TE die channel; (c) evolution of PTE cross-section; (d) evolution of TE cross-section

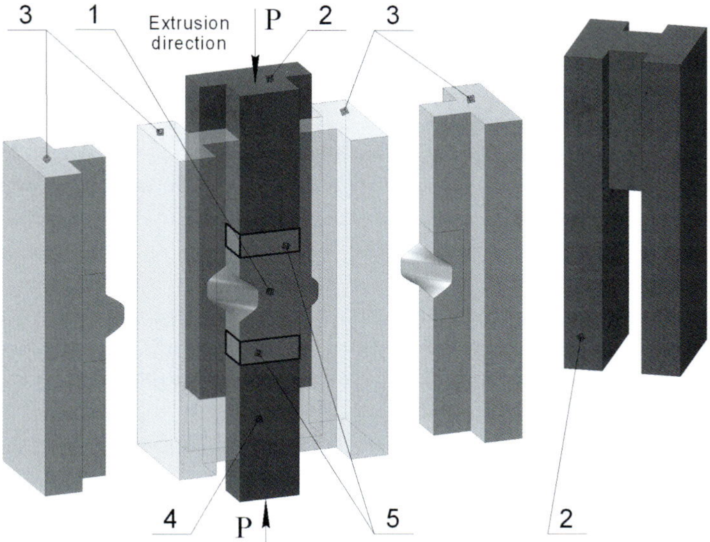

Figure 6.28 PTE tool set with two flat movable walls: 1 - billet, 2 - main plunger, 3 - curved stationary walls, 4 – back pressure plunger, 5 - dummy billets

back to rectangular (Figure 6.27(c)). The non-flat walls of this segment are ruled surfaces formed by moving a straight line along two directrices made by two straight segments conjugated with arcs (Figure 6.27(a)). Further details of this surface geometry can be found in Hilbert and Cohn-Vossen (1999).

Beygelzimer *et al.* (2011) investigated PTE using a tool set which has two flat movable walls (Figure 6.28). In PTE, billet (1) is pushed by plunger (2) through the die channel formed by two curved (3) and two flat walls. The latter is made as a single part together with the plunger. At the loading point, the plunger has an H-beam cross-section with an area larger than the cross-section of the billet being extruded. This improves the stability of the plunger and increases the length of the billet. Load from the main plunger (2) and back-pressure plungers (4) is transferred to the billet through two dummy billets (5). Back pressure is necessary for the billet to completely fill the die channel. This pressure also increases the hydrostatic stress in the deformation zone, thus improving ductility and promoting grain refinement (Beygelzimer, 2005).

Investigations by Beygelzimer *et al.* (2011) revealed that, although on average the equivalent von Mises strain accumulated after one pass can be the same in both PTE and TE processes, its distribution is substantially different depending on processing technique. In PTE, strain contours are elongated along the shorter side of the rectangular billet cross-section, while in TE the contours are arranged into loops around the centre of billet cross-section. PTE has also been successfully applied to processing polymers (Beloshenko *et al.*, 2008, 2015).

6.5.2 Laboratory horizontal tool set for TE

The schematic of a horizontal tool set for TE and its principal components are shown in Figure 6.29. The principal novelty of such a tool set is a split-design TE die Beygelzimer, Tarasov *et al.*, 2012. Its parts are compressed against each other during an extrusion pass by a hydraulic actuator.

When a billet being processed is inside, the TE die can only be split into two parts if a perpendicular dropped from any point of the die surface on the contact plane nowhere crosses the die body. We can obtain a mathematical equation for this condition.

In a cylindrical coordinate system (r,φ, z), where the z-axis is aligned with the TE die axis, $r = R(\varphi)$ describes the contour of the die channel in cross-section $z = 0$. The contact plane is defined by conditions $\varphi = 0$ and $\varphi = \pi$ as shown in Figure 6.30.

The contour is considered smooth (i.e. function $R(\varphi)$ if continuous along with the derivative $dR(\varphi)/d\varphi$ in the entire range $0 \leq \varphi \leq 2\pi$). The contour of the die channel in section z is the same as the contour in section $z = 0$, just rotated by an angle $\alpha = 2\pi\,(z/h)$, where h is the twist line increment. Hence, equation describing the contour of the die channel in section z will be

$$r = R\left(\varphi - 2\pi\frac{z}{h}\right) \tag{6.10}$$

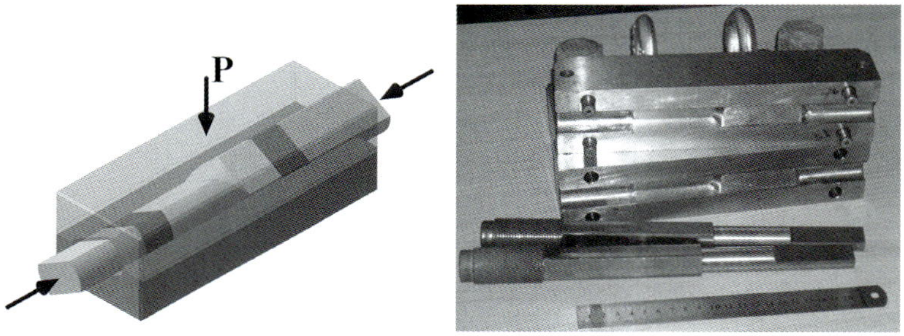

(a) (b)

Figure 6.29 Schematic of a horizontal tool set for TE: (a) schematic; (b) photograph

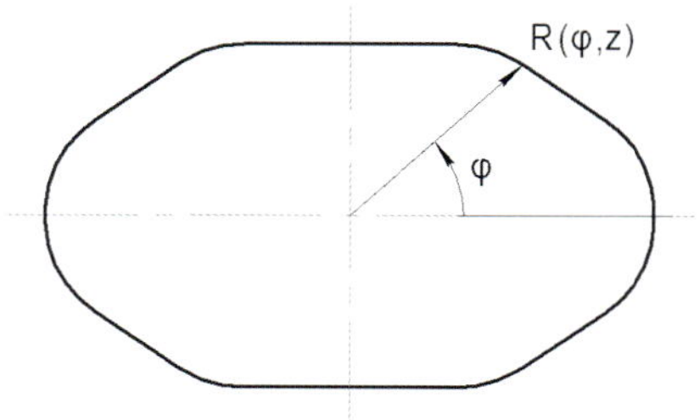

Figure 6.30 Profile of cross-section of split channel of TE die in section $z = 0$

From Figure 6.30, the aforementioned condition for the TE die split is equivalent to

$$\begin{cases} 0 \le \varphi < \pi, & \dfrac{d(R\cos\varphi)}{d\varphi} \le 0 \\ \pi \le \varphi \le 2\pi, & \dfrac{d(R\cos\varphi)}{d\varphi} \ge 0 \end{cases} \tag{6.11}$$

From Equation (6.11) we can obtain

$$\begin{cases} 0 \le \varphi < \pi, & \dfrac{dR}{d\varphi}\cos\varphi \le R\sin\varphi \\ \pi \le \varphi \le 2\pi, & \dfrac{dR}{d\varphi}\cos\varphi \le R\sin\varphi \end{cases} \tag{6.12}$$

From Equation (6.12), on the contact plane we have

$$\frac{dR}{d\varphi} = 0 \tag{6.13}$$

In this case, from Equation (6.10), in the contact section

$$\frac{\partial R}{\partial z} = 0 \tag{6.14}$$

In the tool sets for horizontal TE, a billet being processed is sequentially extruded through the die in forward and backward directions. Two hydro cylinders located on both sides of the die work by turns acting as either the main or the back-pressure cylinder. This allows the billets to be processed under conditions that can be controlled by temperature, back pressure, strain rate etc. A number of tool sets for horizontal TE have been designed based on this scheme, two of which are shown in Figure 6.31.

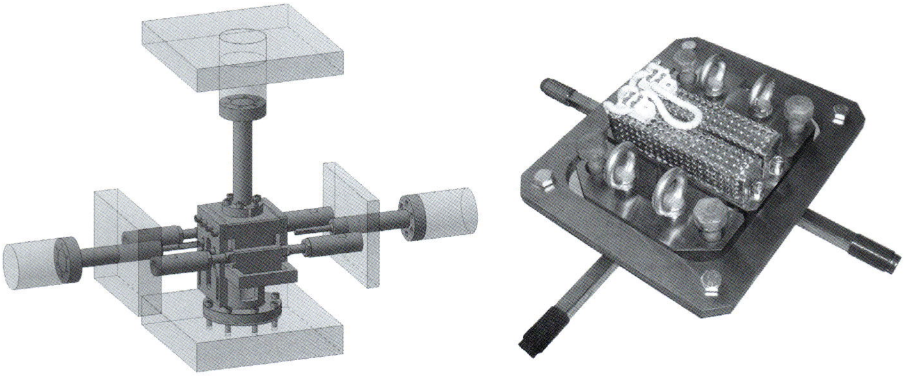

(a) (b)

Figure 6.31 Tool sets for horizontal TE: (a) design for Monash University (Melbourne, Australia); (b) photograph of tooling used by Laboratory of Excellence 'DAMAS' ('Design of Alloy Metals for low-mAss Structures'), University of Lorraine-Metz (Metz, France)

Summary

This chapter has introduced TE as a unique technique for plastic deformation processing. The mechanics of plastic deformation in TE have been thoroughly discussed and simple engineering equations for the calculation of process parameters have been proposed.

Like HPT and ECAE, the most widely used SPD techniques, the dominant deformation mode in TE is simple shear. However, due to the unique properties of TE process, it opens extra options for materials engineering.

The effectiveness of TE processing for microstructure and property tailoring in metallic materials such as: Ti alloys, super alloy Inconel 718, Al–Mg–Sc alloys, and recycled grades of Cu and Al alloys is illustrated. There are three possible areas in which TE processing might be applied.

- manufacturing of bulk nano-structured materials through grain refinement
- manufacturing of bulk nano-structured materials through consolidating powders
- improvement of ductility and microstructure homogenisation in recycled non-ferrous alloys due to vortex-like plastic flow in TE

Most of the recent developments in tool design for TE and its derivatives have been described in this chapter.

References

Akbari Mousavi, S. A. A., Shahab, A. R. and Mastoori, M. (2008) 'Computational study of Ti–6AL–4V flow behaviors during the twist extrusion process', *Materials and Design*, 29: 1316–1329.

Bahadori, S. R. and Akbari Mousavi, S. A. A. (2011) 'Examination of an aluminum alloy behavior under different routes of twist extrusion processing', *Materials Science and Engineering A*, 528: 6527–6534.

Bahadori, S. R., Akbari Mousavi, S. A. A. and Shahab, A. R. (2010) 'Sequence effects of twist extrusion and rolling on microstructure and mechanical properties of aluminum alloy 8112', *Journal of Physics: Conference Series*, 240: 012132.

Bahadori, Sh.R., Dehghania, K. and Akbari Mousavib, S. A. A. (2015) 'Comparison of microstructure and mechanical properties of pure copper processed by twist extrusion and equal channel angular Pressing', *Materials letters*, 152: 48–52.

Beigelzimer, J. E., Efros, B. M., Varyukhin, V. N. and Khokhlov, A. V. (1994) 'Continuum model of the structural-inhomogeneous porous body and its application for the study of stability and viscous fracture of materials deformed under pressure', *Engineering Fracture Mechanics*, 48: 629–640.

Beloshenko, V., Beygelzimer, Y. and Varyukhin, V. (2008) *Solid Phase Extrusion of Polymers*, Naukova Dumka, Kiev, Ukraine.

Beloshenko, V., Beygelzimer, Y. and Voznyak, Y., (2015) 'Solid-State Extrusion', *Encyclopedia of Polymer Science and Technology*, John Wiley & Sons, Inc.

Berta, M., Orlov, D. and Prangnell, P. B. (2007) 'Grain refinement response during twist extrusion of an Al-0.13% Mg alloy', *International Journal of Materials Research*, 98: 200–204.

Beygelzimer, Y. (2005) 'Grain refinement versus voids accumulation during severe plastic deformation', *Mechanics of Materials*, 37: 753–767.
Beygelzimer, Y. (2010a) 'Simple shear of metals: What is it?', *Fizika and Tekhnika Vysokikh Davlenii*, 20(4): 40–52 (in Russian).
Beygelzimer, Y. (2010b) 'Simple shear and turbulence in the metals', *Fizika and Tekhnika Vysokikh Davlenii*, 20(1): 26–32 (in Russian).
Beygelzimer, Y. (2011) 'Vortices and mixing in metals during severe plastic deformation', *Materials Science Forum*, 683: 213–224.
Beygelzimer, Y. E., Mikhailov, O. V., Synkov, A. S., Shtern, M. B. and Olevsky, E. (2008) 'Twist extrusion of powder metallurgy billets. I. Finite element analysis', *Fizika and Tekhnika Vysokikh Davlenii*, 18(1): 69–82 (in Russian).
Beygelzimer, Y. and Orlov, D. (2002) 'Metal plasticity during the twist extrusion', *Defect and Diffusion Forum*, 208/9: 311–314.
Beygelzimer, Y., Orlov, D., Korshunov, A., Synkov, S., Varyukhin, V., Vedernikova, I., Reshetov, A., Synkov, A., Polyakov, L. and Korotchenkova, I. (2006) 'Features of twist extrusion: Method, structures and material properties', *Solid State Phenomena*, 114: 69–78.
Beygelzimer, Y., Orlov, D. and Varyukhin, V. (2002) 'A new severe plastic deformation method: Twist extrusion' in *2002 TMS Annual Meeting and Exhibition*, Zhu, Y. T., Langdon, T. G., Mishra, R. S., Semiatin, S. L., Saran, M. J. and Lowe, T. C. (eds), pp. 297–304, Minerals, Metals and Materials Society, Warrendale, PA.
Beygelzimer, Y., Prilepo, D., Kulagin, R., Grishaev, V., Abramova, O., Varyukhin, V. and Kulakov, M. (2011) 'Planar twist extrusion versus twist extrusion', *Journal of Materials Processing Technology*, 211: 522–529.
Beygelzimer, Y., Prilepo, D. and Synkov, S. (2007) 'Semi-continuous twist extrusion', *Fizika and Tekhnika Vysokikh Davlenii*, 17(2): 100–104 (in Russian).
Beygelzimer, Y., Prokof'eva, O., Kulagin, R., Varyukhin, V. and Synkov, S. (2010) 'Measures of ductility for UFG materials obtained by SPD', *Materials Science Forum*, 633/4: 223–230.
Beygelzimer, Y., Reshetov, A., Synkov, S., Prokof'eva, O. and Kulagin, R. (2009) 'Kinematics of metal flow during twist extrusion investigated with a new experimental method', *Journal of Materials Processing Technology*, 209: 3650–3656.
Beygelzimer, Y., Valiev, R. Z. and Varyukhin, V. (2011) 'Simple shear: Double-stage deformation', *Materials Science Forum*, 667/9: 97–102.
Beygelzimer, Y., Varyukhin, V., Orlov, D., Efros, B., Stolyarov, V. and Salimgareyev, G. H. (2002) 'Microstructural evolution of titanium under twist extrusion' in *2002 TMS Annual Meeting and Exhibition*, Zhu, Y. T., Langdon, T. G., Mishra, R. S., Semiatin, S. L., Saran, M. J. and Lowe, T. C. (eds), pp. 43–46, Minerals, Metals and Materials Society, Warrendale, PA.
Beygelzimer, Y., Varyukhin, V., Orlov, D. and Synkov, S. (2003) *Twist Extrusion – Process For Strain Accumulation*, TEAN, Donetsk, Ukraine (in Russian).
Beygelzimer, Y. Y., Varyukhin, V. N., Synkov, S. G., Sapronov, A. N. and Synkov, V. G. (1999) 'New techniques for accumulating large plastic deformations using hydroextrusion', *Fizika and Tekhnika Vysokikh Davlenii*, 9/3: 109–110 (in Russian).
Beygelzimer, Y. Y., Varyukhin, V. N., Synkov, V. G. and Synkov, S. G. (2000) 'Intensive plastic deformation of materials under torsional extrusion', *Fizika and Tekhnika Vysokikh Davlen*, 10/1: 24–27 (in Russian).
Beygelzimer, Y., Estrin, Y. and Kulagin, R. (2015) 'Synthesis of Hybrid Materials by Severe Plastic Deformation: A New Paradigm of SPD Processing', *Advanced Engineering Materials*, 17: 1853–1861.

Beygelzimer, Y., Kulagin, R., Latypov, M., Varyukhin, V. and Kim, H.-S. (2015) 'Off-Axis Twist Extrusion for Uniform Processing of Round Bars', *Metals and Materials International*, 21: 734–740.

Beygelzimer, Y., Tarasov, A. and Kulagin, R. (2012) 'Sectional die for Twist Extrusion', *Patent Ukraine*, No. 71855.

Bouaziz, O. (2010) 'The 'ductilities' in single phase steels from usual to nano-scale microstructures', *Materials Science Forum*, 633/4: 205–210.

Bridgman, P. W. (1952) *Studies in Large Plastic Flow and Fracture*, McGraw-Hill, New York.

Gubicza, J., Dobatkin, S. V., Khosravi, E., Kuznetsov, A. A. and Lábár, J. L. (2011) 'Microstructural stability of Cu processed by different routes of severe plastic deformation', *Materials Science and Engineering A*, 528: 1828–1832.

Hilbert, D. and Cohn-Vossen, S. (1999) *Geometry and the Imagination*, second edn., AMS Chelsea Publishing, Providence, RI.

Hughes, D. A. and Hansen, N. (1997) 'High angle boundaries formed by grain subdivision mechanisms', *Acta Materialia*, 45: 3871–3886.

Kim, J. G., Latypov, M. I., Pardis, N., Beygelzimer, Y. and Kim H. S. (2015) 'Finite element analysis of the plastic deformation in tandem process of simple shear extrusion and twist extrusion', Materials & Design, 83: 858–865.

Kuhlmann-Wilsdorf, D. and Hansen, N. (1991) 'Geometrically necessary, incidental and subgrain boundaries', *Scripta Metallurgica and Materialia*, 25: 1557–1562.

Kulagin, R. (2014) 'Improving the technology and equipment for the Twist Extrusion process based on the analysis of the billet deformed state', PhD Thesis, Donetsk, Ukraine (in Ukrainian).

Kulagin, R., Latypov, M., Kim, H.-S., Varyukhin, V. and Beygelzimer, Y. (2013) 'Cross Flow During Twist Extrusion: Theory, Experiment, and Application', Metallurgical and Materials Transactions *A*, 44a: 3211–3220.

Latypov, M. I., Alexandrov, I. V., Beygelzimer, Y. E., Lee, S. and Kim, H. S. (2012) 'Finite element analysis of plastic deformation in twist extrusion', *Computational Materials Science*, 60: 194–200.

Latypov, M. I., Beygelzimer, Y. and Kim, H. S. (2013) 'Comparative Analysis of Two Twist-Based SPD Processes: Elliptical Cross-Section Spiral Equal-Channel Extrusion vs. Twist Extrusion', Materials Transaction, 54: 1587–1591.

Latypov, M. I., Lee M. G., Beygelzimer, Y., Kulagin, R. and Kim, H. S. (2015) 'Simple shear model of twist extrusion and its deviations', Metals and Materials International, 21: 569–579.

Latypov, M., Beygelzimer, Y., Kulagin, R., Varyukhin V. and Kim, H.-S. (2015) 'Toward architecturing of metal composites by twist extrusion', *Materials Research Letters*, 3: 161–168.

Latypov, M., Yoon, E. Y., Lee, D. J., Kulagin, R., Beygelzimer, Y., Salehi, M. S. and Kim H.-S. (2014) 'Microstructure and Mechanical Properties of Copper Processed by Twist Extrusion with a Reduced Twist-Line Slope', *Metallurgical and Materials Transactions A*, 45a: 2232–2241.

Latypov, M., Lee, M.-G., Beygelzimer, Y., Prilepo, D., Gusar, Y. and Kim H.-S. (2016) 'Modeling and Characterization of Texture Evolution in Twist Extrusion', *Metallurgical and materials transactions A*, 47A: 1248–1260.

Olshanetskii, V., Stepanova, L., Greshta, V., Pavlenko, D. and Tkach, D. (2014) 'Deformation Behavior of Titanium VT1-0 With Submicrocrystalline Structure Formed by the Method of Helical Extrusion', *Metal Science and Heat Treatment*, 55: 603–607.

Orlov, D. (2003) 'Theory developing and design of twist extrusion technology for titanium and copper', PhD thesis, Donetsk, Ukraine (in Ukrainian).

Orlov, D., Beygelzimer, Y., Synkov, S., Varyukhin, V. and Horita, Z. (2008) 'Evolution of microstructure and hardness in pure Al by twist extrusion', *Materials Transactions*, 49: 2–6.

Orlov, D., Beygelzimer, Y., Synkov, S., Varyukhin, V., Tsuji, N. and Horita, Z. (2009a) 'Refinement of microstructure and hardness in pure Al by twist extrusion', *Materials Transactions*, 50: 96–100.

Orlov, D., Beygelzimer, Y., Synkov, S., Varyukhin, V., Tsuji, N. and Horita, Z. (2009b) 'Plastic flow structure and mechanical properties in pure Al deformed by twist extrusion', *Materials Science and Engineering A*, 519: 105–111.

Orlov, D., Kamikawa, N. and Tsuji, N. (2012) 'High pressure torsion to refine grains in pure aluminum up to saturation: mechanisms of structure evolution and their dependence on strain', *Philosophical Magazine*, 92: 2329–2350.

Orlov, D., Reshetov, A. Synkov, A., Varyukhin, V., Lotsko, D., Sirko,O., Zakharova, N., Sharovsky, A., Voropaiev, V., Milman, Y. and Synkov, S. (2006) 'Twist extrusion as a tool for grain refinement in Al-Mg-Sc-Zr alloys' in *Nanostructured Materials by High-Pressure Severe Plastic Deformation*, Zhu, Y. and Varyukhin, V. (eds), pp. 77–82. Dordrecht, the Netherlands: Springer.

Orlov, D., Stolyarov, V. V., Salimgareyev, H. S., Soshnikova, E. P., Reshetov, A. V., Beygelzimer, Y. Y., Synkov, S. G. and Varyukhin, V. N. (2004) 'Microstructure, mechanical properties and anisotropy of pure Ti processed by twist extrusion and cold rolling' in *2004 TMS Annual Meeting and Exhibition*, Zhu, Y. T., Langdon, T. G., Valiev, R. Z., Semiatin, S. L., Shin, D. H. and Lowe, T. C. (eds), pp. 457–462, Minerals, Metals and Materials Society, Warrendale, PA.

Orlov, D., Todaka, Y., Umemoto, M., Beygelzimer, Y., Horita, Z. and Tsuji, N. (2009) 'Plastic flow and grain refinement under simple shear-based severe plastic deformation processing', *Materials Science Forum*, 604/5: 171–178.

Pardis, N. and Ebrahimi, R. (2009) 'Deformation behavior in simple shear extrusion (SSE) as a new severe plastic deformation technique', *Materials Science and Engineering A*, 527: 355–360.

Pashinska, E., Varyukhin, V. and Dobatkin, S. (2013) 'Mechanisms of structure formation in low-carbon steel at warm twist extrusion', *Emerging materials research*, 2: 138–143.

Pavlenko, D. and Ovchinnikov, A. (2015) 'Effect of Deformation by the Method of Screw Extrusion on the Structure and Properties of VT1-0 Alloy in Different States', *Materials Science*, 51: 52–60.

Prokof'eva, O. (2008) 'Development of methods of design of Twist Extrusion', PhD Thesis, Donetsk, Ukraine (in Ukranian).

Reshetov, A. (2007) 'Developing theory and technology of Twist Extrusion process for obtaining blanks with enhanced mechanical properties', PhD Thesis, Donetsk, Ukraine (in Ukrainian).

Reshetov, A., Korshunov, A., Smolyakov, A., Beygelzimer, Y., Varyukhin, V., Kaganova, I. and Morozov, A. (2011) 'Distribution of mechanical properties by volume in titanium billets processed by twist extrusion', *Materials Science Forum*, 667–669: 851–856.

Rybin, V. V. (1986) *Large Plastic Deformation and Fracture of Metals*, Metallurgia, Moscow.

Segal, V. M. (1996) 'Plastic deformation of crystalline materials', US Patent 5513512.

Shevelev, A., Beygelzimer, Y., Varyukhin, V., Synkov, S., Ryabicheva, L. and Reshetov, A. (2010) *Metal Forming of the Recycled Aluminum and Aluminum-Containing Alloys*, Noulidzh, Donetsk, Ukraine.

Shpak, A. P., Varyukhin, V. N., Tkatch, V. I., Maslov, V. V., Beygelzimer, Y. Y., Synkov, S. G., Nosenko, V. K. and Rassolov, S. G. (2006) 'Nanostructured $Al_{86}Gd_6Ni_6Co_2$ bulk alloy produced by twist extrusion of amorphous melt-spun ribbons', *Materials Science and Engineering A*, 425: 172–177.

Stolyarov, V. V., Beigel'zimer, Y. E., Orlov, D. V. and Valiev, R. Z. (2005) 'Refinement of microstructure and mechanical properties of titanium processed by twist extrusion and subsequent rolling', *Physics of Metals and Metallography*, 99: 204–211.

Stolyarov, V., Pashinskaya, E. and Beigel'zimer, Y. (2010) 'Effect of combined deformation on the structure and propertieas of copper and titanium alloys', *Russian Metallurgy*, 2010/10: 904–909.

Synkov, A. (2011) 'Consolidation of non-compact blanks through Twist Extrusion', PhD Thesis, Donetsk, Ukraine (in Ukrainian).

Titov, V., Zlochevskaya, N., Kachan, A., Savchinskii, I. and Vishnevskii, P. (2014) 'Features of integrated manufacture of blade workpieces from eutectically strengthened titanium alloys', *Metallurgist*, 58: 141–148.

Varyukhin, V., Beygelzimer, Y. and Efros, B. (2008) 'Nanostructured materials by twist extrusion and high-pressure torsion', *Materials Science Forum*, 584/6: 102–107.

Varyukhin, V., Beygelzimer, Y., Kulagin, R., Prokof'eva, O. and Reshetov, A. (2011) 'Twist extrusion: Fundamentals and applications', *Materials Science Forum*, 667/9: 31–37.

Varyukhin, V., Beygelzimer, Y., Synkov, S. and Orlov, D. (2006) 'Applications of twist extrusion', *Materials Science Forum*, 503/4: 335–340.

Varyukhin, V., Pashinska, O., Tkachenko, V. and Myshlyaev, M., (2012) 'Change of structure and regularities of copper hardening under Twist Extrusion', *Metallofizika i noveishie tekhnologii*, 34: 1655–1661.

Voorhes, W. G. (1975) 'Extrusion process' (Wanskuck Company, Providence, RI), US Patent 3922898.

Zavdoveev, A., Belousov, N. and Pashinska, E. (2015) 'Structure and properties of low-carbon steel after twist extrusion', *Emerging materials research*, 4: 89–93.

Zendehdel, H. and Hassani, A. (2012) 'Influence of twist extrusion process on microstructure and mechanical properties of 6063 aluminum alloy', *Materials and Design*, 37: 13–18.

Accumulative roll-bonding (ARB)

CHAPTER

7

Nobuhiro Tsuji

7.1 Introduction

Among various severe plastic deformation (SPD) processes, where a considerable plastic strain is applied to bulky samples in order to realize ultra-fine grained (UFG) or nanostructured materials, accumulative roll-bonding (ARB) is one of the few SPD processes that can be applied to continuous mass production of bulky materials. The principle of ARB has been already used for commercial fabrication of thin-foil strips of stainless steels having UFG structures, as will be discussed in Section 7.4. ARB is a relatively new SPD process, but it has already been recognised as a promising method for fabricating UFG/nanostructured materials which have homogeneous microstructures. It can also be used to fabricate multi-layered materials, metal-matrix composites and non-equilibrium materials. This chapter summarises the background to the development of ARB, the principle and process details, and examples of applications to various kinds of nanostructured bulk sheets.

7.2 Background of development and principles of ARB process

Since it was found that SPD can fabricate bulky UFG materials with a grain size below 1 μm, intensive studies have been conducted on SPD processes and the bulk UFG and nanostructured materials that are obtained (Valiev *et al.*, 2000; Altan *et al.*, 2006; Zehtbauer *et al.*, 2009; Whang, 2011; Tsuji *et al.*, 2003; Azushima *et al.*, 2008; Zhu *et al.*, 2010). Nevertheless, the definition of SPD is still unclear. It may help to compare the plastic strains applied by various conventional metal working processes. Table 7.1 indicates the correspondence between (von Mises) equivalent strain and nominal (engineering) strains in various deformation processes (i.e. tension, compression, rolling, extrusion and wire drawing). Here the equivalent strain (ε) and engineering strains (e, r_c, r_r, r_A) are determined by the following equations in each process.

Tension
$$e = \frac{L - L_0}{L_0} \tag{7.1}$$

$$\varepsilon = \ln \frac{L}{L_0} \tag{7.2}$$

Where L_0 is the initial gauge length and L is the gauge length after deformation.

Compression
$$r_c = \frac{H_0 - H}{H_0} \tag{7.3}$$

$$\varepsilon = -\ln \frac{H}{H_0} \tag{7.4}$$

Where H_0 is the initial height of the specimen and H is the height of the specimen after deformation.

Rolling
$$r_r = \frac{t - t_0}{t_0} \tag{7.5}$$

$$\varepsilon = -\frac{2}{\sqrt{3}} \ln \frac{t}{t_0} \tag{7.6}$$

Where t_0 is the initial thickness of the sheet and t is the thickness of the sheet after rolling.

Extrusion and wire drawing

$$r_A = \frac{A_0 - A}{A_0} \tag{7.7}$$

$$\varepsilon = -\ln \frac{A}{A_0} \tag{7.8}$$

Where A_0 is the initial cross-sectional area and A is the cross-sectional area after deformation. We assume uniform deformation. In tension (tensile test), metallic materials usually exhibit fracture before reaching 100% elongation.

Table 7.1 Correspondence between equivalent strain and engineering strain in various deformation processes

Equivalent strain (ε)	Engineering strain × 100 (%)			
	Tension (e)	Compression (r_c)	Rolling (r_r)	Extrusion and wire drawing (r_A)
0.5	64.90%	39.30%	35.10%	39.30%
1	172%	63.20%	57.90%	63.20%
2	639%	86.50%	82.30%	86.50%
3	1900%	95.00%	92.60%	95.00%
4	5360%	98.20%	96.90%	98.20%
8	298000%	99.97%	99.90%	99.97%
10	2200000%	99.99995%	99.98%	99.99995%

In uniaxial compression, deformation over 90% is substantially difficult due to the friction between the material and dies and their elastic constraint. Similarly, the maximum engineering strains conventionally applied in rolling and extrusion/wire drawing are about 95% and 98%, respectively. That is, the maximum equivalent strain that can be achieved by conventional deformation processes are 3–4 (Table 7.1). It is therefore possible to define the SPD processes as the deformation processes where logarithmic equivalent strain above 4–5 can be applied to the materials (i.e. SPD processes can plastically deform bulky materials up to very high strains that cannot be realized in conventional metal working processes).

Various unique SPD processes have been invented in order to realize SPD, the most popular and best known being equal channel angular extrusion (or pressing) (ECAE or ECAP) and high pressure torsion (HPT) (Valiev *et al.*, 2000; Azushima *et al.*, 2008; Zhu *et al.*, 2010). Since these processes have also been introduced elsewhere in this book, they are not considered in detail in this chapter. They are certainly promising for the fabrication of UFG or nanostructured materials, and a large number of studies have been conducted using ECAE and HPT. Although they are very useful methods, especially in fundamental research at laboratory scale, they also have some disadvantages. First, these processes can only fabricate relatively small samples. HPT has the great advantage that really large plastic strains can be applied without the materials fracturing under high (substantially hydrostatic) pressure, but standard specimens for HPT are consequently tiny discs that are much thinner than coins. ECAE can easily be performed, as only a pressing machine and dies are needed. However, conventional materials processed by ECAE have dimensions similar to pens. Large machines with a huge capacity would be necessary to scale up the dimensions of the samples and the cost of tooling would also increase. Secondly, ECAE and HPT are essentially 'batch' processes and it would be difficult to realize continuous production of SPD materials. Although several methods to make ECAE continuous have been proposed (Saito *et al.*, 2000; Lee *et al.*, 2002; Raab *et al.*, 2004; Huang and Prangnell 2007), they have not yet become popular. These disadvantages of HPT and ECAE have been considerable barriers to the development of SPD studies, especially in terms of future commercialisation of materials production.

ARB was developed by Saito and Tsuji with their colleagues and students in 1997–1998 (Tsuji *et al.*, 1999, 2003; Saito *et al.*, 1998, 1999). At that time, some pioneer works on SPD and UFG materials had already been done using ECAE and HPT processes. The main aim for Saito, who was a metal working professor at Osaka University, and Tsuji, who was an assistant professor in charge of physical metallurgy in Prof. Saito's group, was to develop a new SPD process applicable to continuous mass production of bulk nanostructured metallic materials, overcoming the issues in the existing SPD processes described above. They focused on rolling deformation, which was the most advantageous metal working process for continuous production of bulky materials, such as plates, sheets and bars. However, there is a limit to the strain applicable in conventional rolling, as shown in Table 7.1. Quite heavy rolling above 99% is technically possible, but with increasing the total reduction in thickness (strain) the sheet becomes thinner and thinner in sheet rolling and finally only foils are obtained. Thus, it is substantially impossible to realize SPD of bulky materials by conventional rolling. Saito and Tsuji then considered using a 'roll-bonding' process that is used commercially to produce 'clad-sheet' (Li *et al.*, 2008). The principle of the ARB process is illustrated schematically in Figure 7.1. Here, 50% rolling of sheet metal is shown as a typical example. After 50% rolling, the sheet has half its initial thickness, and the length doubles. The rolled sheet is cut into two pieces, which are stacked to obtain the initial dimensions before the first 50% rolling. The stacked sheets can then be 50% rolled again. In principle, this procedure can be repeated limitlessly, so that the process can apply a huge amount of plastic strain to the material. In order to obtain one-body solid materials, rolling in the ARB is not only a deformation process but also a bonding process (i.e. roll-bonding). Surface treatments are conducted on the contact surfaces of the sheets before stacking in order to obtain good bonding. Two sheets are tightened by steel wires or partly spot-welded after stacking. Sometimes roll-bonding is carried out at elevated temperatures in order to achieve better bonding. The temperature must be below the recrystallisation temperature of the material, so that the accumulating lattice defects in the materials can realize UFG structures. In that case, the stacked sheets are heated in the furnace and then immediately roll-bonded. However, the present author has found that even steels and pure Ti can be successfully roll-bonded at ambient temperature under appropriate conditions and techniques.

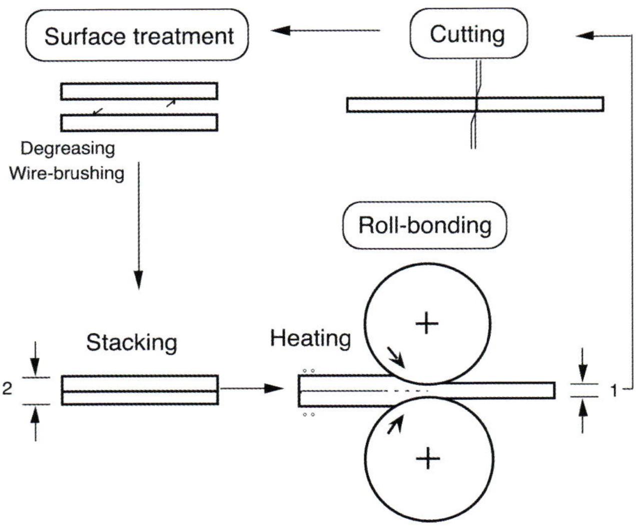

Figure 7.1 Schematic showing principle of ARB process

Table 7.2 Geometry of sheet material in ARB using 50% thickness reduction from 2 mm to 1 mm per cycle

No. of cycles	1	2	3	4	5	6	7	8	9	10	....	n
No. of layers	2	4	8	16	32	64	128	256	512	1024		2^n
No. of bonded boundaries	1	3	7	15	31	63	127	255	511	1023		$2^n - 1$
Layer interval (μm)	500	250	125	62.5	31.3	15.6	7.8	3.9	1.9	0.96		$1000/2^n$
Total reduction (%)	50	75	87.5	93.8	96.9	98.4	99.2	99.6	99.8	99.9		$(1 - 1/2^n) \times 100$
Equivalent strain	0.8	1.6	2.4	3.2	4	4.8	5.6	6.4	7.2	8		$0.8n$

Table 7.2 summarises the change in geometry for ARB using 50% rolling per cycle, assuming that the two pieces of the 1 mm thick sheet are stacked and roll-bonded. As the number of the original sheets within the product sheet 1 mm thick is 2^n after n cycles, more than 1000 original sheets are included within the 1 mm product sheet after 10 cycles of the ARB, which means that the thickness of the original sheet is already below 1 μm. In addition to ARB using 50% rolling per cycle, multi-stack ARB is also possible. For ARB with 3-stack or 4-stack sheets, respectively, 66.7% or 75% rolling reduction in thickness per cycle is necessary to keep the same rolling geometry. The von Mises equivalent strain after n cycles of the ARB using a reduction of r_r (Equation (7.5)) per cycle is

$$\varepsilon = -\frac{2n}{\sqrt{3}}\ln(1 - r_r) \tag{7.9}$$

It should be noted that the strain shown in Equation (7.9) only takes the reduction in thickness in rolling into account. When redundant shear strain is introduced in unlubricated ARB, the substantial plastic strain becomes much larger, as will be described later in this section.

Interestingly, the principle of the ARB process is quite similar to a process of traditional Japanese sword production, although the developers of ARB were not aware of that similarity when they began their research. Figure 7.2 illustrates the processes used to produce traditional Japanese swords, including the repeated folding and forging process (*orikaeshi-tanren* in Japanese) (Sasaki and Momono, 2007), a principle similar to ARB. When producing Japanese swords, the sword-smiths repeat the processes of hot-forging, cutting a notch, and folding the material. Typically, the procedures are repeated for up to 12–15 cycles, so that the Japanese swords are composed of 2^{12}–2^{15} layers of original material (see Table 7.2). In contrast to ARB, the process is carried out at elevated temperatures (in the austenite region of steels) and aims to control the carbon content as well as to increase the toughness of the sword, possibly through grain refinement of austenite. The present author would also like to point out that similar processes, such as 'repeated folding and rolling' and 'repeated pressing and

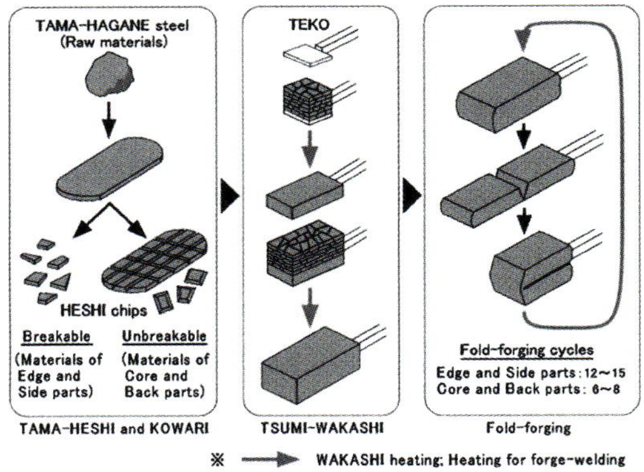

Figure 7.2 Schematics showing processes for traditional Japanese sword production (Sasaki and Momono, 2007)

diffusion bonding' (Atzmon *et al.*, 1985; Battezzati *et al.*, 1995; Yasuna *et al.*, 1997: Sagel *et al.*, 1998; Sieber *et al.*, 1999), have been tried using stacked sheets of different metals, aiming to achieve bulk mechanical alloying, before or nearly in the same period as the ARB process for fabricating bulk UFG materials was introduced.

Initially, the present author was rather doubtful about the success of the ARB process, because the bonding seemed difficult. However, immediately after starting the actual trials of the ARB process, we were surprised to realize that bonding of sheets was not so difficult even at low temperatures. As Tylecote (1968) wrote in his textbook

> ... pressure welding of metals has been used for joining different pieces of metals since ancient ages.

Bonding is affected by temperature and pressure, so that the degree of deformation of the material necessary for good bonding strongly depends on the temperature. An example of the effects of temperature and deformation on the bonding of Kroll Ti sheets by Wilsford and Tylecote (1960) is shown in Figure 7.3 (Tylecote, 1968). There is a critical reduction in thickness (deformation) with increasing temperature. That is, larger deformation is necessary at lower temperatures in order to achieve good bonding. The tendency is the same in roll-bonding. From the present author's experiences, 30–50% rolling reduction in one pass is necessary for good roll-bonding in various metallic materials. Steels or pure titanium can be successfully roll-bonded even at room temperature. When the material is hard, or becomes significantly harder by strain hardening during the process, the repetition of the ARB process becomes difficult as critical deformation (large one-pass rolling) is not achieved. For large one-pass rolling, the rolling mill must have a certain capacity. As is well known, a smaller

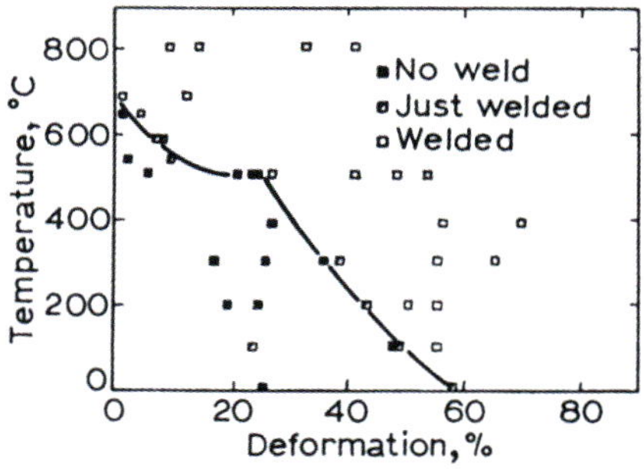

Figure 7.3 Effect of deformation and temperature on bonding (welding) of Kroll titanium in pressure welding (Tylecote, 1968; Wilford and Tylecote, 1960)

diameter work-roll helps to reduce the force needed for rolling (Wagoner and Chenot, 1997; Lenard, 2007). But at the same time, large rolls are necessary to carry out large one-pass rolling. Although the present author's group has only used rolling mills equipped with relatively large work-rolls (i.e. 250–310 mm in diameter) and has not experienced ARB using small rolls, the large rolls seem preferable for achieving good bonding. Figure 7.4 shows a two-high rolling mill equipped with rolls that are 250 mm in diameter, which is used for the ARB process at Kyoto University. Lubrication also affects bonding. Roll-bonding without lubrication greatly increases the rolling force, but from the viewpoint

Figure 7.4 Rolling mill used for ARB process at Kyoto University

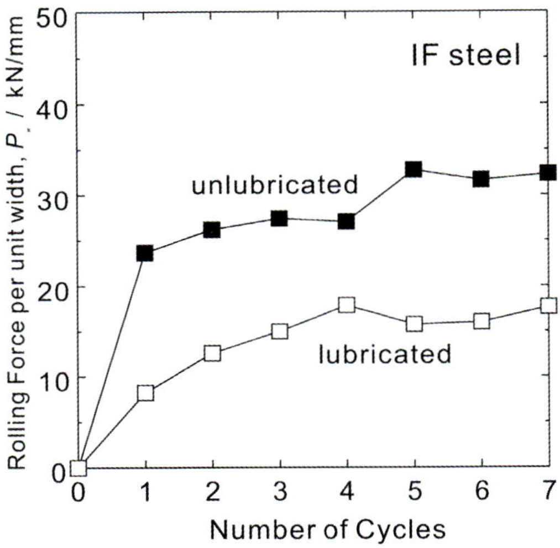

Figure 7.5 Rolling force per unit width of sheet as a function of number of ARB cycles. IF steel sheet 1 mm thick was ARB processed at 500°C using 50% one-pass roll-bonding per cycle (Tsuji, 2006)

of increasing pressure this is preferable for bonding. Figure 7.5 shows the rolling force per unit width of the sheet as a function of the number of ARB cycles, when an ultra-low carbon interstitial-free (IF) steel is ARB processed at 500°C using 50% one-pass roll-bonding per cycle (Tsuji, 2006). It can clearly be seen that, in the case of unlubricated ARB, the rolling force is about twice that in an ARB process with good lubrication. Unlubricated ARB also accelerates the formation of UFG structures, because large redundant shear strains are imposed on the material in addition to the rolling strain (ideally plane-strain compression) (Lee *et al.*, 2002; Kamikawa *et al.*, 2004; Inoue and Tsuji, 2009). However, too high a rolling force makes the large one-pass rolling, which is also necessary for bonding, difficult because of the large elastic deformation of the rolling mill (mill jump). Thus, appropriate lubrication should be chosen, depending on the materials and deformation conditions.

Adequate surface treatments are also essential in order to achieve good bonding. In the present author's group, the contact surfaces of the sheets are degreased by acetone and wire-brushing using a stainless or steel wire brush before stacking. Contamination of the sheet surface severely inhibits bonding. Although the details of the bonding mechanism are still unclear, we have considered an explanation like this. For example, oxide layers quickly form after wire-brushing of Al alloy, even at room temperature. The oxide layer on the Al alloys is a hard and brittle ceramic which is quite thin, of the order of nanometres. The surface area is doubled by 50% rolling. In this deformation, the oxide layers are broken down and fresh metallic atoms come to the surface between the broken oxides. Once such metallic atoms on the upper and lower surfaces of the

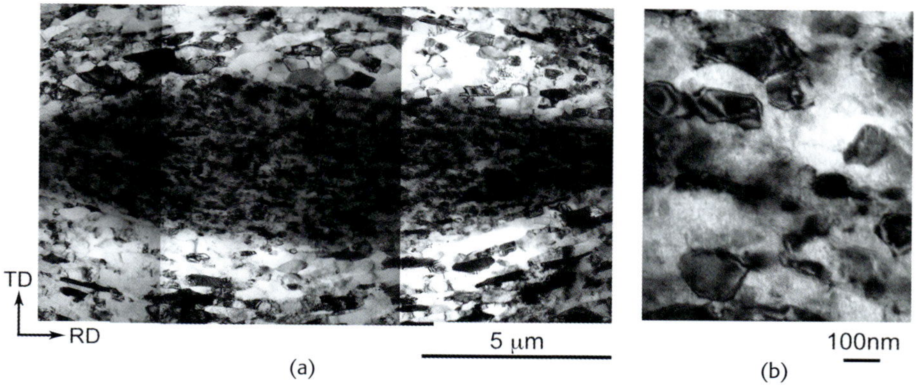

Figure 7.6 (a) TEM microstructure near a bonded surface of commercial purity Al (JIS-1100) ARB processed by six cycles at RT. Observed in TD of sheet; (b) high magnification microstructure of wire-brushed layer in (a) (Tsuji 2006)

sheets approach within a critical distance under the rolling pressure, metallic bonding can occur. Obviously, the bonding obtained by one-pass 50% rolling is not perfect compared with the natural bonding in the bulk of the metal. In fact, the last bonded boundary in the middle of the ARB processed sheet used to be found to be delaminated after fracture in a tensile test. However, the delaminations are rarely observed at another bonded layer position in the previous cycles even after tensile fracture. The bonded boundaries are repeatedly extended during subsequent cycles of roll-bonding, and the bonding improves. This suggests that additional (normal) rolling after the last cycle of the ARB would strengthen the bonding between the sheets.

It has been found that wire-brushing acts as a kind of SPD process for thin surface layers, so that nanocrystalline structures form in the surface layers (Tsuji 2006; Sato *et al.*, 2003, 2004). Figure 7.6(a) shows a transmission electron microscopy (TEM) micrograph near one of the bonded boundaries in commercial purity Al (JIS-1100) sheets that have been ARB processed by six cycles at room temperature (RT) (Tsuji 2006). The wire-brushed layer is a messy band parallel to the rolling direction (RD). A nanocrystalline structure involving oxide particles can be seen at higher magnification (Figure 7.6(b)). It is probable that the grain structures observed in the wire-brushed layers are finer than those in the ARB processed matrix because larger strains are generated at a higher rate in the wire-brushing process while the oxide particles in the matrix material inhibit grain growth. Surface nanocrystalline materials can be fabricated using these phenomena (Sato *et al.*, 2003, 2004).

Cracking of the sheets in heavy one-pass rolling may become a serious problem for the ARB process. Some kinds of materials show edge cracks especially after repeated ARB cycles. Once cracking like that shown in Figure 7.7 occurs, it becomes difficult to proceed to the next cycle of ARB. The material shown

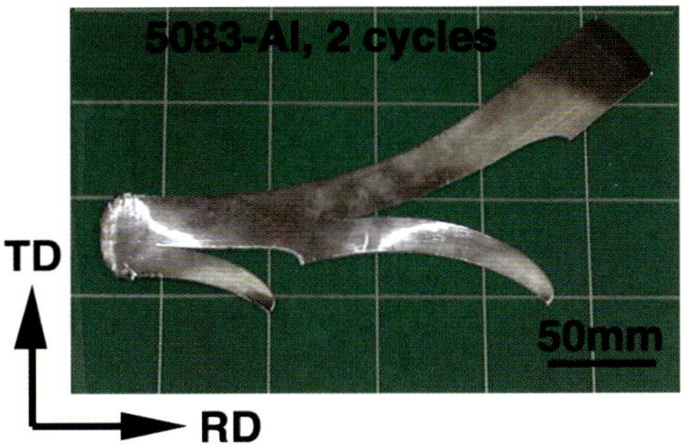

Figure 7.7 Appearance of a JIS-5083 (Al-4.5Mg) alloy sheet showing propagations of edge cracks after two cycles of ARB process at RT

in Figure 7.7 is an Al–Mg alloy (JIS-5083), which shows serious propagation of cracks. Even if the edge cracks stop near the edges, the cracked edges should be trimmed before the next cycle in order to avoid propagating the cracks. Thus, the sheet becomes smaller and smaller especially in laboratory-scale experiments. Such edge cracking is caused by tensile stress along the RD at the edge of the sheet. When the width of the sheet is relatively small compared to its thickness, larger lateral spreading is expected. Metal flow for the lateral spreading is supplied from regions near the edges, while the material near the width of the centre mainly elongates to the RD. As a result, larger tensile stress along the RD at edge parts is induced when larger lateral spreading is caused. Therefore, increasing the width of the sheets helps to reduce the lateral spreading and to minimise the effect of cracking, but it leads to an increase in the rolling force. This can be an important issue for laboratory-scale experiments, since the rolling mills used in laboratories usually have limited capacity. Attaching dummy sheets onto both sides of the specimen is a useful way of avoiding the lateral spreading and edge cracking.

It is sometimes considered that the ARB process results in bad quality interfacial bonding and edge cracks (Wei *et al.*, 2009). However, that is completely wrong. Although there are some issues to be overcome, sound sheets having bulky dimensions and UFG structures can be fabricated in various kinds of metallic materials by the ARB process. As long as the material can be deformed by rolling, the ARB process can be adopted and bulk nanostructured metals could be fabricated. If one claims that the ARB process is inappropriate for fabricating sound nanostructured material, it would be on the basis of unskilled processing. This is demonstrated by the fact that many articles have reported bulk nanostructured metallic materials fabricated by the ARB process in different institutions worldwide and various interesting and reliable data have been accumulated.

It has also been suggested that ARB processed sheets might be significantly contaminated due to the repeated roll-bonding (Valiev, 2013). The present author does not deny that contamination may occur, but considers it to be negligible. The nanostructured area shown in Figure 7.6, which was formed by wire-brushing, is one of the most extreme (thickest) regions under observation. In real materials, the regions are not continuous but fragmented. Assuming that the thickness of the wire-brushed nano-layer is on average 1 μm, the volume fraction of the nano-layer in the 1 mm thick sheet is 0.1% (1 μm/1000 μm). Also, assuming that the thickness of the Al_2O_3 oxide layers on the sheet surface averages 5 nm, the volume fraction of oxide within the 1 μm nano-layer is 1% (10 nm/1000 nm), so that the fraction of oxide in total sheet having 1 mm thickness is 0.001%. It should be noted that, once bonded the surfaces are no longer contaminated during subsequent roll-bonding. Thus, the total volume fraction of oxide after n cycles is $(0.001 \times n)$% while the total number of bonded interfaces is $2^n - 1$. As it is known that 5–6 cycles of ARB are enough to obtain uniform UFG structures, the number of ARB cycles usually used is 5–6. Therefore, the volume fraction of oxide within the 5–6 cycle-ARB processed materials is only 0.005–0.006%. Additionally, the average interval between adjacent bonded interfaces after six cycles is 15.6 μm according to Table 7.2, which is much larger than the grain sizes (of the order of 100 nm) realized in the ARB processed sheets. Of course, there might be contamination by other light elements, but this simple calculation suggests that the effects of contamination and bonded interfaces on the microstructures and properties of the ARB processed materials are not significant.

7.3 Nanostructures obtained by ARB

The ARB process has been applied to various metals and alloys in order to obtain UFGs or nanostructures. This section considers the typical nanostructures obtained by ARB. Figure 7.8 shows TEM micrographs of an ultra-low-carbon

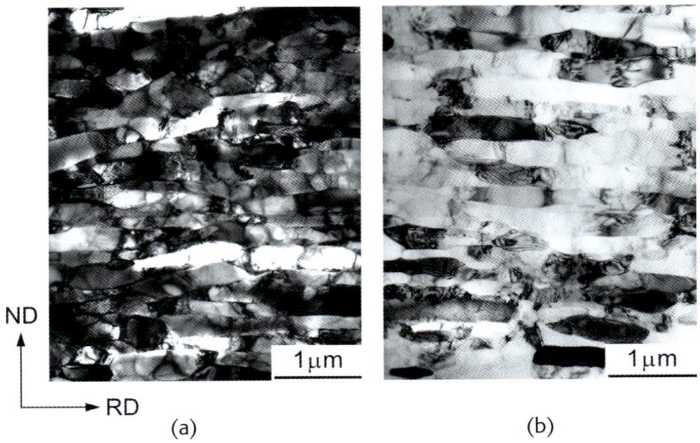

Figure 7.8 TEM microstructures of single-phased materials ARB processed by six cycles without lubrication. Microstructures were observed in TD of sheets. (a) IF steel processed at 500°C; (b) 1100-Al processed at RT

IF steel (a) and a commercial purity aluminum (JIS 1100-Al, 99 wt.% purity), (b) ARB processed by six cycles. The microstructures were observed in the transverse direction (TD) of the sheets. These are the typical nanostructures in single-phase materials that are ARB processed by many cycles. Both materials showed quite similar microstructures elongated along the RD. These are so-called lamellar boundary (LB) structures which have been observed in heavily rolled FCC metals (Hansen and Juul Jensen, 1999). The mean spacing of the LBs was about 200 nm for both materials. Most of the boundaries in these microstructures are already high-angle ones, as has been confirmed by Kikuchi-line analysis in TEM and in field emission scanning microscopy (FE-SEM)/electron backscatter diffraction (EBSD) (Tsuji *et al.*, 2002a, 2009, Tsuji 2011). The boundaries between adjacent regions seem quite sharp. It can therefore be concluded that the elongated regions surrounded by high-angle boundaries are certainly 'grains' from the viewpoint of misorientation. At the same time, however, the morphology of the grains is elongated along the principal direction of deformation and the microstructure involves many low-angle boundaries and dislocations. That is, the ultra-fine grains fabricated by the ARB process (and by other SPD processes) are essentially also deformation microstructures. The microstructural evolution during the ARB (or during SPD in general) can be understood in terms of grain subdivision (Tsuji, 2011; Hansen and Mehl, 2001; Huang *et al.*, 2003).

Subsequent annealing could change the microstructures of the ARB processed materials (Tsuji *et al.*, 2002b, 2004). Figure 7.9 shows the TEM microstructures of the 1100-Al ARB processed at 200°C without lubrication and then annealed at various temperatures (Tsuji *et al.*, 2002b). During low-temperature annealing, recovery decreases the dislocation density inside the elongated ultra-fine grains, and they grow slightly. Equiaxial grains free from dislocations can be observed in a specimen annealed at 225°C for 1.8 ks (Figure 7.9(d)). It is quite difficult to distinguish these grains from conventionally recrystallised grains in morphology. It should be noted, however, that the mean grain size is still around 1 μm, which cannot be achieved through a conventional deformation and recrystallisation process. Further annealing causes grain growth, resulting in equiaxial grain structures having various mean grain sizes. Similar microstructural change has also been observed in the ARB processed IF steel (Tsuji *et al.*, 2004). The changes in microstructures during the annealing of the ARB processed pure Al and IF steel are fairly homogeneous and continuous. On the other hand, discontinuous changes in the microstructure, which are similar to conventional recrystallisation by nucleation and growth, can also occur during annealing in materials such as Cu (Takata *et al.*, 2007) and Ti (Terada *et al.*, 2008).

ARB can be used as a bulk mechanical alloying process. After many repetitions of the ARB cycles for the stacks of different metals, non-equilibrium materials such as supersaturated solid solutions and even amorphous ones are obtained. An example of the formation of non-equilibrium nanostructures by the ARB process is shown in Ohsaki *et al.* (2007). Sheets of pure Cu (99.96% purity) and pure Zr (99.2% purity), 200 mm in length, 50 mm in width, and

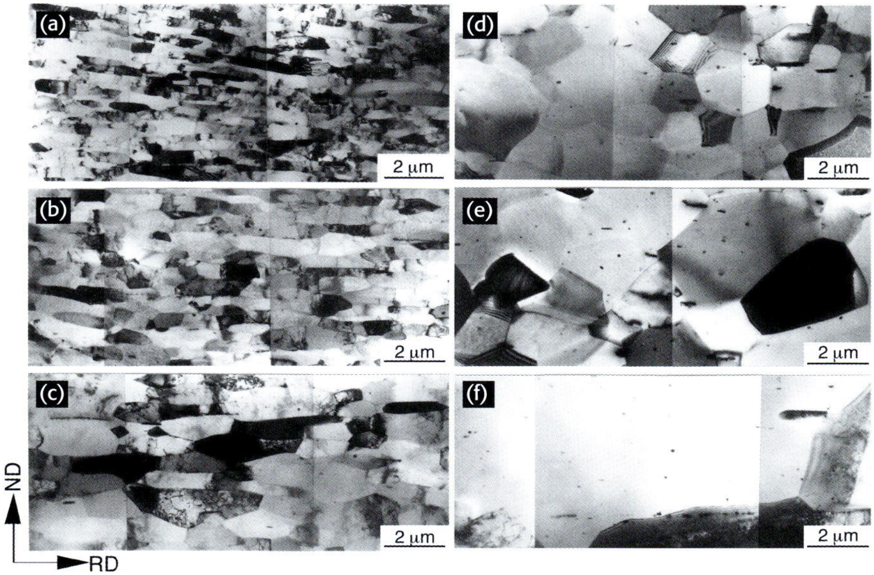

Figure 7.9 TEM microstructures of 1100-Al ARB processed by six cycles at 200°C without lubrication and then annealed at: (a) 100°C; (b) 150°C; (c) 200°C; (d) 225°C; (e) 250°C; (f) 300°C for 1.8 ks. Observed in TD

0.2 mm in thickness, were mutually stacked so that the total thickness became 1 mm. The overall composition of the multi-stacks was Cu-29at%Zr. The ARB process using 75% reduction per cycle was carried out at RT with lubrication. The ARB process was repeated for up to 10 cycles, equivalent strain of 15.8, which corresponds to conventional rolling of a huge plate with a thickness of 568 m down to 1 mm. Figure 7.10 shows a high-resolution TEM microstructure

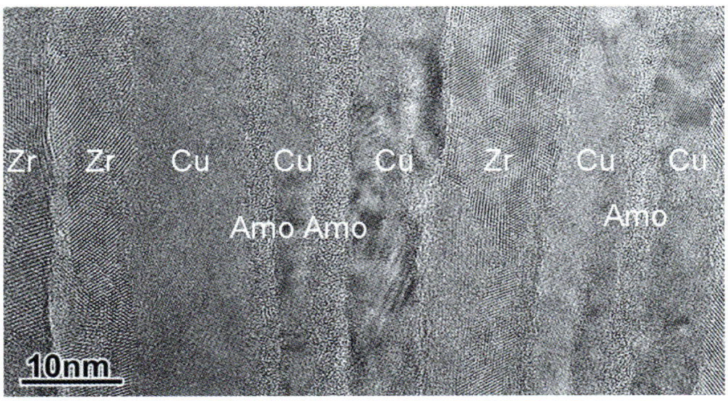

Figure 7.10 High-resolution TEM microstructure of Cu + Zr multi-stacks ARB processed by nine cycles (equivalent strain of 13.7) at RT. Observed in TD

of the Cu/Zr multi-stacks ARB processed by nine cycles at RT. A nano-lamellar structure composed of mutually aligned Cu and Zr layers is observed. The thickness of each Cu or Zr layer is below 10 nm. In some thinner Zr layers, no periodic contrast can be recognised, indicating that an amorphous phase has formed (indicated as 'Amo'). Thin amorphous regions are also observed at the interfaces between the Cu and Zr layers. Energy-dispersive X-ray analysis using a nano-beam has clarified that atomistic scale mixing of Cu and Zr occurs at interface regions during the ARB and an amorphous phase is eventually formed (Ohsaki *et al.*, 2007). It has also been found that subsequent heat treatment of the ARB processed stacks causes thermally induced amorphisation to form sheets of nearly 100% metallic glass (Sun *et al.*, 2008). Attempts to fabricate a metal-matrix composite using the ARB process have been also made (Alizadeh *et al.*, 2012).

UFG or nanostructured materials fabricated by the ARB process show excellent mechanical properties, such as high strength, considerable toughness and low-temperature superplasticity. It should be noted that they often show peculiar mechanical behaviours that never occur in conventional coarse-grained materials. Yield-point phenomena in pure Al (Tsuji *et al.*, 2002b) and 'hardening by annealing and softening by deformation' phenomena (Huang *et al.*, 2006) are examples of such unique behaviours. The mechanical properties of bulk nanostructured metals obtained by the ARB process have been summarised elsewhere (Tsuji 2007, 2011).

7.4 Summary

The principle and details of the ARB process, especially its processing details, have all been shown without any concealment, in this chapter. Although the ARB process is a promising SPD technique which can be applied to continuous production of large bulky materials, adopting the ARB process to industrial application is still not easy. One of the reasons for such difficulty is that materials industries such as large steel companies use huge facilities for mass production. They cannot replace the facilities flexibly. The need for several repetitions of roll-bonding is also a big hurdle that production would need to overcome before it could be efficient. However, it would be possible to adopt the ARB to small-scale mills. Actually, thin strips of stainless steels having a thickness of about 0.1 mm have been already produced by a small-scale Japanese company, partly using the principle of the ARB process (Tsuji, 2010). Figure 7.11 shows the appearance and microstructure of the UFG thin strip of SUS 304 austenitic stainless steel. Recrystallised UFGs with an average grain size of 1.1 μm are observed in the EBSD orientation colour maps, and typical dimensions of the UFG stainless steel strips are 0.1 mm in thickness, 100 mm in width, and 3000 m in length. They could be used for precise machines and their parts. For fundamental, laboratory-scale experiments, it is certain that sound bulky nanostructured sheets can be fabricated by the ARB process without special equipment. It is to be hoped that the ARB process could contribute to the progress of materials science and engineering as an effective fabrication process for nanostructured metals and alloys.

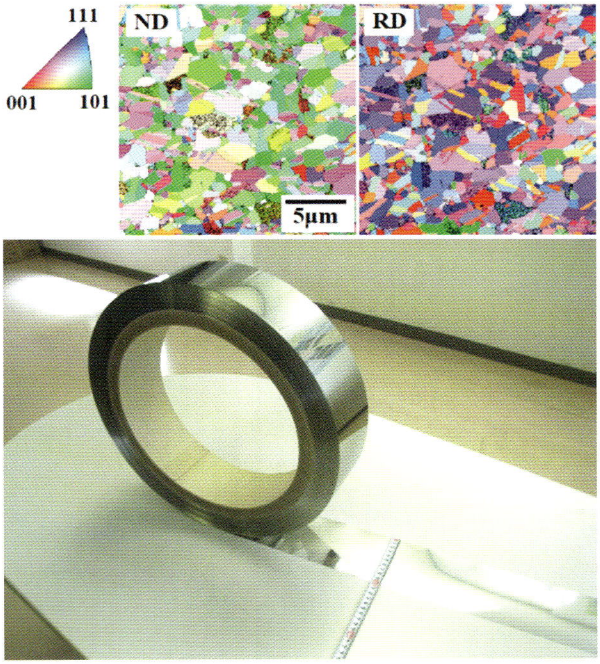

Figure 7.11 Appearance and EBSD colour maps of UFG strip of austenitic stainless steel (SUS 304) fabricated through repetition of deformation induced martensitic transformation in cold-rolling and reverse transformation to austenite (courtesy of Tokushu Kinzoku Excel Co., Ltd)

References

Alizadeh, M., Paydar, M. H., Terada, D. and Tsuji, N. (2012) 'Effect of SiC particles on the microstructure evolution and mechanical properties of aluminum during ARB process', *Materials Science and Engineering A*, 540: 13–23.

Altan, B. S., Miskioglu, I., Purcek, G., Mulyukov, R. R. and Artan, R. (eds) (2006) *Severe Plastic Deformation: Towards Bulk Production of Nanostructured Materials*, Nova Science Publishers, Hauppauge, NY.

Atzmon, M., Unruh, K. M. and Johnson, W. L. (1985) 'Formation and characterization of amorphous erbium-based alloys prepared by near-isothermal cold-rolling of elemental composites', *Journal of Applied Physics*, 58: 3865–3870.

Azushima, A., Kopp, R., Korhonen, A., Yang, D. Y., Micari, F., Lahoti, G. D., Groche, P., Yanagimoto, J., Tsuji, N., Rosochowski, A. and Yanagida, A. (2008) 'Severe plastic deformation (SPD) processes for metals', *CIRP Annals-Manufacturing Technology*, 57(2): 716–735.

Battezzati, L., Antonione, C. and Fracchia, F. (1995) 'Ni-Al intermetallics produced by cold-rolling elemental sheets', *Intermetallics*, 3: 67–71.

Hansen, N. and Juul Jensen, D. (1999) 'Development of microstructure in FCC metals during cold work', *Philosophical Transactions of the Royal Society London A*, 357: 1447–1469.

Hansen, N. and Mehl, R. F. (2001) 'New discoveries in deformed metals', *Metallurgical and Materials Transactions A*, 32: 2917–2935.

Huang, X., Hansen, N. and Tsuji, N. (2006) 'Hardening by annealing and softening by deformation in nanostructured metals', *Science*, 312(5771): 249–251.

Huang, X., Tsuji, N., Hansen, N. and Minamino, Y. (2003) 'Microstructural evolution during accumulative roll-bonding of commercial purity aluminum', *Materials Science and Engineering A*, 340: 265–271.

Huang, Y. and Prangnell, P. B. (2007) 'Continuous frictional angular extrusion and its application in the production of ultrafine grained sheet metals', *Scripta Materialia*, 56: 333–336.

Inoue, T. and Tsuji, N. (2009) 'Quantification of strain in accumulative roll-bonding under unlubricated condition by finite element analysis', *Computational Materials Science*, 46: 261–266.

Kamikawa, N., Tsuji, N. and Minamino, Y. (2004) 'Microstructure and texture through thickness of ultralow carbon IF steel sheet severely deformed by accumulative roll-bonding', *Science and Technology of Advanced Materials*, 5(1/2): 163–172.

Lee, J. C., Seok, H. K. and Suh, J. Y. (2002) 'Microstructural evolutions of the Al strip prepared by cold rolling and continuous equal channel angular pressing', *Acta Materialia*, 50: 4005–4019.

Lee, S. H., Saito, Y., Tsuji, N., Utsunomiya, H. and Sakai, T. (2002) 'Role of shear strain in ultragrain refinement by accumulative roll-bonding (ARB) process', *Scripta Materialia*, 46: 281–285.

Lenard, J. G. (2007) *Primer on Flat Rolling*, Elsevier, Oxford, UK.

Li, L., Nagai, K. and Yin, F. (2008) 'Progress in cold roll bonding of metals', *Science and Technology of Advanced Materials*, 9(2): 023001.

Ohsaki, S., Kato, S., Tsuji, N., Ohkubo, T. and Hono, K. (2007) 'Bulk mechanical alloying of Cu–Ag and Cu/Zr two-phase microstructures by accumulative roll-bonding process', *Acta Materialia*, 55: 2885–2895.

Raab, G. J., Valiev, R. Z., Lowe, T. C. and Zhu, Y. T. (2004) 'Continuous processing of ultrafine grained Al by ECAP-Conform', *Materials Science and Engineering A*, 382: 30–34.

Sagel, A., Sieber, H., Fecht, H. J. and Perepezko, J. H. (1998) 'Synthesis of an amorphous Zr–Al–Ni–Cu alloy with large supercooled liquid region by cold-rolling of elemental foils', *Acta Materialia*, 46: 4233–4241.

Saito, Y., Tsuji, N., Utsunomiya, H., Sakai, T. and Hong, R. G. (1998) 'Ultra-fine grained bulk aluminum produced by accumulative roll-bearing (ARB) process', *Scripta Materialia*, 39: 1221–1227.

Saito, Y., Utsunomiya, H., Suzuki, H. and Sakai, T. (2000) 'Improvement in the *r*-value of aluminum strip by a continuous shear deformation process', *Scripta Materialia*, 42: 1139–1144.

Saito, Y., Utsunomiya, H., Tsuji, N. and Sakai, T. (1999) 'Novel ultra-high straining process for bulk materials–development of the accumulative roll-bonding (ARB) process', *Acta Materialia*, 47: 579–583.

Sasaki, N. and Momono, T. (2007) 'Changes in carbon content of materials of the Japanese sword under traditional forging process', *Tetsu-to-Hagané*, 93: 792–798 (in Japanese).

Sato, M., Tsuji, N., Minamino, Y. and Koizumi, Y. (2003) 'Fabrication of surface nanocrystalline aluminum alloys', *Materials Science Forum*, 426/432: 2753–2758.

Sato, M., Tsuji, N., Minamino, Y. and Koizumi, Y. (2004) 'Formation of nanocrystalline surface layers in various metallic materials by near surface severe plastic deformation', *Science and Technology of Advanced Materials*, 5: 145–152.

Sieber, H., Wilde, G. and Perepezko, J. H. (1999) 'Thermally activated amorphous phase formation in cold-rolled multilayers of Al–Ni, Al–Ta, Al–Fe and Zr–Cu', *Journal of Non-Crystalline Solids*, 250/252: 611–615.

Sun, Y. F., Todaka, Y., Umemoto, M. and Tsuji, N. (2008) 'Solid-state amorphization of Cu + Zr multi-stacks by ARB and HPT techniques', *Journal of Materials Science*, 43: 7457–7464.

Takata, N., Yamada, K., Ikeda, K., Yoshida, F., Nakashima, H. and Tsuji, N. (2007) 'Changes in microstructure and texture during annealing of pure copper deformed by accumulative roll bonding', *Materials Transactions*, 48: 2043–2048.

Terada, D., Inoue, M., Kitahara, H. and Tsuji, N. (2008) 'Change in mechanical properties and microstructure of ARB processed Ti during annealing', *Materials Transactions*, 49(1): 41–46.

Tsuji, N. (2006) 'Production of bulk nanostructured metals by accumulative roll bonding (ARB) process' in *Severe Plastic Deformation: Towards Bulk Production of Nanostructured Materials*, Altan, B. S., Miskioglu, I., Purcek, G., Mulyukov, R. R. and Artan, R. (eds), pp. 545–565, Nova Science Publishers, Hauppauge, NY.

Tsuji, N. (2007) 'Unique mechanical properties of nanostructured metals', *Journal of Nanoscience and Nanotechnology*, 7(11): 3765–3770.

Tsuji, N. (2010) 'New routes for fabricating ultrafine-grained microstructures in bulky steels without very-high strains', *Advanced Engineering Materials*, 12(8): 701–707.

Tsuji, N. (2011) 'Bulk nanostructured metals and alloys produced by accumulative roll-bonding' in *Nanostructured Metals and Alloys: Processing, Microstructure, Mechanical Properties and Applications*, Whang, S. H. (ed.), pp. 40–58. Woodhead Publishing, Cambridge, UK.

Tsuji, N., Ito, Y., Saito, Y. and Minamino, Y. (2002) 'Strength and ductility of ultrafine grained aluminum and iron produced by ARB and annealing', *Scripta Materialia*, 47(12): 893–899.

Tsuji, N., Okuno, S., Koizumi, Y. and Minamino, Y. (2004) 'Toughness of ultrafine grained ferritic steels fabricated by ARB and annealing process', *Materials Transactions*, 45: 2272–2281.

Tsuji, N., Saito, Y., Lee, S. H. and Minamino, Y. (2003) 'ARB (accumulative roll-bonding) and other new techniques to produce bulk ultrafine grained materials', *Advanced Engineering Materials*, 5: 338–344.

Tsuji, N., Saito, Y., Utsunomiya, H. and Tanigawa, S. (1999) 'Ultra-fine grained bulk steel produced by accumulative roll-bonding (ARB) process', *Scripta Materialia*, 40: 795–800.

Tsuji, N., Ueji, R., Kamikawa, N. and Terada, D. (2009) 'Microstructures of ultra low-carbon IF steel highly deformed by ARB process', *Ferrum*, 14(1): 16–17.

Tsuji, N., Ueji, R. and Minamino, Y. (2002) 'Nanoscale crystallographic analysis of ultrafine grained IF steel fabricated by ARB process', *Scripta Materialia*, 47(2): 69–76.

Tylecote, R. F. (1968) *The Solid Phase Welding of Metals*, Edward Arnold, London.

Valiev, R. Z. (2013) Private communication.

Valiev, R. Z., Islamgaliev, R. K. and Alexandrov, I. V. (2000) 'Bulk nanostructured materials from severe plastic deformation', *Progress in Materials Science*, 45: 103–189.

Wagoner, R. H. and Chenot, J. L. (1997) *Fundamentals of Metal Forming*, Wiley, New York.

Wei, W., Zhang, W., Wei, K. X., Zhong, Y., Cheng, G. and Hu, J. (2009) 'Finite element analysis of deformation behavior in continuous ECAP process', *Materials Science and Engineering A*, 516: 111–118.

Whang, S. H. (ed.) (2011) *Nanostructured Metals and Alloys: Processing, Microstructure, Mechanical Properties and Applications*, Woodhead Publishing, Cambridge, UK.

Wilford, C. F. and Tylecote, R. F. (1960) 'Hot pressure welding of titanium', *British Welding Journal*, 7: 708–712.

Yasuna, K., Terauchi, M., Otsuki, A., Ishihara, K. N. and Shingu, P. H. (1997) 'Bulk metallic multilayers produced by repeated press-rolling and their perpendicular magnetoresisitance', *Journal of Applied Physics*, 82: 2435–2438.

Zehtbauer, M. J. and Zhu, Y. T. (eds) (2009) *Bulk Nanostructured Materials*, Wiley-VCH, Weinheim, Germany.

Zhu, Y., Valiev, R. Z., Langdon, T. G., Tsuji, N. and Lu, K. (2010) 'Processing of nanostructured metals and alloys via plastic deformation', *MRS Bulletin*, 35(12): 977–981.

Index